De Gruyter Graduate

Celine Carstensen
Benjamin Fine
Gerhard Rosenberger

Abstract Algebra

Applications to Galois Theory, Algebraic Geometry and Cryptography

De Gruyter

Mathematics Subject Classification 2010: Primary: 12-01, 13-01, 16-01, 20-01; Secondary: 01-01, 08-01, 11-01, 14-01, 94-01.

This book is Volume 11 of the *Sigma Series in Pure Mathematics*, Heldermann Verlag.

ISBN 978-3-11-025008-4
e-ISBN 978-3-11-025009-1

Library of Congress Cataloging-in-Publication Data

Carstensen, Celine.
Abstract algebra : applications to Galois theory, algebraic geometry, and cryptography / by Celine Carstensen, Benjamin Fine, and Gerhard Rosenberger.
p. cm. – (Sigma series in pure mathematics ; 11)
Includes bibliographical references and index.
ISBN 978-3-11-025008-4 (alk. paper)
1. Algebra, Abstract. 2. Galois theory. 3. Geometry, Algebraic. 4. Crytography. I. Fine, Benjamin, 1948– II. Rosenberger, Gerhard. III. Title.
QA162.C375 2011
515'.02–dc22
2010038153

Bibliographic information published by the Deutsche Nationalbibliothek

The Deutsche Nationalbibliothek lists this publication in the Deutsche Nationalbibliografie; detailed bibliographic data are available in the Internet at http://dnb.d-nb.de.

Typesetting: Da-TeX Gerd Blumenstein, Leipzig, www.da-tex.de
Printing and binding: AZ Druck und Datentechnik GmbH, Kempten
∞ Printed on acid-free paper

Printed in Germany

www.degruyter.com

Preface

Traditionally, mathematics has been separated into three main areas; algebra, analysis and geometry. Of course there is a great deal of overlap between these areas. For example, topology, which is geometric in nature, owes its origins and problems as much to analysis as to geometry. Further the basic techniques in studying topology are predominantly algebraic. In general, algebraic methods and symbolism pervade all of mathematics and it is essential for anyone learning any advanced mathematics to be familiar with the concepts and methods in abstract algebra.

This is an introductory text on abstract algebra. It grew out of courses given to advanced undergraduates and beginning graduate students in the United States and to mathematics students and teachers in Germany. We assume that the students are familiar with Calculus and with some linear algebra, primarily matrix algebra and the basic concepts of vector spaces, bases and dimensions. All other necessary material is introduced and explained in the book. We assume however that the students have some, but not a great deal, of mathematical sophistication. Our experience is that the material in this can be completed in a full years course. We presented the material sequentially so that polynomials and field extensions preceded an in depth look at group theory. We feel that a student who goes through the material in these notes will attain a solid background in abstract algebra and be able to move on to more advanced topics.

The centerpiece of these notes is the development of Galois theory and its important applications, especially the insolvability of the quintic. After introducing the basic algebraic structures, groups, rings and fields, we begin the theory of polynomials and polynomial equations over fields. We then develop the main ideas of field extensions and adjoining elements to fields. After this we present the necessary material from group theory needed to complete both the insolvability of the quintic and solvability by radicals in general. Hence the middle part of the book, Chapters 9 through 14 are concerned with group theory including permutation groups, solvable groups, abelian groups and group actions. Chapter 14 is somewhat off to the side of the main theme of the book. Here we give a brief introduction to free groups, group presentations and combinatorial group theory. With the group theory material in hand we return to Galois theory and study general normal and separable extensions and the fundamental theorem of Galois theory. Using this we present several major applications of the theory including solvability by radicals and the insolvability of the quintic, the fundamental theorem of algebra, the construction of regular n-gons and the famous impossibilities; squaring the circling, doubling the cube and trisecting an angle. We finish in a slightly different direction giving an introduction to algebraic and group based cryptography.

October 2010

Celine Carstensen
Benjamin Fine
Gerhard Rosenberger

Contents

Chapter 1
Groups, Rings and Fields

1.1 Abstract Algebra

Abstract algebra or *modern algebra* can be best described as the theory of *algebraic structures*. Briefly, an *algebraic structure* is a set S together with one or more binary operations on it satisfying axioms governing the operations. There are many algebraic structures but the most commonly studied structures are *groups, rings, fields* and *vector spaces*. Also widely used are *modules* and *algebras*. In this first chapter we will look at some basic preliminaries concerning groups, rings and fields. We will only briefly touch on groups here, a more extensive treatment will be done later in the book.

Mathematics traditionally has been subdivided into three main areas – *analysis, algebra and geometry*. These areas overlap in many places so that it is often difficult to determine whether a topic is one in geometry say or in analysis. Algebra and algebraic methods permeate all these disciplines and most of mathematics has been algebraicized – that is uses the methods and language of algebra. Groups, rings and fields play a major role in the modern study of analysis, topology, geometry and even applied mathematics. We will see these connections in examples throughout the book.

Abstract algebra has its origins in two main areas and questions that arose in these areas – the *theory of numbers* and the *theory of equations*. The theory of numbers deals with the properties of the basic number systems – integers, rationals and reals while the theory of equations, as the name indicates, deals with solving equations, in particular polynomial equations. Both are subjects that date back to classical times. A whole section of Euclid's elements is dedicated to number theory. The foundations for the modern study of number theory were laid by Fermat in the 1600s and then by Gauss in the 1800s. In an attempt to prove Fermat's big theorem Gauss introduced the complex integers $a + bi$ where a and b are integers and showed that this set has unique factorization. These ideas were extended by Dedekind and Kronecker who developed a wide ranging theory of algebraic number fields and algebraic integers. A large portion of the terminology used in abstract algebra, rings, ideals, factorization comes from the study of algebraic number fields. This has evolved into the modern discipline of algebraic number theory.

The second origin of modern *abstract algebra* was the problem of trying to determine a formula for finding the solutions in terms of radicals of a fifth degree polynomial. It was proved first by Ruffini in 1800 and then by Abel that it is impossible to find a formula in terms of radicals for such a solution. Galois in 1820 extended this and showed that such a formula is impossible for any degree five or greater. In proving this he laid the groundwork for much of the development of modern abstract algebra especially field theory and finite group theory. Earlier, in 1800, Gauss proved the *fundamental theorem of algebra* which says that any nonconstant complex polynomial equation must have a solution. One of the goals of this book is to present a comprehensive treatment of Galois theory and a proof of the results mentioned above.

The locus of real points (x, y) which satisfy a polynomial equation $f(x, y) = 0$ is called an algebraic plane curve. *Algebraic geometry* deals with the study of algebraic plane curves and extensions to loci in a higher number of variables. Algebraic geometry is intricately tied to abstract algebra and especially commutative algebra. We will touch on this in the book also.

Finally *linear algebra*, although a part of abstract algebra, arose in a somewhat different context. Historically it grew out of the study of solution sets of systems of linear equations and the study of the geometry of real n-dimensional spaces. It began to be developed formally in the early 1800s with work of Jordan and Gauss and then later in the century by Cayley, Hamilton and Sylvester.

1.2 Rings

The primary motivating examples for algebraic structures are the basic number systems; the integers $\mathbb{Z}$, the rational numbers $\mathbb{Q}$, the real numbers $\mathbb{R}$ and the complex numbers $\mathbb{C}$. Each of these has two basic operations, addition and multiplication and form what is called a *ring*. We formally define this.

Definition 1.2.1. A *ring* is a set R with two binary operations defined on it, addition, denoted by $+$, and multiplication, denoted by $\cdot$, or just by juxtaposition, satisfying the following six axioms:

(1) Addition is commutative: $a + b = b + a$ for each pair a, b in R.

(2) Addition is associative: $a + (b + c) = (a + b) + c$ for $a, b, c \in R$.

(3) There exists an additive identity, denoted by 0, such that $a + 0 = a$ for each $a \in R$.

(4) For each $a \in R$ there exists an additive inverse, denoted by $-a$, such that $a + (-a) = 0$.

(5) Multiplication is associative: $a(bc) = (ab)c$ for $a, b, c \in R$.

(6) Multiplication is left and right distributive over addition: $a(b + c) = ab + ac$ and $(b + c)a = ba + ca$ for $a, b, c \in R$.

If in addition

(7) Multiplication is commutative: $ab = ba$ for each pair a, b in R.

then R is a *commutative ring*.

Further if

(8) There exists a multiplicative identity denoted by 1 such that $a \cdot 1 = a$ and $1 \cdot a = a$ for each a in R.

then R is a *ring with identity*.

If R satisfies (1) through (8) then R is a commutative ring with an identity.

A set G with one operation, $+$, on it satisfying axioms (1) through (4) is called an *abelian group*. We will discuss these further later in the chapter.

The numbers systems $\mathbb{Z}, \mathbb{Q}, \mathbb{R}, \mathbb{C}$ are all commutative rings with identity.

A ring R with only one element is called *trivial*. A ring R with identity is trivial if and only if $0 = 1$.

A *finite ring* is a ring R with only finitely many elements in it. Otherwise R is an infinite ring. $\mathbb{Z}, \mathbb{Q}, \mathbb{R}, \mathbb{C}$ are all infinite rings. Examples of finite rings are given by the integers modulo n, $\mathbb{Z}_n$, with $n > 1$. The ring $\mathbb{Z}_n$ consists of the elements $0, 1, 2, \dots, n-1$ with addition and multiplication done modulo n. That is, for example $4 \cdot 3 = 12 = 2$ modulo 5. Hence in $\mathbb{Z}_5$ we have $4 \cdot 3 = 2$. The rings $\mathbb{Z}_n$ are all finite commutative rings with identity.

To give examples of rings without an identity consider the set $n\mathbb{Z} = \{nz : z \in \mathbb{Z}\}$ consisting of all multiples of the fixed integer n. It is an easy verification (see exercises) that this forms a ring under the same addition and multiplication as in $\mathbb{Z}$ but that there is no identity for multiplication. Hence for each $n \in \mathbb{Z}$ with $n > 1$ we get an infinite commutative ring without an identity.

To obtain examples of noncommutative rings we consider matrices. Let $M_2(\mathbb{Z})$ be the set of 2×2 matrices with integral entries. Addition of matrices is done componentwise, that is

$$\begin{pmatrix} a_1 & b_1 \\ c_1 & d_1 \end{pmatrix} + \begin{pmatrix} a_2 & b_2 \\ c_2 & d_2 \end{pmatrix} = \begin{pmatrix} a_1 + a_2 & b_1 + b_2 \\ c_1 + c_2 & d_1 + d_2 \end{pmatrix}$$

while multiplication is matrix multiplication

$$\begin{pmatrix} a_1 & b_1 \\ c_1 & d_1 \end{pmatrix} \cdot \begin{pmatrix} a_2 & b_2 \\ c_2 & d_2 \end{pmatrix} = \begin{pmatrix} a_1a_2 + b_1c_2 & a_1b_2 + b_1d_2 \\ c_1a_2 + d_1c_2 & c_1b_2 + d_1d_2 \end{pmatrix}.$$

Then again it is an easy verification (see exercises) that $M_2(\mathbb{Z})$ forms a ring. Further since matrix multiplication is noncommutative this forms a noncommutative ring. However the identity matrix does form a multiplicative identity for it. $M_2(n\mathbb{Z})$ with $n > 1$ provides an example of an infinite noncommutative ring without an identity.

Finally $M_2(\mathbb{Z}_n)$ for $n > 1$ will give an example of a finite noncommutative ring.

1.3 Integral Domains and Fields

Our basic number systems have the property that if $ab = 0$ then either $a = 0$ or $b = 0$. However this is not necessarily true in the modular rings. For example $2 \cdot 3 = 0$ in $\mathbb{Z}_6$.

Definition 1.3.1. A *zero divisor* in a ring R is an element $a \in R$ with $a \neq 0$ such that there exists an element $b \neq 0$ with $ab = 0$. A commutative ring with an identity $1 \neq 0$ and with no zero divisors is called an *integral domain*. Notice that having no zero divisors is equivalent to the fact that if $ab = 0$ in R then either $a = 0$ or $b = 0$.

Hence $\mathbb{Z}, \mathbb{Q}, \mathbb{R}, \mathbb{C}$ are all integral domains but from the example above $\mathbb{Z}_6$ is not. In general we have the following.

Theorem 1.3.2. *$\mathbb{Z}_n$ is an integral domain if and only if n is a prime.*

Proof. First of all notice that under multiplication modulo n an element m is 0 if and only if n divides m. We will make this precise shortly. Recall further Euclid's lemma which says that if a prime p divides a product ab then p divides a or p divides b.

Now suppose that n is a prime and $ab = 0$ in $\mathbb{Z}_n$. Then n divides ab. From Euclid's lemma it follows that n divides a or n divides b. In the first case $a = 0$ in $\mathbb{Z}_n$ while in the second $b = 0$ in $\mathbb{Z}_n$. It follows that there are no zero divisors in $\mathbb{Z}_n$ and since $\mathbb{Z}_n$ is a commutative ring with an identity it is an integral domain.

Conversely suppose $\mathbb{Z}_n$ is an integral domain. Suppose that n is not prime. Then $n = ab$ with $1 < a < n$, $1 < b < n$. It follows that $ab = 0$ in $\mathbb{Z}_n$ with neither a nor b being zero. Therefore they are zero divisors which is a contradiction. Hence n must be prime. □

In $\mathbb{Q}$ every nonzero element has a multiplicative inverse. This is not true in $\mathbb{Z}$ where only the elements $-1, 1$ have multiplicative inverses within $\mathbb{Z}$.

Definition 1.3.3. A *unit* in a ring R with identity is an element a which has a multiplicative inverse, that is an element b such that $ab = ba = 1$. If a is a unit in R we denote its inverse by a^{-1}.

Hence every nonzero element of $\mathbb{Q}$ and of $\mathbb{R}$ and of $\mathbb{C}$ is a unit but in $\mathbb{Z}$ the only units are ± 1. In $M_2(\mathbb{R})$ the units are precisely those matrices that have nonzero determinant while in $M_2(\mathbb{Z})$ the units are those integral matrices that have determinant ± 1.

Definition 1.3.4. A *field* F is a commutative ring with an identity $1 \neq 0$ where every nonzero element is a unit.

The rationals $\mathbb{Q}$, the reals $\mathbb{R}$ and the complexes $\mathbb{C}$ are all fields. If we relax the commutativity requirement and just require that in the ring R with identity each nonzero element is a unit then we get a *skew field* or *division ring*.

Lemma 1.3.5. *If F is a field then F is an integral domain.*

Proof. Since a field F is already a commutative ring with an identity we must only show that there are no zero divisors in F.

Suppose that $ab = 0$ with $a \neq 0$. Since F is a field and a is nonzero it has an inverse a^{-1}. Hence

$$a^{-1}(ab) = a^{-1}0 = 0 \implies (a^{-1}a)b = 0 \implies b = 0.$$

Therefore F has no zero divisors and must be an integral domain. □

Recall that $\mathbb{Z}_n$ was an integral domain only when n was a prime. This turns out to also be necessary and sufficient for $\mathbb{Z}_n$ to be a field.

Theorem 1.3.6. *$\mathbb{Z}_n$ is a field if and only if n is a prime.*

Proof. First suppose that $\mathbb{Z}_n$ is a field. Then from Lemma 1.3.5 it is an integral domain, so from Theorem 1.3.2 n must be a prime.

Conversely suppose that n is a prime. We must show that $\mathbb{Z}_n$ is a field. Since we already know that $\mathbb{Z}_n$ is an integral domain we must only show that each nonzero element of $\mathbb{Z}_n$ is a unit. Here we need some elementary facts from number theory. If a, b are integers we use the notation $a|b$ to indicate that a divides b.

Recall that given nonzero integers a, b their *greatest common divisor* or GCD $d > 0$ is a positive integer which is a common divisor, that is $d|a$ and $d|b$, and if d_1 is any other common divisor then $d_1|d$. We denote the greatest common divisor of a, b by either $\gcd(a, b)$ or (a, b). It can be proved that given nonzero integers a, b their GCD exists, is unique and can be characterized as the least positive linear combination of a and b. If the GCD of a and b is 1 then we say that a and b are *relatively prime* or *coprime*. This is equivalent to being able to express 1 as a linear combination of a and b.

Now let $a \in \mathbb{Z}_n$ with n prime and $a \neq 0$. Since $a \neq 0$ we have that n does not divide a. Since n is prime it follows that a and n must be relatively prime, $(a, n) = 1$. From the number theoretic remarks above we then have that there exist x, y with

$$ax + ny = 1.$$

However in $\mathbb{Z}_n$ the element $ny = 0$ and so in $\mathbb{Z}_n$ we have

$$ax = 1.$$

Therefore a has a multiplicative inverse in $\mathbb{Z}_n$ and is hence a unit. Since a was an arbitrary nonzero element we conclude that $\mathbb{Z}_n$ is a field. □

The theorem above is actually a special case of a more general result from which Theorem 1.3.6 could also be obtained.

Theorem 1.3.7. *Each finite integral domain is a field.*

Proof. Let F be a finite integral domain. We must show that F is a field. It is clearly sufficient to show that each nonzero element of F is a unit. Let

$$\{0, 1, r_1, \ldots, r_n\}$$

be the elements of F. Let r_i be a fixed nonzero element and multiply each element of F by r_i on the left. Now

$$\text{if } r_i r_j = r_i r_k \text{ then } r_i(r_j - r_k) = 0.$$

Since $r_i \neq 0$ it follows that $r_j - r_k = 0$ or $r_j = r_k$. Therefore all the products $r_i r_j$ are distinct. Hence

$$R = \{0, 1, r_1, \ldots, r_n\} = r_i R = \{0, r_i, r_i r_1, \ldots, r_i r_n\}.$$

Hence the identity element 1 must be in the right-hand list, that is there is an r_j such that $r_i r_j = 1$. Therefore r_i has a multiplicative inverse and is hence a unit. Therefore F is a field. □

1.4 Subrings and Ideals

A very important concept in algebra is that of a *substructure* that is a subset having the same structure as the superset.

Definition 1.4.1. A *subring* of a ring R is a nonempty subset S that is also a ring under the same operations as R. If R is a field and S also a field then its a *subfield.*

If $S \subset R$ then S satisfies the same basic axioms, associativity and commutativity of addition for example. Therefore S will be a subring if it is nonempty and closed under the operations, that is closed under addition, multiplication and taking additive inverses.

Lemma 1.4.2. *A subset S of a ring R is a subring if and only if S is nonempty and whenever $a, b \in S$ we have $a + b \in S$, $a - b \in S$ and $ab \in S$.*

Example 1.4.3. Show that if $n > 1$ the set $n\mathbb{Z}$ is a subring of $\mathbb{Z}$. Here clearly $n\mathbb{Z}$ is nonempty. Suppose $a = nz_1, b = nz_2$ are two element of $n\mathbb{Z}$. Then

$$\begin{aligned} a + b &= nz_1 + nz_2 = n(z_1 + z_2) \in n\mathbb{Z} \\ a - b &= nz_1 - nz_2 = n(z_1 - z_2) \in n\mathbb{Z} \\ ab &= nz_1 \cdot nz_2 = n(nz_1 z_2) \in n\mathbb{Z}. \end{aligned}$$

Therefore $n\mathbb{Z}$ is a subring.

Example 1.4.4. Show that the set of real numbers of the form

$$S = \{u + v\sqrt{2} : u, v \in \mathbb{Q}\}$$

is a subring of $\mathbb{R}$.

Here $1 + \sqrt{2} \in S$, so S is nonempty. Suppose $a = u_1 + v_1\sqrt{2}, b = u_2 + v_2\sqrt{2}$ are two element of S. Then

$$\begin{aligned} a + b &= (u_1 + v_1\sqrt{2}) + (u_2 + v_2\sqrt{2}) = u_1 + u_2 + (v_1 + v_2)\sqrt{2} \in S \\ a - b &= (u_1 + v_1\sqrt{2}) - (u_2 + v_2\sqrt{2}) = u_1 - u_2 + (v_1 - v_2)\sqrt{2} \in S \\ a \cdot b &= (u_1 + v_1\sqrt{2}) \cdot (u_2 + v_2\sqrt{2}) = (u_1u_2 + 2v_1v_2) + (u_1v_2 + v_1u_2)\sqrt{2} \in S. \end{aligned}$$

Therefore S is a subring.

We will see this example later as an algebraic number field.

In the following we are especially interested in special types of subrings called *ideals*.

Definition 1.4.5. Let R be a ring and $I \subset R$. Then I is a (two-sided) *ideal* if the following properties holds:

(1) I is nonempty.

(2) If $a, b \in I$ then $a \pm b \in I$.

(3) If $a \in I$ and r is any element of R then $ra \in I$ and $ar \in I$.

We denote the fact that I forms an ideal in R by $I \triangleleft R$.

Notice that if $a, b \in I$, then from (3) we have $ab \in I$ and $ba \in I$. Hence I forms a subring, that is each ideal is also a subring. $\{0\}$ and the whole ring R are trivial ideals of R.

If we assume that in (3) only $ra \in I$ then I is called a left ideal. Analogously we define a right ideal.

Lemma 1.4.6. *Let R be a commutative ring and $a \in R$. Then the set*

$$\langle a \rangle = aR = \{ar : r \in R\}$$

is an ideal of R.

This ideal is called the *principal ideal generated by a.*

Proof. We must verify the three properties of the definition. Since $a \in R$ we have that aR is nonempty. If $u = ar_1, v = ar_2$ are two elements of aR then

$$u \pm v = ar_1 \pm ar_2 = a(r_1 \pm r_2) \in aR$$

so (2) is satisfied.

Finally let $u = ar_1 \in aR$ and $r \in R$. Then

$$ru = rar_1 = a(rr_1) \in aR \quad \text{and} \quad ur = ar_1r = a(r_1r) \in aR.$$ □

Recall that $a \in \langle a \rangle$ if R has an identity.

Notice that if $n \in \mathbb{Z}$ then the principal ideal generated by n is precisely the ring $n\mathbb{Z}$, that we have already examined. Hence for each $n > 1$ the subring $n\mathbb{Z}$ is actually an ideal. We can show more.

Theorem 1.4.7. *Any subring of $\mathbb{Z}$ is of the form $n\mathbb{Z}$ for some n. Hence each subring of $\mathbb{Z}$ is actually a principal ideal.*

Proof. Let S be a subring of $\mathbb{Z}$. If $S = \{0\}$ then $S = 0\mathbb{Z}$ so we may assume that S has nonzero elements. Since S is a subring if it has nonzero elements it must have positive elements (since it has the additive inverse of any element in it).

Let S^+ be the set of positive elements in S. From the remarks above this is a nonempty set and so there must be a least positive element n. We claim that $S = n\mathbb{Z}$.

Let m be a positive element in S. By the division algorithm

$$m = qn + r,$$

where either $r = 0$ or $0 < r < n$. Suppose that $r \neq 0$. Then

$$r = m - qn.$$

Now $m \in S$ and $n \in S$. Since S is a subring it is closed under addition so that $qn \in S$. But S is a subring so $m - qn \in S$. It follows that $r \in S$. But this is a contradiction since n was the least positive element in S. Therefore $r = 0$ and $m = qn$. Hence each positive element in S is a multiple of n.

Now let m be a negative element of S. Then $-m \in S$ and $-m$ is positive. Hence $-m = qn$ and thus $m = (-q)n$. Therefore every element of S is a multiple of n and so $S = n\mathbb{Z}$.

It follows that every subring of $\mathbb{Z}$ is of this form and therefore every subring of $\mathbb{Z}$ is an ideal. □

We mention that this is true in $\mathbb{Z}$ but not always true. For example $\mathbb{Z}$ is a subring of $\mathbb{Q}$ but not an ideal.

An extension of the proof of Lemma 1.4.2 gives the following. We leave the proof as an exercise.

Lemma 1.4.8. *Let R be a commutative ring and $a_1, \dots, a_n \in R$ be a finite set of elements in R. Then the set*

$$\langle a_1, \dots, a_n \rangle = \{r_1 a_1 + r_2 a_2 + \dots + r_n a_n : r_i \in R\}$$

is an ideal of R.

This ideal is called the *ideal generated by* $a_1, \dots, a_n$.

Recall that $a_1, \dots, a_n$ are in $\langle a_1, \dots, a_n \rangle$ if R has an identity.

Theorem 1.4.9. *Let R be a commutative ring with an identity $1 \neq 0$. Then R is a field if and only if the only ideals in R are $\{0\}$ and R.*

Proof. Suppose that R is a field and $I \lhd R$ is an ideal. We must show that either $I = \{0\}$ or $I = R$. Suppose that $I \neq \{0\}$ then we must show that $I = R$.

Since $I \neq \{0\}$ there exists an element $a \in I$ with $a \neq 0$. Since R is a field this element a has an inverse a^{-1}. Since I is an ideal it follows that $a^{-1}a = 1 \in I$. Let $r \in R$ then, since $1 \in I$, we have $r \cdot 1 = r \in I$. Hence $R \subset I$ and hence $R = I$.

Conversely suppose that R is a commutative ring with an identity whose only ideals are $\{0\}$ and R. We must show that R is a field or equivalently that every nonzero element of R has a multiplicative inverse.

Let $a \in R$ with $a \neq 0$. Since R is a commutative ring and $a \neq 0$, the principal ideal aR is a nontrivial ideal in R. Hence $aR = R$. Therefore the multiplicative identity $1 \in aR$. It follows that there exists an $r \in R$ with $ar = 1$. Hence a has a multiplicative inverse and R must be a field. □

1.5 Factor Rings and Ring Homomorphisms

Given an ideal I in a ring R we can build a new ring called the *factor ring* or *quotient ring* of R modulo I. The special condition on the subring I that $rI \subset I$ and $Ir \subset I$ for all $r \in R$, that makes it an ideal, is specifically to allow this construction to be a ring.

Definition 1.5.1. Let I be an ideal in a ring R. Then a *coset* of I is a subset of R of the form

$$r + I = \{r + i : i \in I\}$$

with r a fixed element of R.

Lemma 1.5.2. *Let I be an ideal in a ring R. Then the cosets of I partition R, that is any two cosets are either coincide or disjoint.*

We leave the proof to the exercises.

Now on the set of all cosets of an ideal we will build a new ring.

Theorem 1.5.3. *Let I be an ideal in a ring R. Let R/I be the set of all cosets of I in R, that is*

$$R/I = \{r + I : r \in R\}.$$

We define addition and multiplication on R/I in the following manner:

$$(r_1 + I) + (r_2 + I) = (r_1 + r_2) + I$$
$$(r_1 + I) \cdot (r_2 + I) = (r_1 \cdot r_2) + I.$$

Then R/I forms a ring called the factor ring of R modulo I. The zero element of R/I is $0 + I$ and the additive inverse of $r + I$ is $-r + I$.

Further if R is commutative then R/I is commutative and if R has an identity then R/I has an identity $1 + I$.

Proof. The proofs that R/I satisfies the ring axioms under the definitions above is straightforward. For example

$$(r_1 + I) + (r_2 + I) = (r_1 + r_2) + I = (r_2 + r_1) + I = (r_2 + I) + (r_1 + I)$$

and so addition is commutative.

What must be shown is that both addition and multiplication are well-defined. That is, if

$$r_1 + I = r_1' + I \quad \text{and} \quad r_2 + I = r_2' + I$$

then

$$(r_1 + I) + (r_2 + I) = (r_1' + I) + (r_2' + I)$$

and

$$(r_1 + I) \cdot (r_2 + I) = (r_1' + I) \cdot (r_2' + I).$$

Now if $r_1 + I = r_1' + I$ then $r_1 \in r_1' + I$ and so $r_1 = r_1' + i_1$ for some $i_1 \in I$. Similarly if $r_2 + I = r_2' + I$ then $r_2 \in r_2' + I$ and so $r_2 = r_2' + i_2$ for some $i_2 \in I$. Then

$$(r_1 + I) + (r_2 + I) = (r_1' + i_1 + I) + (r_2' + i_2 + I) = (r_1' + I) + (r_2' + I)$$

since $i_1 + I = I$ and $i_2 + I = I$. Similarly

$$\begin{aligned}(r_1 + I) \cdot (r_2 + I) &= (r_1' + i_1 + I) \cdot (r_2' + i_2 + I) \\ &= r_1' \cdot r_2' + r_1' i_2 + r_2' i_1 + r_1' I + r_2' I + I \cdot I \\ &= (r_1' \cdot r_2') + I\end{aligned}$$

since all the other products are in the ideal I.

This shows that addition and multiplication are well-defined. It also shows why the ideal property is necessary. □

As an example let R be the integers $\mathbb{Z}$. As we have seen each subring is an ideal and of the form $n\mathbb{Z}$ for some natural number n. The factor ring $\mathbb{Z}/n\mathbb{Z}$ is called the *residue class ring modulo n* denoted $\mathbb{Z}_n$. Notice that we can take as cosets

$$0 + n\mathbb{Z}, 1 + n\mathbb{Z}, \ldots, (n-1) + n\mathbb{Z}.$$

Addition and multiplication of cosets is then just addition and multiplication modulo n, as we can see, that this is just a formalization of the ring $\mathbb{Z}_n$, that we have already looked at. Recall that $\mathbb{Z}_n$ is an integral domain if and only if n is prime and $\mathbb{Z}_n$ is a field for precisely the same n. If $n = 0$ then $\mathbb{Z}/n\mathbb{Z}$ is the same as $\mathbb{Z}$.

We now show that ideals and factor rings are closely related to certain mappings between rings.

Definition 1.5.4. Let R and S be rings. Then a mapping $f : R \to S$ is a *ring homomorphism* if

$$f(r_1 + r_2) = f(r_1) + f(r_2) \quad \text{for any } r_1, r_2 \in R$$
$$f(r_1 \cdot r_2) = f(r_1) \cdot f(r_2) \quad \text{for any } r_1, r_2 \in R.$$

In addition,

(1) f is an *epimorphism* if it is surjective.

(2) f is an *monomorphism* if it is injective.

(3) f is an *isomorphism* if it is bijective, that is both surjective and injective. In this case R and S are said to be *isomorphic rings* which we denote by $R \cong S$.

(4) f is an *endomorphism* if $R = S$, that is a ring homomorphism from a ring to itself.

(5) f is an *automorphism* if $R = S$ and f is an isomorphism.

Lemma 1.5.5. *Let R and S be rings and let $f : R \to S$ be a ring homomorphism. Then*

(1) $f(0) = 0$ *where the first* 0 *is the zero element of* R *and the second is the zero element of* S.

(2) $f(-r) = -f(r)$ *for any* $r \in R$.

Proof. We obtain $f(0) = 0$ from the equation $f(0) = f(0+0) = f(0) + f(0)$. Hence $0 = f(0) = f(r - r) = f(r + (-r)) = f(r) + f(-r)$, that is $f(-r) = -f(r)$. □

Definition 1.5.6. Let R and S be rings and let $f : R \to S$ be a *ring homomorphism*. Then the *kernel* of f is

$$\ker(f) = \{r \in R : f(r) = 0\}.$$

The *image* of f, denoted $\mathrm{im}(f)$, is the range of f within S. That is

$$\mathrm{im}(f) = \{s \in S : \text{there exists } r \in R \text{ with } f(r) = s\}.$$

Theorem 1.5.7 (ring isomorphism theorem). *Let R and S be rings and let*

$$f : R \to S$$

be a ring homomorphism. Then

(1) $\ker(f)$ *is an ideal in R,* $\mathrm{im}(f)$ *is a subring of S and*

$$R/\ker(f) \cong \mathrm{im}(f).$$

(2) *Conversely suppose that I is an ideal in a ring R. Then the map $f : R \to R/I$ given by $f(r) = r + I$ for $r \in R$ is a ring homomorphism whose kernel is I and whose image is R/I.*

The theorem says that the concepts of ideal of a ring and kernel of a ring homomorphism coincide, that is each ideal is the kernel of a homomorphism and the kernel of each ring homomorphism is an ideal.

Proof. Let $f : R \to S$ be a ring homomorphism and let $I = \ker(f)$. We show first that I is an ideal. If $r_1, r_2 \in I$ then $f(r_1) = f(r_2) = 0$. It follows from the homomorphism property that

$$f(r_1 \pm r_2) = f(r_1) \pm f(r_2) = 0 + 0 = 0$$
$$f(r_1 \cdot r_2) = f(r_1) \cdot f(r_2) = 0 \cdot 0 = 0.$$

Therefore I is a subring.

Now let $i \in I$ and $r \in R$. Then

$$f(r \cdot i) = f(r) \cdot f(i) = f(r) \cdot 0 = 0 \quad \text{and} \quad f(i \cdot r) = f(i) \cdot f(r) = 0 \cdot f(r) = 0$$

and hence I is an ideal.

Consider the factor ring R/I. Let $f^* : R/I \to \mathrm{im}(f)$ by $f^*(r + I) = f(r)$. We show that f^* is an isomorphism.

First we show that it is well-defined. Suppose that $r_1 + I = r_2 + I$ then $r_1 - r_2 \in I = \ker(f)$. It follows that $f(r_1 - r_2) = 0$ so $f(r_1) = f(r_2)$. Hence $f^*(r_1 + I) = f^*(r_2 + I)$ and the map f^* is well-defined.

Now

$$f^*((r_1 + I) + (r_2 + I)) = f^*((r_1 + r_2) + I) = f(r_1 + r_2)$$
$$= f(r_1) + f(r_2) = f^*(r_1 + I) + f^*(r_2 + I)$$

and

$$f^*((r_1 + I) \cdot (r_2 + I)) = f^*((r_1 \cdot r_2) + I) = f(r_1 \cdot r_2)$$
$$= f(r_1) \cdot f(r_2) = f^*(r_1 + I) \cdot f^*(r_2 + I).$$

Hence f^* is a homomorphism. We must now show that it is injective and surjective.

Suppose that $f^*(r_1 + I) = f^*(r_2 + I)$. Then $f(r_1) = f(r_2)$ so that $f(r_1 - r_2) = 0$. Hence $r_1 - r_2 \in \ker(f) = I$. Therefore $r_1 \in r_2 + I$ and thus $r_1 + I = r_2 + I$ and the map f^* is injective.

Finally let $s \in \mathrm{im}(f)$. Then there exists and $r \in R$ such that $f(r) = s$. Then $f^*(r + I) = s$ and the map f^* is surjective and hence an isomorphism. This proves the first part of the theorem.

To prove the second part let I be an ideal in R and R/I the factor ring. Consider the map $f : R \to R/I$ given by $f(r) = r + I$. From the definition of addition and multiplication in the factor ring R/I it is clear that this is a homomorphism. Consider the kernel of f. If $r \in \ker(f)$ then $f(r) = r + I = 0 = 0 + I$. This implies that $r \in I$ and hence the kernel of this map is exactly the ideal I completing the theorem. □

Theorem 1.5.7 is called the *ring isomorphism theorem* or the *first ring isomorphism theorem*. We mention that there is an analogous theorem for each algebraic structure. In particular for groups and vector spaces. We will mention the result for groups in Section 1.8.

1.6 Fields of Fractions

The integers are an integral domain and the rationals $\mathbb{Q}$ are a field that contains the integers. First we show that $\mathbb{Q}$ is the smallest field containing $\mathbb{Z}$.

Theorem 1.6.1. *The rationals $\mathbb{Q}$ are the smallest field containing the integers $\mathbb{Z}$. That is if $\mathbb{Z} \subset F \subset \mathbb{Q}$ with F a subfield of $\mathbb{Q}$ then $F = \mathbb{Q}$.*

Proof. Since $\mathbb{Z} \subset F$ we have $m, n \in F$ for any two integers m, n. Since F is a subfield, it is closed under taking division, that is taking multiplicative inverses and hence the fraction $\frac{m}{n} \in F$. Since each element of $\mathbb{Q}$ is such a fraction it follows that $\mathbb{Q} \subset F$. Since $F \subset \mathbb{Q}$ it follows that $F = \mathbb{Q}$. □

Notice that to construct the rationals from the integers we form all the fractions $\frac{m}{n}$ with $n \neq 0$ and where $\frac{m_1}{n_1} = \frac{m_2}{n_2}$ if $m_1 n_2 = n_1 m_2$. We then do the standard operations on fractions. If we start with any integral domain D we can mimic this construction to build a *field of fractions* from D that is the smallest field containing D.

Theorem 1.6.2. *Let D be an integral domain. Then there is a field F containing D, called the field of fractions for D, such that each element of F is a fraction from D, that is an element of the form $d_1 d_2^{-1}$ with $d_1, d_2 \in D$. Further F is unique up to isomorphism and is the smallest field containing D.*

Proof. The proof is just the mimicking of the construction of the rationals from the integers. Let

$$F^\star = \{(d_1, d_2) : d_1, d_2 \neq 0, d_1, d_2 \in D\}.$$

Define on $F^\star$ the equivalence relation

$$(d_1, d_2) = (d_1', d_2') \quad \text{if } d_1 d_2' = d_2 d_1'.$$

Let F be the set of equivalence classes and define addition and multiplication in the usual manner as for fractions where the result is the equivalence class.

$$(d_1, d_2) + (d_3, d_4) = (d_1 d_4 + d_2 d_3, d_2 d_4)$$
$$(d_1, d_2) \cdot (d_3, d_4) = (d_1 d_3, d_2 d_4).$$

It is now straightforward to verify the ring axioms for F. The inverse of $(d_1, 1)$ is $(1, d_1)$ for $d_1 \neq 0$ in D.

As with $\mathbb{Z}$ we identify the elements of F as fractions $\frac{d_1}{d_2}$.

The proof that F is the smallest field containing D is the same as for $\mathbb{Q}$ from $\mathbb{Z}$. □

As examples we have that $\mathbb{Q}$ is the field of fractions for $\mathbb{Z}$. A familiar but less common example is the following.

Let $\mathbb{R}[x]$ be the set of polynomials over the real numbers $\mathbb{R}$. It can be shown that $\mathbb{R}[x]$ forms an integral domain. The field of fractions consists of all formal functions $\frac{f(x)}{g(x)}$ where $f(x), g(x)$ are real polynomials with $g(x) \neq 0$. The corresponding field of fractions is called the *field of rational functions* over $\mathbb{R}$ and is denoted $\mathbb{R}(x)$.

1.7 Characteristic and Prime Rings

We saw in the last section that $\mathbb{Q}$ is the smallest field containing the integers. Since any subfield of $\mathbb{Q}$ must contain the identity, it follows that any nontrivial subfield of $\mathbb{Q}$ must contain the integers and hence be all of $\mathbb{Q}$. Therefore $\mathbb{Q}$ has no nontrivial subfields. We say that $\mathbb{Q}$ is a *prime field*.

Definition 1.7.1. A field F is a prime field if F contains no nontrivial subfields.

Lemma 1.7.2. *Let K be any field. Then K contains a prime field F as a subfield.*

Proof. Let K_1, K_2 be subfields of K. If $k_1, k_2 \in K_1 \cap K_2$ then $k_1 \pm k_2 \in K_1$ since K_1 is a subfield and $k_1 \pm k_2 \in K_2$ since K_2 is a subfield. Therefore $k_1 \pm k_2 \in K_1 \cap K_2$. Similarly $k_1 k_2^{-1} \in K_1 \cap K_2$. It follows that $K_1 \cap K_2$ is again a subfield.

Now let F be the intersection of all subfields of K. From the argument above F is a subfield and the only nontrivial subfield of F is itself. Hence F is a prime field. □

Definition 1.7.3. Let R be a commutative ring with an identity $1 \neq 0$. The smallest positive integer n such that $n \cdot 1 = 1 + 1 + \cdots + 1 = 0$ is called the *characteristic* of R. If there is no such n, then R has characteristic 0. We denote the characteristic by $\operatorname{char}(R)$.

Notice first that the characteristic of $\mathbb{Z}, \mathbb{Q}, \mathbb{R}$ are all zero. Further the characteristic of $\mathbb{Z}_n$ is n.

Theorem 1.7.4. *Let R be an integral domain. Then the characteristic of R is either 0 or a prime. In particular the characteristic of a field is zero or a prime.*

Proof. Suppose that R is an integral domain and $\operatorname{char}(R) = n \neq 0$. Suppose that $n = mk$ with $1 < m < n$, $1 < k < n$. Then $n \cdot 1 = 0 = (m \cdot 1)(k \cdot 1)$. Since R is an integral domain we have no zero divisors and hence $m \cdot 1 = 0$ or $k \cdot 1 = 0$. However this is a contradiction since n is the least positive integer such that $n \cdot 1 = 0$. Therefore n must be a prime. □

We have seen that every field contains a prime field. We extend this.

Definition 1.7.5. A commutative ring R with an identity $1 \neq 0$ is a *prime ring* if the only subring containing the identity is the whole ring.

Clearly both the integers $\mathbb{Z}$ and the modular integers $\mathbb{Z}_n$ are prime rings. In fact up to isomorphism they are the only prime rings.

Theorem 1.7.6. *Let R be a prime ring. If $\operatorname{char}(R) = 0$ then $R \cong \mathbb{Z}$, while if $\operatorname{char}(R) = n > 0$ then $R \cong \mathbb{Z}_n$.*

Proof. Suppose that $\operatorname{char}(R) = 0$. Let $S = \{r = m \cdot 1 : r \in R, m \in \mathbb{Z}\}$. Then S is a subring of R containing the identity (see the exercises) and hence $S = R$. However the map $m \cdot 1 \to m$ gives an isomorphism from S to $\mathbb{Z}$. It follows that R is isomorphic to $\mathbb{Z}$.

If $\operatorname{char}(R) = n > 0$ the proof is identical. Since $n \cdot 1 = 0$ the subring S of R defined above is all of R and isomorphic to $\mathbb{Z}_n$. □

Theorem 1.7.6 can be extended to fields with $\mathbb{Q}$ taking the place of $\mathbb{Z}$ and $\mathbb{Z}_p$, with p a prime, taking the place of $\mathbb{Z}_n$.

Theorem 1.7.7. *Let K be a prime field. If K has characteristic 0 then $K \cong \mathbb{Q}$ while if K has characteristic p then $K \cong \mathbb{Z}_p$.*

Proof. The proof is identical to that of Theorem 1.7.6; however we consider the smallest subfield K_1 of K containing S. □

We mention that there can be infinite fields of characteristic p. Consider for example the field of fractions of the polynomial ring $\mathbb{Z}_p[x]$. This is the field of rational functions with coefficients in $\mathbb{Z}_p$.

We give a theorem on fields of characteristic p that will be important much later when we look at Galois theory.

Theorem 1.7.8. *Let K be a field of characteristic p. Then the mapping $\phi : K \to K$ given by $\phi(k) = k^p$ is an injective endomorphism of K. In particular $(a+b)^p = a^p + b^p$ for any $a, b \in K$.*

This mapping is called the Frobenius homomorphism of K.

Further if K is finite, ϕ is an automorphism.

Proof. We first show that ϕ is a homomorphism. Now

$$\phi(ab) = (ab)^p = a^p b^p = \phi(a)\phi(b).$$

We need a little more work for addition.

$$\phi(a+b) = (a+b)^p = \sum_{i=0}^{p} \binom{p}{i} a^i b^{p-i} = a^p + \sum_{i=1}^{p-1} \binom{p}{i} a^i b^{p-i} + b^p$$

by the binomial expansion which holds in any commutative ring. However

$$\binom{p}{i} = \frac{p(p_1)\cdots(p-i+1)}{i\cdot(i-1)\cdots 1}$$

and it is clear that $p | \binom{p}{i}$ for $1 \le i \le p-1$. Hence in K we have $\binom{p}{i}\cdot 1 = 0$ and so we have

$$\phi(a+b) = (a+b)^p = a^p + b^p = \phi(a) + \phi(b).$$

Therefore ϕ is a homomorphism.

Further ϕ is always injective. To see this suppose that $\phi(x) = \phi(y)$. Then

$$\phi(x-y) = 0 \implies (x-y)^p = 0.$$

But K is a field so there are no zero divisors so we must have $x - y = 0$ or $x = y$.

If K is finite and ϕ is injective it must also be surjective and hence an automorphism of K. □

1.8 Groups

We close this first chapter by introducing some basic definitions and results from group theory, that mirror the results, that were presented for rings and fields. We will look at group theory in more detail later in the book. Proofs will be given at that point.

Definition 1.8.1. A *group* G is a set with one binary operation (which we will denote by multiplication) such that

(1) The operation is associative.

(2) There exists an identity for this operation.

(3) Each $g \in G$ has an inverse for this operation.

If, in addition, the operation is commutative, the group G is called an *abelian group*. The *order* of G is the number of elements in G, denoted by $|G|$. If $|G| < \infty$, G is a *finite group* otherwise G is an *infinite group*.

Groups most often arise from invertible mappings of a set onto itself. Such mappings are called *permutations*.

Theorem 1.8.2. *The group of all permutations on a set A forms a group called the symmetric group on A which we denote by S_A. If A has more than 2 elements then S_A is nonabelian.*

Definition 1.8.3. Let G_1 and G_2 be groups. Then a mapping $f : G_1 \to G_2$ is a *(group) homomorphism* if

$$f(g_1 g_2) = f(g_1) f(g_2) \quad \text{for any } g_1, g_2 \in G_1.$$

As with rings we have further

(1) f is an *epimorphism* if it is surjective.

(2) f is an *monomorphism* if it is injective.

(3) f is an *isomorphism* if it is bijective, that is both surjective and injective. In this case G_1 and G_2 are said to be *isomorphic groups*, which we denote by $G_1 \cong G_2$.

(4) f is an *endomorphism* if $G_1 = G_2$, that is a homomorphism from a group to itself.

(5) f is an *automorphism* if $G_1 = G_2$ and f is an isomorphism.

Lemma 1.8.4. *Let G_1 and G_2 be groups and let $f : G_1 \to G_2$ be a homomorphism. Then*

(a) $f(1) = 1$ *where the first* 1 *is the identity element of G_1 and the second is the identity element of G_2.*

(b) $f(g^{-1}) = (f(g))^{-1}$ *for any $g \in G_1$.*

If A is a set, $|A|$ denotes the size of A.

Theorem 1.8.5. *If A_1 and A_2 are sets with $|A_1| = |A_2|$ then $S_{A_1} \cong S_{A_2}$. If $|A| = n$ with n finite we call S_A the symmetric group on n elements which we denote by S_n. Further we have $|S_n| = n!$.*

Subgroups are defined in an analogous manner to subrings. Special types of subgroups called *normal subgroups* take the place in group theory that ideals play in ring theory.

Definition 1.8.6. A subset H of a group G is a *subgroup* if $H \neq \emptyset$ and H forms a group under the same operation as G. Equivalently, H is a subgroup if $H \neq \emptyset$ and H is closed under the operation and inverses.

Definition 1.8.7. If H is a subgroup of a group G, then a *left coset* of H is a subset of G of the form $gH = \{gh : h \in H\}$. A *right coset* of H is a subset of G of the form $Hg = \{hg : h \in H\}$.

As with rings the cosets of a subgroup partition a group. We call the number of right cosets of a subgroup H in a group G then *index* of H in G, denoted $|G : H|$. One can prove that the number of right cosets is equal to the number of left cosets. For finite groups we have the following beautiful result called *Lagrange's theorem*.

Theorem 1.8.8 (Lagrange's theorem). *Let G be a finite group and H a subgroup. Then the order of H divides the order of G. In particular*

$$|G| = |H||G : H|.$$

Normal subgroups take the place of ideals in group theory.

Definition 1.8.9. A subgroup H of a group G is a *normal subgroup*, denoted $H \lhd G$, if every left coset of H is also a right coset, that is $gH = Hg$ for each $g \in G$. Note that this does not say that g and H commute elementwise, just that the subsets gH and Hg are the same. Equivalently H is normal if $g^{-1}Hg = H$ for any $g \in G$.

Normal subgroups allow us to construct factor groups just as ideals allowed us to construct factor rings.

Theorem 1.8.10. *Let H be a normal subgroup of a group G. Let G/H be the set of all cosets of H in G, that is*

$$G/H = \{gH : g \in G\}.$$

We define multiplication on G/H in the following manner

$$(g_1H)(g_2H) = g_1g_2H.$$

Then G/H forms a group called the factor group or quotient group of G modulo H. The identity element of G/H is $1H$ and the inverse of gH is $g^{-1}H$.

Further if G is abelian then G/H is also abelian.

Finally as with rings normal subgroups, factor groups are closely tied to homomorphisms.

Definition 1.8.11. Let G_1 and G_2 be groups and let $f : G_1 \to G_2$ be a homomorphism. Then the *kernel* of f, denoted $\ker(f)$, is

$$\ker(f) = \{g \in G_1 : f(g) = 1\}.$$

The *image* of f, denoted $\mathrm{im}(f)$, is the range of f within G_2. That is

$$\mathrm{im}(f) = \{h \in G_2 : \text{there exists } g \in G_1 \text{ with } f(g) = h\}.$$

Theorem 1.8.12 (group isomorphism theorem). *Let G_1 and G_2 be groups and let $f : G_1 \to G_2$ be a homomorphism. Then*

(1) $\ker(f)$ *is a normal subgroup in G_1.* $\mathrm{im}(f)$ *is a subgroup of G_2 and*

$$G_1/\ker(f) \cong \mathrm{im}(f).$$

(2) *Conversely suppose that H is a normal subgroup of a group G. Then the map $f : G \to G/H$ given by $f(g) = gH$ for $g \in G$ is a homomorphism whose kernel is H and whose image is G/H.*

1.9 Exercises

1. Let $\phi : K \to R$ be a homomorphism from a field K to a ring R. Show: Either $\phi(a) = 0$ for all $a \in K$ or ϕ is a monomorphism.
2. Let R be a ring and $M \neq \emptyset$ an arbitrary set. Show that the following are equivalent:

   (i) The ring of all mappings from M to R is a field.

   (ii) M contains only one element and R is a field.

3. Let π be a set of prime numbers. Define

$$\mathbb{Q}_\pi = \left\{ \frac{a}{b} : \text{ all prime divisors of } b \text{ are in } \pi \right\}.$$

   (i) Show that $\mathbb{Q}_\pi$ is a subring of $\mathbb{Q}$.
   (ii) Let R be a subring of $\mathbb{Q}$ and let $\frac{a}{b} \in R$ with coprime integers a, b. Show that $\frac{1}{b} \in R$.
   (iii) Determine all subrings R of $\mathbb{Q}$. (Hint: Consider the set of all prime divisors of denominators of reduced elements of R.)

4. Prove Lemma 1.5.2.

5. Let R be a commutative ring with an identity $1 \in R$. Let $\mathcal{A}$, $\mathcal{B}$ and $\mathcal{C}$ be ideals in R. $\mathcal{A} + \mathcal{B} := \{a + b : a \in \mathcal{A}, b \in \mathcal{B}\}$ and $\mathcal{A}\mathcal{B} := (\{ab : a \in \mathcal{A}, b \in \mathcal{B}\})$. Show:

   (i) $\mathcal{A} + \mathcal{B} \lhd R$, $\mathcal{A} + \mathcal{B} = (\mathcal{A} \cup \mathcal{B})$
   (ii) $\mathcal{A}\mathcal{B} = \{a_1b_1 + \cdots + a_nb_n : n \in \mathbb{N}, a_i \in \mathcal{A}, b_i \in \mathcal{B}\}$, $\mathcal{A}\mathcal{B} \subset \mathcal{A} \cap \mathcal{B}$
   (iii) $\mathcal{A}(\mathcal{B} + \mathcal{C}) = \mathcal{A}\mathcal{B} + \mathcal{A}\mathcal{C}$, $(\mathcal{A} + \mathcal{B})\mathcal{C} = \mathcal{A}\mathcal{B} + \mathcal{B}\mathcal{C}$, $(\mathcal{A}\mathcal{B})\mathcal{C} = \mathcal{A}(\mathcal{B}\mathcal{C})$
   (iv) $\mathcal{A} = R \Leftrightarrow \mathcal{A} \cap R^* \neq \emptyset$
   (v) $a, b \in R \Rightarrow \langle a \rangle + \langle b \rangle = \{xa + yb : x, y \in R\}$
   (vi) $a, b \in R \Rightarrow \langle a \rangle \langle b \rangle = \langle ab \rangle$. Here $\langle a \rangle = Ra = \{xa : x \in R\}$.

6. Solve the following congruence:

$$3x \equiv 5 \mod 7.$$

   Is this congruence also solvable mod 17?

7. Show that the set of 2×2 matrices over a ring R forms a ring.

8. Prove Lemma 1.4.8.

9. Prove that if R is a ring with identity and $S = \{r = m \cdot 1 : r \in R, m \in \mathbb{Z}\}$ then S is a subring of R containing the identity.

Chapter 2
Maximal and Prime Ideals

2.1 Maximal and Prime Ideals

In the first chapter we defined ideals I in a ring R and then the factor ring R/I of R modulo the ideal I. We saw further that if R is commutative then R/I is also commutative and if R has an identity then so does R/I. This raises further questions concerning the structure of factor rings. In particular we can ask under what conditions does R/I form an integral domain and under what conditions does R/I form a field. These questions lead us to define certain special properties of ideals, called prime ideals and maximal ideals.

For motivation let us look back at the integers $\mathbb{Z}$. Recall that each proper ideal in $\mathbb{Z}$ has the form $n\mathbb{Z}$ for some $n > 1$ and the resulting factor ring $\mathbb{Z}/n\mathbb{Z}$ is isomorphic to $\mathbb{Z}_n$. We proved the following result.

Theorem 2.1.1. *$\mathbb{Z}_n = \mathbb{Z}/n\mathbb{Z}$ is an integral domain if and only if $n = p$ a prime. Further $\mathbb{Z}_n$ is a field again if and only if $n = p$ is a prime.*

Hence for the integers $\mathbb{Z}$ a factor ring is a field if and only if it is an integral domain. We will see later that this is not true in general. However what is clear is that the special ideals $n\mathbb{Z}$ leading to integral domains and fields are precisely when n is a prime. We look at the ideals $p\mathbb{Z}$ with p a prime in two different ways and then use these in subsequent sections to give the general definitions. We first need a famous result, Euclid's lemma, from number theory. For integers a, b the notation $a|b$ means that a divides b.

Lemma 2.1.2 (Euclid). *If p is a prime and $p|ab$ then $p|a$ or $p|b$.*

Proof. Recall that the greatest common divisor or GCD of two integers a, b is an integer $d > 0$ such that d is a common divisor of both a and b and if d_1 is another common divisor of a and b then $d_1|d$. We express the GCD of a, b by $d = (a, b)$. It is known that for any two integers a, b their GCD exists and is unique and further is the least positive linear combination of a and b, that is the least positive integer of the form $ax + by$ for integers x, y. The integers a, b are *relatively prime* if their GCD is 1, $(a, b) = 1$. In this case 1 is a linear combination of a and b.

Now suppose $p|ab$ where p is a prime. If p does not divide a then since the only positive divisors of p are 1 and p it follows that $(a, p) = 1$. Hence 1 is expressible as a linear combination of a and p. That is $ax + py = 1$ for some integers x, y. Multiply through by b, so that

$$abx + pby = b.$$

Now $p|ab$ so $p|abx$ and $p|pby$. Therefore $p|abx + pby$, that is, $p|b$. □

We now recast this lemma in two different ways in terms of the ideal $p\mathbb{Z}$. Notice that $p\mathbb{Z}$ consists precisely of all the multiples of p. Hence $p|ab$ is equivalent to $ab \in p\mathbb{Z}$.

Lemma 2.1.3. *If p is a prime and $ab \in p\mathbb{Z}$ then $a \in p\mathbb{Z}$ or $b \in p\mathbb{Z}$.*

This conclusion will be taken as the definition of a *prime ideal* in the next section.

Lemma 2.1.4. *If p is a prime and $p\mathbb{Z} \subset n\mathbb{Z}$ then $n = 1$ or $n = p$. That is, every ideal in $\mathbb{Z}$ containing $p\mathbb{Z}$ with p a prime is either all of $\mathbb{Z}$ or $p\mathbb{Z}$.*

Proof. Suppose that $p\mathbb{Z} \subset n\mathbb{Z}$. Then $p \in n\mathbb{Z}$ so p is a multiple of n. Since p is a prime it follows easily that either $n = 1$ or $n = p$. □

In Section 2.3 the conclusion of this lemma will be taken as the definition of a *maximal ideal.*

2.2 Prime Ideals and Integral Domains

Motivated by Lemma 2.1.3 we make the following general definition for commutative rings R with identity.

Definition 2.2.1. Let R be a commutative ring. An ideal P in R with $P \neq R$ is a *prime ideal* if whenever $ab \in P$ with $a, b \in R$ then either $a \in P$ or $b \in P$.

This property of an ideal is precisely what is necessary and sufficient to make the factor ring R/I an integral domain.

Theorem 2.2.2. *Let R be a commutative ring with an identity $1 \neq 0$ and let P be a nontrivial ideal in R. Then P is a prime ideal if and only if the factor ring R/P is an integral domain.*

Proof. Let R be a commutative ring with an identity $1 \neq 0$ and let P be a prime ideal. We show that R/P is an integral domain. From the results in the last chapter we have that R/P is again a commutative ring with an identity. Therefore we must show that there are no zero divisors in R/P. Suppose that $(a + I)(b + I) = 0$ in R/P. The zero element in R/P is $0 + P$ and hence

$$(a + P)(b + P) = 0 = 0 + P \implies ab + P = 0 + P \implies ab \in P.$$

However P is a prime ideal so we must then have $a \in P$ or $b \in P$. If $a \in P$ then $a + P = P = 0 + P$ so $a + P = 0$ in R/P. The identical argument works if $b \in P$. Therefore there are no zero divisors in R/P and hence R/P is an integral domain.

Conversely suppose that R/P is an integral domain. We must show that P is a prime ideal. Suppose that $ab \in P$. Then $(a + P)(b + P) = ab + P = 0 + P$. Hence in R/P we have

$$(a + P)(b + P) = 0.$$

However R/P is an integral domain so it has no zero divisors. It follows that either $a + P = 0$ and hence $a \in P$ or $b + P = 0$ and $b \in P$. Therefore either $a \in P$ or $b \in P$ so P is a prime ideal. □

In a commutative ring R we can define a multiplication of ideals. We then obtain an exact analog of Euclid's lemma. Since R is commutative each ideal is 2-sided.

Definition 2.2.3. Let R be a commutative ring with an identity $1 \neq 0$ and let A and B be ideals in R. Define

$$AB = \{a_1 b_1 + \cdots + a_n b_n : a_i \in A, b_i \in B, n \in \mathbb{N}\}.$$

That is AB is the set of finite sums of products ab with $a \in A$ and $b \in B$.

Lemma 2.2.4. *Let R be a commutative ring with an identity $1 \neq 0$ and let A and B be ideals in R. Then AB is an ideal.*

Proof. We must verify that AB is a subring and that it is closed under multiplication from R. Le $r_1, r_2 \in AB$. Then

$$r_1 = a_1 b_1 + \cdots + a_n b_n \quad \text{for some } a_i \in A,\ b_i \in B$$

and

$$r_2 = a_1' b_1' + \cdots + a_m' b_m' \quad \text{for some } a_i' \in A,\ b_i' \in B.$$

Then

$$r_1 \pm r_2 = a_1 b_1 + \cdots + a_n b_n \pm a_1' b_1' \pm \cdots \pm a_m' b_m'$$

which is clearly in AB. Further

$$r_1 \cdot r_2 = a_1 b_1 a_1' b_1' + \cdots + a_n b_n a_m' b_m'.$$

Consider for example the first term $a_1 b_1 a_1' b_1'$. Since R is commutative this is equal to

$$(a_1 a_1')(b_1 b_1').$$

Now $a_1 a_1' \in A$ since A is a subring and $b_1 b_1' \in B$ since B is a subring. Hence this term is in AB. Similarly for each of the other terms. Therefore $r_1 r_2 \in AB$ and hence AB is a subring.

Now let $r \in R$ and consider rr_1. This is then

$$rr_1 = ra_1b_1 + \cdots + ra_nb_n.$$

Now $ra_i \in A$ for each i since A is an ideal. Hence each summand is in AB and then $rr_1 \in AB$. Therefore AB is an ideal. □

Lemma 2.2.5. *Let R be a commutative ring with an identity $1 \neq 0$ and let A and B be ideals in R. If P is a prime ideal in R then $AB \subset P$ implies that $A \subset P$ or $B \subset P$.*

Proof. Suppose that $AB \subset P$ with P a prime ideal and suppose that B is not contained in P. We show that $A \subset P$. Since $AB \subset P$ each product $a_ib_j \in P$. Choose a $b \in B$ with $b \notin P$ and let a be an arbitrary element of A. Then $ab \in P$. Since P is a prime ideal this implies either $a \in P$ or $b \in P$. But by assumption $b \notin P$ so $a \in P$. Since a was arbitrary we have $A \subset P$. □

2.3 Maximal Ideals and Fields

Now, motivated by Lemma 2.1.4 we define a maximal ideal.

Definition 2.3.1. Let R be a ring and I an ideal in R. Then I is a *maximal ideal* if $I \neq R$ and if J is an ideal in R with $I \subset J$ then $I = J$ or $J = R$.

If R is a commutative ring with an identity this property of an ideal I is precisely what is necessary and sufficient so that R/I is a field.

Theorem 2.3.2. *Let R be a commutative ring with an identity $1 \neq 0$ and let I be an ideal in R. Then I is a maximal ideal if and only if the factor ring R/I is a field.*

Proof. Suppose that R is a commutative ring with an identity $1 \neq 0$ and let I be an ideal in R. Suppose first that I is a maximal ideal and we show that the factor ring R/I is a field.

Since R is a commutative ring with an identity the factor ring R/I is also a commutative ring with an identity. We must show then that each nonzero element of R/I has a multiplicative inverse. Suppose then that $\overline{r} = r + I \in R/I$ is a nonzero element of R/I. It follows that $r \notin I$. Consider the set $\langle r, I \rangle = \{rx + i : x \in R, i \in I\}$. This is also an ideal (see exercises) called the ideal generated by r and I, denoted $\langle r, I \rangle$. Clearly $I \subset \langle r, I \rangle$ and since $r \notin I$ and $r = r \cdot 1 + 0 \in \langle r, I \rangle$ it follows that $\langle r, I \rangle \neq I$. Since I is a maximal ideal it follows that $\langle r, I \rangle = R$ the whole ring. Hence the identity element $1 \in \langle r, I \rangle$ and so there exist elements $x \in R$ and $i \in I$ such that $1 = rx + i$. But then $1 \in (r + I)(x + I)$ and so $1 + I = (r + I)(x + I)$.

Since $1 + I$ is the multiplicative identity of R/I is follows that $x + I$ is the multiplicative inverse of $r + I$ in R/I. Since $r + I$ was an arbitrary nonzero element of R/I it follows that R/I is a field.

Now suppose that R/I is a field for an ideal I. We show that I must be maximal. Suppose then that I_1 is an ideal with $I \subset I_1$ and $I \neq I_1$. We must show that I_1 is all of R. Since $I \neq I_1$ there exists an $r \in I_1$ with $r \notin I$. Therefore the element $r + I$ is nonzero in the factor ring R/I and since R/I is a field it must have a multiplicative inverse $x + I$. Hence $(r + I)(x + I) = rx + I = 1 + I$ and therefore there is an $i \in I$ with $1 = rx + i$. Since $r \in I_1$ and I_1 is an ideal we get that $rx \in I_1$. Further since $I \subset I_1$ it follows that $rx + i \in I_1$ and so $1 \in I_1$. If r_1 is an arbitrary element of R then $r_1 \cdot 1 = r_1 \in I_1$. Hence $R \subset I_1$ and so $R = I_1$. Therefore I is a maximal ideal. □

Recall that a field is already an integral domain. Combining this with the ideas of prime and maximal ideals we obtain:

Theorem 2.3.3. *Let R be a commutative ring with an identity $1 \neq 0$. Then each maximal ideal is a prime ideal*

Proof. Suppose that R is a commutative ring with an identity and I is a maximal ideal in R. Then from Theorem 2.3.2 we have that the factor ring R/I is a field. But a field is an integral domain so R/I is an integral domain. Therefore from Theorem 2.2.2 we have that I must be a prime ideal. □

The converse is not true in general. That is there are prime ideals that are not maximal. Consider for example $R = \mathbb{Z}$ the integers and $I = \{0\}$. Then I is an ideal and $R/I = \mathbb{Z}/\{0\} \cong \mathbb{Z}$ is an integral domain. Hence $\{0\}$ is a prime ideal. However $\mathbb{Z}$ is not a field so $\{0\}$ is not maximal. Note however that in the integers $\mathbb{Z}$ a proper ideal is maximal if and only if it is a prime ideal.

2.4 The Existence of Maximal Ideals

In this section we prove that in any ring R with an identity there do exist maximal ideals. Further given an ideal $I \neq R$ then there exists a maximal ideal I_0 such that $I \subset I_0$. To prove this we need three important equivalent results from logic and set theory.

First recall that a *partial order* $\leq$ on a set S is a reflexive, transitive relation on S. That is $a \leq a$ for all $a \in S$ and if $a \leq b, b \leq c$ then $a \leq c$. This is a "partial" order since there may exist elements $a \in S$ where neither $a \leq b$ nor $b \leq a$. If A is any set then it is clear that containment of subsets is a partial order on the power set $\mathcal{P}(A)$.

If $\leq$ is a partial order on a set M, then a *chain* on M is a subset $K \subset M$ such that $a, b \in K$ implies that $a \leq b$ or $b \leq a$. A chain on M is *bounded* if there exists an

$m \in M$ such that $k \leq m$ for all $k \in K$. The element m is called an *upper bound* for K. An element $m_0 \in M$ is *maximal* if whenever $m \in M$ with $m_0 \leq m$ then $m = m_0$. We now state the three important results from logic.

Zorn's lemma. *If each chain of M has an upper bound in M then there is at least one maximal element in M.*

Axiom of well-ordering. *Each set M can be well-ordered, such that each nonempty subset of M contains a least element.*

Axiom of choice. *Let $\{M_i : i \in I\}$ be a nonempty collection of nonempty sets. Then there is a mapping $f : I \to \bigcup_{i \in I} M_i$ with $f(i) \in M_i$ for all $i \in I$.*

The following can be proved.

Theorem 2.4.1. *Zorn's lemma, the axiom of well-ordering and the axiom of choice are all equivalent.*

We now show the existence of maximal ideals in commutative rings with identity.

Theorem 2.4.2. *Let R be a commutative ring with an identity $1 \neq 0$ and let I be an ideal in R with $I \neq R$. Then there exists a maximal ideal I_0 in R with $I \subset I_0$. In particular a ring with an identity contains maximal ideals.*

Proof. Let I be an ideal in the commutative ring R. We must show that there exists a maximal ideal I_0 in R with $I \subset I_0$.

Let
$$M = \{X : X \text{ is an ideal with } I \subset X \neq R\}.$$
Then M is partially ordered by containment. We want to show first that each chain in M has a maximal element. If $K = \{X_j : X_j \in M, j \in J\}$ is a chain let
$$X' = \bigcup_{j \in J} X_j.$$

If $a, b \in X'$ then there exists an $i, j \in J$ with $a \in X_i, b \in X_j$. Since K is a chain either $X_i \subset X_j$ or $X_j \subset X_i$. Without loss of generality suppose that $X_i \subset X_j$ so that $a, b \in X_j$. Then $a \pm b \in X_j \subset X'$ and $ab \in X_j \subset X'$ since X_j is an ideal. Further if $r \in R$ then $ra \in X_j \subset X'$ since X_j is an ideal. Therefore X' is an ideal in R.

Since $X_j \neq R$ it follows that $1 \notin X_j$ for all $j \in J$. Therefore $1 \notin X'$ and so $X' \neq R$. It follows that under the partial order of containment X' is an upper bound for K.

We now use Zorn's lemma. From the argument above we have that each chain has a maximal element. Hence for an ideal I the set M above has a maximal element. This maximal element I_0 is then a maximal ideal containing I. □

2.5 Principal Ideals and Principal Ideal Domains

Recall again that in the integers $\mathbb{Z}$ each ideal I is of the form $n\mathbb{Z}$ for some integer n. Hence in $\mathbb{Z}$ each ideal can be generated by a single element.

Lemma 2.5.1. *Let R be a commutative ring and $a_1, \ldots, a_n$ be elements of R. Then the set*

$$\langle a_1, \ldots, a_n \rangle = \{r_1 a_1 + \cdots + r_n a_n : r_i \in R\}$$

forms an ideal in R called the ideal generated by $a_1, \ldots, a_n$.

Proof. The proof is straightforward. Let

$$a = r_1 a_1 + \cdots + r_n a_n, \quad b = s_1 a_1 + \cdots + s_n a_n$$

with $r_1, \ldots, r_n, s_1, \ldots, s_n$ elements of R, be two elements of $\langle a_1, \ldots, a_n \rangle$. Then

$$a \pm b = (r_1 \pm s_1) a_1 + \cdots + (r_n \pm s_n) a_n \in \langle a_1, \ldots, a_n \rangle$$
$$ab = (r_1 s_1 a_1) a_1 + (r_1 s_2 a_1) a_2 + \cdots + (r_n s_n a_n) a_n \in \langle a_1, \ldots, a_n \rangle$$

so $\langle a_1, \ldots, a_n \rangle$ forms a subring. Further if $r \in R$ we have

$$ra = (r r_1) a_1 + \cdots + (r r_n) a_n \in \langle a_1, \ldots, a_n \rangle$$

and so $\langle a_1, \ldots, a_n \rangle$ is an ideal. □

Definition 2.5.2. Let R be a commutative ring. An ideal $I \subset R$ is a *principal ideal* if it has a single generator. That is

$$I = \langle a \rangle = aR \quad \text{for some } a \in R.$$

We now restate Theorem 1.4.7 of Chapter 1.

Theorem 2.5.3. *Every nonzero ideal in $\mathbb{Z}$ is a principal ideal.*

Proof. Every ideal I in $\mathbb{Z}$ is of the form $n\mathbb{Z}$. This is the principal ideal generated by n. □

Definition 2.5.4. A *principal ideal domain* or PID is an integral domain in which every ideal is principal.

Corollary 2.5.5. *The integers $\mathbb{Z}$ are a principal ideal domain.*

We mention that the set of polynomials $K[x]$ with coefficients from a field K is also a principal ideal domain. We will return to this in the next chapter.

Not every integral domain is a PID. Consider $K[x, y]$ the set of polynomials over K in two variables x, y. Let I consist of all the polynomials with zero constant term.

Lemma 2.5.6. *The set I in $K[x, y]$ as defined above is an ideal but not a principal ideal.*

Proof. We leave the proof that I forms an ideal to the exercises. To show that it is not a principal ideal suppose $I = \langle p(x, y) \rangle$. Now the polynomial $q(x) = x$ has zero constant term so $q(x) \in I$. Hence $p(x, y)$ cannot be a constant polynomial. Further if $p(x, y)$ had any terms with y in them there would be no way to multiply $p(x, y)$ by a polynomial $h(x, y)$ and obtain just x. Therefore $p(x, y)$ can contain no terms with y in them. But the same argument using $s(y) = y$ shows that $p(x, y)$ cannot have any terms with x in them. Therefore there can be no such $p(x, y)$ generating I and so I is not principal and $K[x, y]$ is not a principal ideal domain. □

2.6 Exercises

1. Consider the set $\langle r, I \rangle = \{rx + i : x \in R, i \in I\}$ where I is an ideal. Prove that this is also an ideal called the ideal generated by r and I, denoted $\langle r, I \rangle$.
2. Let R and S be commutative rings and let $\phi : R \to S$ be a ring epimorphism. Let $\mathcal{M}$ be a maximal ideal in R. Show:

   $\phi(\mathcal{M})$ is a maximal ideal in S if and only if $\ker(\phi) \subset \mathcal{M}$. Is $\phi(\mathcal{M})$ always a prime ideal of S?
3. Let $\mathcal{A}_1, \dots, \mathcal{A}_t$ be ideals of a commutative ring R. Let $\mathcal{P}$ be a prime ideal of R. Show:
   (i) $\bigcap_{i=1}^{t} \mathcal{A}_i \subset \mathcal{P} \Rightarrow \mathcal{A}_j \subset \mathcal{P}$ for at least one index j.
   (ii) $\bigcap_{i=1}^{t} \mathcal{A}_i = \mathcal{P} \Rightarrow \mathcal{A}_j = \mathcal{P}$ for at least one index j.
4. Which of the following ideals $\mathcal{A}$ are prime ideals of R? Which are maximal ideals?
   (i) $\mathcal{A} = (x)$, $R = \mathbb{Z}[x]$.
   (ii) $\mathcal{A} = (x^2)$, $R = \mathbb{Z}[x]$.
   (iii) $\mathcal{A} = (1 + \sqrt{5})$, $R = \mathbb{Z}[\sqrt{5}]$.
   (iv) $\mathcal{A} = (x, y)$, $R = \mathbb{Q}[x, y]$.
5. Let $w = \frac{1}{2}(1 + \sqrt{-3})$. Show that $\langle 2 \rangle$ is a prime ideal and even a maximal ideal of $\mathbb{Z}[w]$, but $\langle 2 \rangle$ is neither a prime ideal nor a maximal ideal of $\mathbb{Z}[i]$.
6. Let $R = \{\frac{a}{b} : a, b \in \mathbb{Z}, b \text{ odd}\}$. Show that R is a subring of $\mathbb{Q}$ and that there is only one maximal ideal $\mathcal{M}$ in R.
7. Let R be a ring with an identity. Let $x, y \in R$ and $x \neq 0$ not be a zero divisor. Further let $\langle x \rangle$ be a prime ideal with $\langle x \rangle \subset \langle y \rangle \neq R$. Show that $\langle x \rangle = \langle y \rangle$.
8. Consider $K[x, y]$ the set of polynomials over K in two variables x, y. Let I consist of all the polynomials with zero constant term. Prove that the set I is an ideal.

Chapter 3
Prime Elements and Unique Factorization Domains

3.1 The Fundamental Theorem of Arithmetic

The integers $\mathbb{Z}$ have served as much of our motivation for properties of integral domains. In the last chapter we saw that $\mathbb{Z}$ is a principal ideal domain and furthermore that prime ideals $\neq \{0\}$ are maximal. From the viewpoint of the multiplicative structure of $\mathbb{Z}$ and the viewpoint of classical number theory the most important property of $\mathbb{Z}$ is the *fundamental theorem of arithmetic*. This states that any integer $n \neq 0$ is uniquely expressible as a product of primes where uniqueness is up to ordering and the introduction of ± 1, that is units. In this chapter we show that this property is not unique to the integers and there are many other integral domains where this also holds. These are called unique factorization domains and we will present several examples. First we review the fundamental theorem of arithmetic, its proof and several other ideas from classical number theory.

Theorem 3.1.1 (fundamental theorem of arithmetic). *Given any integer $n \neq 0$ there is a factorization*

$$n = cp_1 p_2 \cdots p_k$$

where $c = \pm 1$ and $p_1, \dots, p_n$ are primes. Further this factorization is unique up to the ordering of the factors.

There are two main ingredients that go into the proof; induction and Euclid's lemma. We presented this in the last chapter. In turn however Euclid's lemma depends upon the existence of greatest common divisors and their linear expressibility. Therefore to begin we present several basic ideas from number theory.

The starting point for the theory of numbers is *divisibility*.

Definition 3.1.2. If a, b are integers we say that a *divides* b, or that a is a *factor* or *divisor* of b, if there exists an integer q such that $b = aq$. We denote this by $a|b$. b is then a *multiple* of a. If $b > 1$ is an integer whose only factors are $\pm 1, \pm b$ then b is a *prime*, otherwise $b > 1$ is *composite*.

The following properties of divisibility are straightforward consequences of the definition.

Lemma 3.1.3. *The following properties hold:*

(1) $a|b \Rightarrow a|bc$ *for any integer* c.

(2) $a|b$ *and* $b|c$ *implies* $a|c$.

(3) $a|b$ *and* $a|c$ *implies that* $a|(bx + cy)$ *for any integers* x, y.

(4) $a|b$ *and* $b|a$ *implies that* $a = \pm b$.

(5) *If* $a|b$ *and* $a > 0, b > 0$ *then* $a \le b$.

(6) $a|b$ *if and only if* $ca|cb$ *for any integer* $c \ne 0$.

(7) $a|0$ *for all* $a \in \mathbb{Z}$ *and* $0|a$ *only for* $a = 0$.

(8) $a| \pm 1$ *only for* $a = \pm 1$.

(9) $a_1|b_1$ *and* $a_2|b_2$ *implies that* $a_1a_2|b_1b_2$.

If b, c, x, y are integers then an integer $bx + cy$ is called a *linear combination* of b, c. Thus part (3) of Lemma 3.1.3 says that if a is a *common divisor* of b, c then a divides any linear combination of b and c.

Further, note that if $b > 1$ is a composite then there exists $x > 0$ and $y > 0$ such that $b = xy$ and from part (5) we must have $1 < x < b$, $1 < y < b$.

In ordinary arithmetic, given a, b we can always attempt to divide a into b. The next result called the *division algorithm* says that if $a > 0$ either a will divide b or the *remainder* of the division of b by a will be less than a.

Theorem 3.1.4 (division algorithm). *Given integers* a, b *with* $a > 0$ *then there exist unique integers* q *and* r *such that* $b = qa + r$ *where either* $r = 0$ *or* $0 < r < a$.

One may think of q and r as the *quotient* and *remainder* respectively when dividing b by a.

Proof. Given a, b with $a > 0$ consider the set

$$S = \{b - qa \ge 0 : q \in \mathbb{Z}\}.$$

If $b > 0$ then $b + a \ge 0$ and the sum is in S. If $b \le 0$ then there exists a $q > 0$ with $-qa < b$. Then $b + qa > 0$ and is in S. Therefore in either case S is nonempty. Hence S is a nonempty subset of $\mathbb{N} \cup \{0\}$ and therefore has a least element r. If $r \ne 0$ we must show that $0 < r < a$. Suppose $r \ge a$, then $r = a + x$ with $x \ge 0$ and $x < r$ since $a > 0$. Then $b - qa = r = a + x \Rightarrow b - (q+1)a = x$. This means that $x \in S$. Since $x < r$ this contradicts the minimality of r which is a contradiction. Therefore if $r \ne 0$ it follows that $0 < r < a$.

The only thing left is to show the uniqueness of q and r. Suppose $b = q_1a + r_1$ also. By the construction above r_1 must also be the minimal element of S. Hence $r_1 \le r$ and $r \le r_1$ so $r = r_1$. Now

$$b - qa = b - q_1a \implies (q_1 - q)a = 0$$

but since $a > 0$ it follows that $q_1 - q = 0$ so that $q = q_1$. □

The next idea that is necessary is the concept of *greatest common divisor.*

Definition 3.1.5. Given nonzero integers a, b their *greatest common divisor* or GCD $d > 0$ is a positive integer which is a common divisor, that is $d|a$ and $d|b$, and if d_1 is any other common divisor then $d_1|d$. We denote the greatest common divisor of a, b by either $\gcd(a, b)$ or (a, b).

Certainly, if a, b are nonzero integers with $a > 0$ and $a|b$ then $a = \gcd(a, b)$.

The next result says that given any nonzero integers they do have a greatest common divisor and it is unique.

Theorem 3.1.6. *Given nonzero integers a, b their GCD exists, is unique and can be characterized as the least positive linear combination of a and b.*

Proof. Given nonzero a, b consider the set

$$S = \{ax + by > 0 : x, y \in \mathbb{Z}\}.$$

Now $a^2 + b^2 > 0$ so S is a nonempty subset of $\mathbb{N}$ and hence has a least element $d > 0$. We show that d is the GCD.

First we must show that d is a common divisor. Now $d = ax + by$ and is the least such positive linear combination. By the division algorithm $a = qd + r$ with $0 \leq r < d$. Suppose $r \neq 0$. Then $r = a - qd = a - q(ax + by) = (1 - qx)a - qby > 0$. Hence r is a positive linear combination of a and b and therefore is in S. But then $r < d$ contradicting the minimality of d in S. It follows that $r = 0$ and so $a = qd$ and $d|a$. An identical argument shows that $d|b$ and so d is a common divisor of a and b. Let d_1 be any other common divisor of a and b. Then d_1 divides any linear combination of a and b and so $d_1|d$. Therefore d is the GCD of a and b.

Finally we must show that d is unique. Suppose d_1 is another GCD of a and b. Then $d_1 > 0$ and d_1 is a common divisor of a, b. Then $d_1|d$ since d is a GCD. Identically $d|d_1$ since d_1 is a GCD. Therefore $d = \pm d_1$ and then $d = d_1$ since they are both positive. □

If $(a, b) = 1$ then we say that a, b are *relatively prime*. It follows that a and b are relatively prime if and only if 1 is expressible as a linear combination of a and b. We need the following three results.

Lemma 3.1.7. *If $d = (a, b)$ then $a = a_1 d$ and $b = b_1 d$ with $(a_1, b_1) = 1$.*

Proof. If $d = (a, b)$ then $d|a$ and $d|b$. Hence $a = a_1 d$ and $b = b_1 d$. We have

$$d = ax + by = a_1 dx + b_1 dy.$$

Dividing both sides of the equation by d we obtain

$$1 = a_1 x + b_1 y.$$

Therefore $(a_1, b_1) = 1$. □

Lemma 3.1.8. *For any integer c we have that $(a, b) = (a, b + ac)$.*

Proof. Suppose $(a, b) = d$ and $(a, b + ac) = d_1$. Now d is the least positive linear combination of a and b. Suppose $d = ax + by$. d_1 is a linear combination of $a, b + ac$ so that

$$d_1 = ar + (b + ac)s = a(cs + r) + bs.$$

Hence d_1 is also a linear combination of a and b and therefore $d_1 \geq d$. On the other hand $d_1 | a$ and $d_1 | (b + ac)$ and so $d_1 | b$. Therefore $d_1 | d$ so $d_1 \leq d$. Combining these we must have $d_1 = d$. □

The next result, called the *Euclidean algorithm*, provides a technique for both finding the GCD of two integers and expressing the GCD as a linear combination.

Theorem 3.1.9 (Euclidean algorithm). *Given integers b and $a > 0$ with $a \nmid b$, form the repeated divisions*

$$\begin{aligned}
b &= q_1 a + r_1, \quad 0 < r_1 < a \\
a &= q_2 r_1 + r_2, \quad 0 < r_2 < r_1 \\
&\vdots \\
r_{n-2} &= q_n r_{n-1} + r_n, \quad 0 < r_n < r_{n-1} \\
r_{n-1} &= q_{n+1} r_n.
\end{aligned}$$

The last nonzero remainder r_n is the GCD of a, b. Further r_n can be expressed as a linear combination of a and b by successively eliminating the r_i's in the intermediate equations.

Proof. In taking the successive divisions as outlined in the statement of the theorem each remainder r_i gets strictly smaller and still nonnegative. Hence it must finally end with a zero remainder. Therefore there is a last nonzero remainder r_n. We must show that this is the GCD.

Now from Lemma 3.1.7 the gcd $(a, b) = (a, b - q_1 a) = (a, r_1) = (r_1, a - q_2 r_1) = (r_1, r_2)$. Continuing in this manner we have then that $(a, b) = (r_{n-1}, r_n) = r_n$ since r_n divides r_{n-1}. This shows that r_n is the GCD.

To express r_n as a linear combination of a and b notice first that

$$r_n = r_{n-2} - q_n r_{n-1}.$$

Substituting this in the immediately preceding division we get

$$r_n = r_{n-2} - q_n(r_{n-3} - q_{n-1}r_{n-2}) = (1 + q_n q_{n-1})r_{n-2} - q_n r_{n-3}.$$

Doing this successively we ultimately express r_n as a linear combination of a and b. □

Example 3.1.10. Find the GCD of 270 and 2412 and express it as a linear combination of 270 and 2412.

We apply the Euclidean algorithm

$$\begin{aligned} 2412 &= (8)(270) + 252 \\ 270 &= (1)(252) + 18 \\ 252 &= (14)(18). \end{aligned}$$

Therefore the last nonzero remainder is 18 which is the GCD. We now must express 18 as a linear combination of 270 and 2412.

From the first equation

$$252 = 2412 - (8)(270)$$

which gives in the second equation

$$270 = 2412 - (8)(270) + 18 \implies 18 = (-1)(2412) + (9)(270)$$

which is the desired linear combination.

The next result that we need is *Euclid's lemma*. We stated and proved this in the last chapter but we restate it here.

Lemma 3.1.11 (Euclid's lemma). *If p is a prime and $p|ab$ then $p|a$ or $p|b$.*

We can now prove the fundamental theorem of arithmetic. Induction suffices to show that there always exists such a decomposition into prime factors.

Lemma 3.1.12. *Any integer $n > 1$ can be expressed as a product of primes, perhaps with only one factor.*

Proof. The proof is by induction. $n = 2$ is prime so its true at the lowest level. Suppose that any integer $2 \le k < n$ can be decomposed into prime factors, we must show that n then also has a prime factorization.

If n is prime then we are done. Suppose then that n is composite. Hence $n = m_1 m_2$ with $1 < m_1 < n$, $1 < m_2 < n$. By the inductive hypothesis both m_1 and m_2 can be expressed as products of primes. Therefore n can, also using the primes from m_1 and m_2, completing the proof. □

Before we continue to the fundamental theorem we mention that the existence of a prime decomposition, unique or otherwise, can be used to prove that the set of primes is infinite. The proof we give goes back to Euclid and is quite straightforward.

Theorem 3.1.13. *There are infinitely many primes.*

Proof. Suppose that there are only finitely many primes $p_1, \dots, p_n$. Each of these is positive so we can form the positive integer

$$N = p_1 p_2 \cdots p_n + 1.$$

From Lemma 3.1.12 N has a prime decomposition. In particular there is a prime p which divides N. Then

$$p \mid (p_1 p_2 \cdots p_n + 1).$$

Since the only primes are assumed $p_1, p_2, \dots, p_n$ it follows that $p = p_i$ for some $i = 1, \dots, n$. But then $p \mid p_1 p_2 \cdots p_i \dots p_n c$ so p cannot divide $p_1 \cdots p_n + 1$ which is a contradiction. Therefore p is not one of the given primes showing that the list of primes must be endless. □

We can now prove the fundamental theorem of arithmetic.

Proof. We assume that $n \geq 1$. If $n \leq -1$ we use $c = -n$ and the proof is the same. The statement certainly holds for $n = 1$ with $k = 0$. Now suppose $n > 1$. From Lemma 3.1.12, n has a prime decomposition

$$n = p_1 p_2 \cdots p_m.$$

We must show that this is unique up to the ordering of the factors. Suppose then that n has another such factorization $n = q_1 q_2 \cdots q_k$ with the q_i all prime. We must show that $m = k$ and that the primes are the same. Now we have

$$n = p_1 p_2 \cdots p_m = q_1 \cdots q_k.$$

Assume that $k \geq m$. From

$$n = p_1 p_2 \cdots p_m = q_1 \cdots q_k$$

it follows that $p_1 \mid q_1 q_2 \cdots q_k$. From Lemma 3.1.11 then we must have that $p_1 \mid q_i$ for some i. But q_i is prime and $p_1 > 1$ so it follows that $p_1 = q_i$. Therefore we can eliminate p_1 and q_i from both sides of the factorization to obtain

$$p_2 \cdots p_m = q_1 \cdots q_{i-1} q_{i+1} \cdots q_k.$$

Continuing in this manner we can eliminate all the p_i from the left side of the factorization to obtain

$$1 = q_{m+1} \cdots q_k.$$

If $q_{m+1}, \dots, q_k$ were primes this would be impossible. Therefore $m = k$ and each prime p_i was included in the primes $q_1, \dots, q_m$. Therefore the factorizations differ only in the order of the factors, proving the theorem. □

3.2 Prime Elements, Units and Irreducibles

We now let R be an arbitrary integral domain and attempt to mimic the divisibility definitions and properties.

Definition 3.2.1. Let R be an integral domain.

(1) Suppose that $a, b \in R$. Then a is a *factor* or *divisor* of b if there exists a $c \in R$ with $b = ac$. We denote this, as in the integers, by $a|b$. If a is a factor of b then b is called a *multiple* of a.

(2) An element $a \in R$ is a *unit* if a has a multiplicative inverse within R, that is there exists an element $a^{-1} \in R$ with $aa^{-1} = 1$.

(3) A *prime element* of R is an element $p \neq 0$ such that p is not a unit and if $p|ab$ then $p|a$ or $p|b$.

(4) An *irreducible* in R is an element $c \neq 0$ such that c is not a unit and if $c = ab$ then a or b must be a unit.

(5) a and b in R are *associates* if there exists a unit $e \in R$ with $a = eb$.

Notice that in the integers $\mathbb{Z}$ the units are just ± 1. The set of prime elements coincides with the set of irreducible elements. In $\mathbb{Z}$ this are precisely the set of prime numbers. On the other hand if K is a field every nonzero element is a unit so in K there are no prime elements and no irreducible elements.

Recall that the modular rings $\mathbb{Z}_n$ are fields (and integral domains) when n is a prime. In general if n is not a prime then $\mathbb{Z}_n$ is a commutative ring with an identity and a unit is still an invertible element. We can characterize the units within $\mathbb{Z}_n$.

Lemma 3.2.2. *$a \in \mathbb{Z}_n$ is a unit if and only if $(a, n) = 1$.*

Proof. Suppose $(a, n) = 1$. Then there exist $x, y \in \mathbb{Z}$ such that $ax + ny = 1$. This implies that $ax \equiv 1 \mod n$ which in turn implies that $ax = 1$ in $\mathbb{Z}_n$ and therefore a is a unit.

Conversely suppose a is a unit in $\mathbb{Z}_n$. Then there is an $x \in \mathbb{Z}_n$ with $ax = 1$. In terms of congruence then

$$ax \equiv 1 \mod n \implies n|ax - 1 \implies ax - 1 = ny \implies ax - ny = 1.$$

Therefore 1 is a linear combination of a and n and so $(a, n) = 1$. □

If R is an integral domain then the set of units within R will form a group.

Lemma 3.2.3. *If R is a commutative ring with an identity then the set of units in R form an abelian group under ring multiplication. This is called the unit group of R denoted $U(R)$.*

Proof. The commutativity and associativity of $U(R)$ follow from the ring properties. The identity of $U(R)$ is the multiplicative identity of R while the ring multiplicative inverse for each unit is the group inverse. We must show that $U(R)$ is closed under ring multiplication. If $a \in R$ is a unit we denote its multiplicative inverse by a^{-1}. Now suppose $a, b \in U(R)$. Then a^{-1}, b^{-1} exist. It follows that

$$(ab)(b^{-1}a^{-1}) = a(bb^{-1})a^{-1} = aa^{-1} = 1.$$

Hence ab has an inverse, namely $b^{-1}a^{-1}$ ($= a^{-1}b^{-1}$ in a commutative ring) and hence ab is also a unit. Therefore $U(R)$ is closed under ring multiplication. □

In general irreducible elements are not prime. Consider for example the subring of the complex numbers (see exercises) given by

$$R = \mathbb{Z}[i\sqrt{5}] = \{x + i\sqrt{5}y : x, y \in \mathbb{Z}\}.$$

This is a subring of the complex numbers $\mathbb{C}$ and hence can have no zero divisors. Therefore R is an integral domain.

For an element $x + iy\sqrt{5} \in R$ define its *norm* by

$$N(x + iy\sqrt{5}) = |x + iy\sqrt{5}| = x^2 + 5y^2.$$

Since $x, y \in \mathbb{Z}$ it is clear that the norm of an element in R is a nonnegative integer. Further if $a \in R$ with $N(a) = 0$ then $a = 0$.

We have the following result concerning the norm.

Lemma 3.2.4. *Let R and N be as above. Then*

(1) $N(ab) = N(a)N(b)$ *for any elements* $a, b \in R$.

(2) *The units of R are those $a \in R$ with $N(a) = 1$. In R the only units are ± 1.*

Proof. The fact that the norm is multiplicative is straightforward and left to the exercises. If $a \in R$ is a unit then there exists a multiplicative inverse $b \in R$ with $ab = 1$. Then $N(ab) = N(a)N(b) = 1$. Since both $N(a)$ and $N(b)$ are nonnegative integers we must have $N(a) = N(b) = 1$.

Conversely suppose that $N(a) = 1$. If $a = x + iy\sqrt{5}$ then $x^2 + 5y^2 = 1$. Since $x, y \in \mathbb{Z}$ we must have $y = 0$ and $x^2 = 1$. Then $a = x = \pm 1$. □

Using this lemma we can show that R possesses irreducible elements that are not prime.

Lemma 3.2.5. *Let R be as above. Then $3 = 3 + i0\sqrt{5}$ is an irreducible element in R but 3 is not prime.*

Proof. Suppose that $3 = ab$ with $a, b \in R$ and a, b nonunits. Then $N(3) = 9 = N(a)N(b)$ with neither $N(a) = 1$ nor $N(b) = 1$. Hence both $N(a) = 3$ and $N(b) = 3$. Let $a = x + iy\sqrt{5}$. It follows that $x^2 + 5y^2 = 3$. Since $x, y \in \mathbb{Z}$ this is impossible. Therefore one of a or b must be a unit and 3 is an irreducible element.

We show that 3 is not prime in R. Let $a = 2 + i\sqrt{5}$ and $b = 2 - i\sqrt{5}$. Then $ab = 9$ and hence $3|ab$. Suppose $3|a$ so that $a = 3c$ for some $c \in R$. Then

$$9 = N(a) = N(3)N(c) = 9N(c) \implies N(c) = 1.$$

Therefore c is a unit in R and from Lemma 3.2.4 we get $c = \pm 1$. Hence $a = \pm 3$. This is a contradiction, so 3 does not divide a. An identical argument shows that 3 does not divide b. Therefore 3 is not a prime element in R. □

We now examine the relationship between prime elements and irreducibles.

Theorem 3.2.6. *Let R be an integral domain. Then*

(1) *Each prime element of R is irreducible.*

(2) *$p \in R$ is a prime element if and only if $p \neq 0$ and $\langle p \rangle = pR$ is a prime ideal.*

(3) *$p \in R$ is irreducible if and only if $p \neq 0$ and $\langle p \rangle = pR$ is maximal in the set of all principal ideals of R which are not equal to R.*

Proof. (1) Suppose that $p \in R$ is a prime element and $p = ab$. We must show that either a or b must be a unit. Now $p|ab$ so either $p|a$ or $p|b$. Without loss of generality we may assume that $p|a$, so $a = pr$ for some $r \in R$. Hence $p = ab = (pr)b = p(rb)$. However R is an integral domain so $p - prb = p(1 - rb) = 0$ implies that $1 - rb = 0$ and hence $rb = 1$. Therefore b is a unit and hence p is irreducible.

(2) Suppose that p is a prime element. Then $p \neq 0$. Consider the ideal pR and suppose that $ab \in pR$. Then ab is a multiple of p and hence $p|ab$. Since p is prime it follows that $p|a$ or $p|b$. If $p|a$ then $a \in pR$ while if $p|b$ then $b \in pR$. Therefore pR is a prime ideal.

Conversely suppose that pR is a prime ideal and suppose that $p = ab$. Then $ab \in pR$ so $a \in pR$ or $b \in pR$. If $a \in pR$ then $p|a$ and if $b \in pR$ then $p|b$ and so p is prime.

(3) Let p be irreducible then $p \neq 0$. Suppose that $pR \subset aR$ where $a \in R$. Then $p = ra$ for some $r \in R$. Since p is irreducible it follows that either a is a unit or r is a unit. If r is a unit we have $pR = raR = aR \neq R$ since p is not a unit. If a is a unit then $aR = R$ and $pR = rR \neq R$. Therefore pR is maximal in the set of principal ideals not equal to R.

Conversely suppose $p \neq 0$ and pR is a maximal ideal in the set of principal ideals $\neq R$. Let $p = ab$ with a not a unit. We must show that b is a unit. Since $aR \neq R$ and $pR \subset aR$ from the maximality we must have $pR = aR$. Hence $a = rp$ for some $r \in R$. Then $p = ab = rpb$ and as before we must have $rb = 1$ and b a unit. □

Theorem 3.2.7. *Let R be a principle ideal domain. Then:*

(1) *An element $p \in R$ is irreducible if and only if it is a prime element.*

(2) *A nonzero ideal of R is a maximal ideal if and only if it is a prime ideal.*

(3) *The maximal ideals of R are precisely those ideals pR where p is a prime element.*

Proof. First note that $\{0\}$ is a prime ideal but not maximal.

(1) We already know that prime elements are irreducible. To show the converse suppose that p is irreducible. Since R is a principal ideal domain from Theorem 3.2.6 we have that pR is a maximal ideal, and each maximal ideal is also a prime ideal. Therefore from Theorem 3.2.6 we have that p is a prime element.

(2) We already know that each maximal ideal is a prime ideal. To show the converse suppose that $I \neq \{0\}$ is a prime ideal. Then $I = pR$ where p is a prime element with $p \neq 0$. Therefore p is irreducible from part (1) and hence pR is a maximal ideal from Theorem 3.2.6.

(3) This follows directly from the proof in part (2) and Theorem 3.2.6. □

3.3 Unique Factorization Domains

We now consider integral domains where there is unique factorization into primes. If R is an integral domain and $a, b \in R$ then we say that a and b are *associates* if there exists a unit $\epsilon \in R$ with $a = \epsilon b$.

Definition 3.3.1. An integral domain D is a *unique factorization domain* or UFD if for each $d \in D$ then either $d = 0$, d is a unit or d has a factorization into primes which is unique up to ordering and unit factors. This means that if

$$r = p_1 \cdots p_m = q_1 \cdots q_k$$

then $m = k$ and each p_i is an associate of some q_j.

There are several relationships in integral domains that are equivalent to unique factorization.

Definition 3.3.2. Let R be an integral domain.

(1) R has property (A) if and only if for each nonunit $a \neq 0$ there are irreducible elements $q_1, \dots, q_r \in R$ satisfying $a = q_1 \cdots q_r$.

(2) R has property (A$'$) if and only if for each nonunit $a \neq 0$ there are prime elements $p_1, \dots, p_r \in R$ satisfying $a = p_1 \cdots p_r$.

(3) R has property (B) if and only if whenever $q_1, \dots, q_r$ and $q'_1, \dots, q'_s$ are irreducible elements of R with

$$q_1 \cdots q_r = q'_1 \cdots q'_s$$

then $r = s$ and there is a permutation $\pi \in S_r$ such that for each $i \in \{1, \dots, r\}$ the elements q_i and $q'_{\pi(i)}$ are associates (uniqueness up to ordering and unit factors).

(4) R has property (C) if and only if each irreducible element of R is a prime element.

Notice that properties (A) and (C) together are equivalent to what we defined as *unique factorization*. Hence an integral domain satisfying (A) and (C) is a UFD. We show next, that there are other equivalent formulations.

Theorem 3.3.3. *In an integral domain R the following are equivalent:*

(1) *R is a UFD.*

(2) *R satisfies properties* (A) *and* (B).

(3) *R satisfies properties* (A) *and* (C).

(4) *R satisfies property* (A′).

Proof. As remarked before the statement of the theorem by definition (A) and (C) are equivalent to unique factorization. We show here that (2), (3) and (4) are equivalent.

First we show that (2) implies (3).

Suppose that R satisfies properties (A) and (B). We must show that it also satisfies (C), that is we must show that if $q \in R$ is irreducible then q is prime. Suppose that $q \in R$ is irreducible and $q|ab$ with $a, b \in R$. Then we have $ab = cq$ for some $c \in R$. If a is a unit from $ab = cq$ we get that $b = a^{-1}cq$ and $q|b$. Identically if b is a unit. Therefore we may assume that neither a nor b are units.

If $c = 0$ then since R is an integral domain either $a = 0$ or $b = 0$ and $q|a$ or $q|b$. We may assume then that $c \neq 0$.

If c is a unit then $q = c^{-1}ab$ and since q is irreducible either $c^{-1}a$ or b are units. If $c^{-1}a$ is a unit then a is also a unit so if c is a unit either a or b are units contrary to our assumption.

Therefore we may assume that $c \neq 0$ and c is not a unit. From property (A) we have

$$\begin{aligned} a &= q_1 \cdots q_r \\ b &= q'_1 \cdots q'_s \\ c &= q''_1 \cdots q''_t \end{aligned}$$

where $q_1, \dots q_r, q_1', \dots, q_s', q_1'', \dots q_t''$ are all irreducibles. Hence

$$q_1 \cdots q_r q_1' \cdots q_s' = q_1'' \cdots q_t'' \cdot q.$$

From property (B) q is an associate of some q_i or q_j'. Hence $q|q_i$ or $q|q_j'$. It follows that $q|a$ or $q|b$ and therefore q is a prime element.

That (3) implies (4) is direct.

We show that (4) implies (2).

Suppose that R satisfies property (A′). We must show that it satisfies both (A) and (B). We show first that (A) follows from (A′) by showing that irreducible elements are prime.

Suppose that q is irreducible. Then from (A′) we have

$$q = p_1 \cdots p_r$$

with each p_i prime. It follows without loss of generality that $p_2 \cdots p_r$ is a unit and p_1 is a nonunit and hence $p_i|1$ for $i = 2, \dots, r$. Thus $q = p_1$ and q is prime. Therefore (A) holds.

We now show that (B) holds. Let

$$q_1 \cdots q_r = q_1' \cdots q_s'$$

where q_i, q_j' are all irreducibles and hence primes. Then

$$q_1' | q_1 \cdots q_r$$

and so $q_1'|q_i$ for some i. Without loss of generality suppose $q_1'|q_1$. Then $q_1 = aq_1'$. Since q_1 is irreducible it follows that a is a unit and q_1 and q_1' are associates. It follows then that

$$aq_2 \cdots q_r = q_2' \cdot q_s'$$

since R has no zero divisors. Property (B) holds then by induction and the theorem is proved. □

Note that in our new terminology $\mathbb{Z}$ is a UFD. In the next section we will present other examples of UFD's however not every integral domain is a unique factorization domain.

As we defined in the last section let R be the following subring of $\mathbb{C}$.

$$R = \mathbb{Z}[i\sqrt{5}] = \{x + i\sqrt{5}y : x, y \in \mathbb{Z}\}.$$

R is an integral domain and we showed using the norm that 3 is an irreducible in R. Analogously we can show that the elements $2 + i\sqrt{5}, 2 - i\sqrt{5}$ are also irreducibles in R and further 3 is not an associate of either $2 + i\sqrt{5}$ or $2 - i\sqrt{5}$. Then

$$9 = 3 \cdot 3 = (2 + i\sqrt{5})(2 - i\sqrt{5})$$

give two different decompositions for an element in terms of irreducible elements. The fact that R is not a UFD also follows from the fact that 3 is an irreducible element which is not prime.

Notice also that Euclid's proof, that there are infinitely many primes in $\mathbb{Z}$, works in any UFD.

Theorem 3.3.4. *Let R be a UFD. Then R has infinitely many prime elements.*

Unique factorization is tied to the famous solution of Fermat's big theorem. Wiles and Taylor in 1995 proved the following.

Theorem 3.3.5. *The equation $x^p + y^p = z^p$ has no integral solutions with $xyz \neq 0$ for any prime $p \geq 3$.*

Kummer tried to prove this theorem by attempting to factor $x^p = z^p - y^p$. Call the conclusion of Theorem 3.3.4 in an integral domain R property (F_p). Let $\epsilon = e^{\frac{2\pi i}{p}}$. Then

$$z^p - y^p = \prod_{j=0}^{p-1} (z - \epsilon^j y).$$

View this equation in the ring

$$R = \mathbb{Z}[\epsilon] = \left\{ \sum_{j=0}^{p-1} a_j \epsilon^j : a_j \in \mathbb{Z} \right\}.$$

Kummer proved that if R is a UFD then property (F_p) holds. However, independently, from Uchida and Montgomery (1971) R is a UFD only if $p \leq 19$ (see [41]).

3.4 Principal Ideal Domains and Unique Factorization

In this section we prove that every principal ideal domain (PID) is a unique factorization domain (UFD). We say that an ascending chain of ideals in R

$$I_1 \subset I_2 \subset \cdots \subset I_n \subset \cdots$$

becomes *stationary* if there exists an m such that $I_r = I_m$ for all $r \geq m$.

Theorem 3.4.1. *Let R be an integral domain. If each ascending chain of principal ideals in R becomes stationary, then R satisfies property* (A).

Proof. Suppose that $a \neq 0$ is a not a unit in R. Suppose that a is not a product of irreducible elements. Clearly then a cannot itself be irreducible. Hence $a = a_1 b_1$ with $a_1, b_1 \in R$ and a_1, b_1 are not units. If both a_1 or b_1 can be expressed as a product of irreducible elements then so can a. Without loss of generality then suppose that a_1 is not a product of irreducible elements.

Since $a_1|a$ we have the inclusion of ideals $aR \subseteq a_1R$. If $a_1R = aR$ then $a_1 \in aR$ and $a_1 = ar = a_1b_1r$ which implies that b_1 is a unit contrary to our assumption. Therefore $aR \neq a_1R$ and the inclusion is proper. By iteration then we obtain a strictly increasing chain of ideals

$$aR \subset a_1R \subset \cdots \subset a_nR \subset \cdots.$$

From our hypothesis on R this must become stationary contradicting the argument above that the inclusion is proper. Therefore a must be a product of irreducibles. □

Theorem 3.4.2. *Each principal ideal domain R is a unique factorization domain.*

Proof. Suppose that R is a principal ideal domain. R satisfies property (C) by Theorem 3.2.7(1), so to show that it is a unique factorization domain we must show that it also satisfies property (A). From the previous theorem it suffices to show that each ascending chain of principal ideals becomes stationary. Consider such an ascending chain

$$a_1R \subset a_2R \subset \cdots \subset a_nR \subset \cdots.$$

Now let

$$I = \bigcup_{i=1}^{\infty} a_iR.$$

Now I is an ideal in R and hence a principal ideal. Therefore $I = aR$ for some $a \in R$. Since I is a union there exists an m such that $a \in a_mR$. Therefore $I = aR \subset a_mR$ and hence $I = a_mR$ and $a_iR \subset a_mR$ for all $i \geq m$. Therefore the chain becomes stationary and from Theorem 3.4.1 R satisfies property (A). □

Since we showed that the integers $\mathbb{Z}$ are a PID we can recover the fundamental theorem of arithmetic from Theorem 3.4.2. We now present another important example of a PID and hence a UFD. In the next chapter we will look in detail at polynomials with coefficients in an integral domain. Below we consider polynomials with coefficients in a field and for the present leave out many of the details.

If F is a field and n is a nonnegative integer, then a *polynomial of degree n over F* is a formal sum of the form

$$P(x) = a_0 + a_1x + \cdots + a_nx^n$$

with $a_i \in F$ for $i = 0, \ldots, n$, $a_n \neq 0$, and x an indeterminate. A *polynomial* $P(x)$ over F is either a polynomial of some degree or the expression $P(x) = 0$, which is called the *zero polynomial* and has degree $-\infty$. We denote the degree of $P(x)$ by $\deg P(x)$. A polynomial of zero degree has the form $P(x) = a_0$ and is called a *constant polynomial* and can be identified with the corresponding element of F. The elements $a_i \in F$ are called the *coefficients of* $P(x)$; a_n is the *leading coefficient*. If $a_n = 1$, $P(x)$ is called a *monic polynomial*. Two nonzero polynomials are equal if and only if they have the same degree and exactly the same coefficients. A polynomial of degree 1 is called a *linear polynomial* while one of degree two is a *quadratic polynomial*.

We denote by $F[x]$ the set of all polynomials over F and we will show that $F[x]$ becomes a principal ideal domain and hence a unique factorization domain. We first define addition, subtraction, and multiplication on $F[x]$ by algebraic manipulation. That is, suppose $P(x) = a_0 + a_1x + \cdots + a_nx^n$, $Q(x) = b_0 + b_1x + \cdots + b_mx^m$ then

$$P(x) \pm Q(x) = (a_0 \pm b_0) + (a_1 \pm b_1)x + \cdots,$$

that is, the coefficient of x^i in $P(x) \pm Q(x)$ is $a_i \pm b_i$, where $a_i = 0$ for $i > n$ and $b_j = 0$ for $j > m$. Multiplication is given by

$$P(x)Q(x) = (a_0b_0) + (a_1b_0 + a_0b_1)x + (a_0b_2 + a_1b_1 + a_2b_0)x^2 + \cdots + (a_nb_m)x^{n+m},$$

that is, the coefficient of x^i in $P(x)Q(x)$ is $(a_0b_i + a_1b_{i-1} + \cdots + a_ib_0)$.

Example 3.4.3. Let $P(x) = 3x^2 + 4x - 6$ and $Q(x) = 2x + 7$ be in $\mathbb{Q}[x]$. Then

$$P(x) + Q(x) = 3x^2 + 6x + 1$$

and

$$P(x)Q(x) = (3x^2 + 4x - 6)(2x + 7) = 6x^3 + 29x^2 + 16x - 42.$$

From the definitions the following degree relationships are clear. The proofs are in the exercises.

Lemma 3.4.4. *Let* $0 \neq P(x), 0 \neq Q(x)$ *in* $F[x]$. *Then:*

(1) $\deg P(x)Q(x) = \deg P(x) + \deg Q(x)$.

(2) $\deg(P(x) \pm Q(x)) \leq \text{Max}(\deg P(x), \deg Q(x))$ *if* $P(x) \pm Q(x) \neq 0$.

We next obtain the following.

Theorem 3.4.5. *If* F *is a field, then* $F[x]$ *forms an integral domain.* F *can be naturally embedded into* $F[x]$ *by identifying each element of* F *with the corresponding constant polynomial. The only units in* $F[x]$ *are the nonzero elements of* F.

Proof. Verification of the basic ring properties is solely computational and is left to the exercises. Since $\deg P(x)Q(x) = \deg P(x) + \deg Q(x)$, it follows that if neither $P(x) \neq 0$ nor $Q(x) \neq 0$ then $P(x)Q(x) \neq 0$ and therefore $F[x]$ is an integral domain.

If $G(x)$ is a unit in $F[x]$, then there exists an $H(x) \in F[x]$ with $G(x)H(x) = 1$. From the degrees we have $\deg G(x) + \deg H(x) = 0$ and since $\deg G(x) \geq 0$, $\deg H(x) \geq 0$. This is possible only if $\deg G(x) = \deg H(x) = 0$. Therefore $G(x) \in F$. □

Now that we have $F[x]$ as an integral domain we proceed to show that $F[x]$ is a principal ideal domain and hence there is unique factorization into primes. We first repeat the definition of a prime in $F[x]$. If $0 \neq f(x)$ has no nontrivial, nonunit factors (it cannot be factorized into polynomials of lower degree) then $f(x)$ is a *prime* in $F[x]$ or a *prime polynomial*. A prime polynomial is also called an *irreducible polynomial*. Clearly, if $\deg g(x) = 1$ then $g(x)$ is irreducible.

The fact that $F[x]$ is a principal ideal domain follows from the division algorithm for polynomials, which is entirely analogous to the division algorithm for integers.

Lemma 3.4.6 (division algorithm in $F[x]$). *If $0 \neq f(x)$, $0 \neq g(x) \in F[x]$ then there exist unique polynomials $q(x), r(x) \in F[x]$ such that $f(x) = q(x)g(x) + r(x)$ where $r(x) = 0$ or $\deg r(x) < \deg g(x)$. (The polynomials $q(x)$ and $r(x)$ are called respectively the quotient and remainder.)*

This theorem is essentially long division of polynomials. A formal proof is based on induction on the degree of $g(x)$. We omit this but give some examples from $\mathbb{Q}[x]$.

Example 3.4.7. (a) Let $f(x) = 3x^4 - 6x^2 + 8x - 6$, $g(x) = 2x^2 + 4$. Then

$$\frac{3x^4 - 6x^2 + 8x - 6}{2x^2 + 4} = \frac{3}{2}x^2 - 6 \quad \text{with remainder } 8x + 18.$$

Thus here $q(x) = \frac{3}{2}x^2 - 6$, $r(x) = 8x + 18$.

(b) Let $f(x) = 2x^5 + 2x^4 + 6x^3 + 10x^2 + 4x$, $g(x) = x^2 + x$. Then

$$\frac{2x^5 + 2x^4 + 6x^3 + 10x^2 + 4x}{x^2 + x} = 2x^3 + 6x + 4.$$

Thus here $q(x) = 2x^3 + 6x + 4$ and $r(x) = 0$.

Theorem 3.4.8. *Let F be a field. Then the polynomial ring $F[x]$ is a principal ideal domain and hence a unique factorization domain.*

Proof. The proof is essentially analogous to the proof in the integers. Let I be an ideal in $F[x]$ with $I \neq F[x]$. Let $f(x)$ be a polynomial in I of minimal degree. We claim that $I = \langle f(x) \rangle$ the principal ideal generated by $f(x)$. Let $g(x) \in I$. We must show that $g(x)$ is a multiple of $f(x)$. By the division algorithm in $F[x]$ we have

$$g(x) = q(x)f(x) + r(x)$$

where $r(x) = 0$ or $\deg(r(x)) < \deg(f(x))$. If $r(x) \neq 0$ then $\deg(r(x)) < \deg(f(x))$. However $r(x) = g(x) - q(x)f(x) \in I$ since I is an ideal and $g(x), f(x) \in I$. This is a contradiction since $f(x)$ was assumed to be a polynomial in I of minimal degree. Therefore $r(x) = 0$ and hence $g(x) = q(x)f(x)$ is a multiple of $f(x)$. Therefore each element of I is a multiple of $f(x)$ and hence $I = \langle f(x) \rangle$.

Therefore $F[x]$ is a principal ideal domain and from Theorem 3.4.2 a unique factorization domain. □

We proved that in a principal ideal domain every ascending chain of ideals becomes stationary. In general a ring R (commutative or not) satisfies the *ascending chain condition* or ACC if every ascending chain of left (or right) ideals in R becomes stationary. A ring satisfying the ACC is called a *Noetherian ring*.

3.5 Euclidean Domains

In analyzing the proof of unique factorization in both $\mathbb{Z}$ and $F[x]$, it is clear that it depends primarily on the division algorithm. In $\mathbb{Z}$ the division algorithm depended on the fact that the positive integers could be ordered and in $F[x]$ on the fact that the degrees of nonzero polynomials are nonnegative integers and hence could be ordered. This basic idea can be generalized in the following way.

Definition 3.5.1. An integral domain D is a *Euclidean domain* if there exists a function N from $D^\star = D \setminus \{0\}$ to the nonnegative integers such that

(1) $N(r_1) \leq N(r_1 r_2)$ for any $r_1, r_2 \in D^\star$.

(2) For all $r_1, r_2 \in D$ with $r_1 \neq 0$ there exist $q, r \in D$ such that

$$r_2 = q r_1 + r$$

where either $r = 0$ or $N(r) < N(r_1)$.

The function N is called a *Euclidean norm* on D.

Therefore Euclidean domains are precisely those integral domains which allow division algorithms. In the integers $\mathbb{Z}$ define $N(z) = |z|$. Then N is a Euclidean norm on $\mathbb{Z}$ and hence $\mathbb{Z}$ is a Euclidean domain. On $F[x]$ define $N(p(x)) = \deg(p(x))$ if $p(x) \neq 0$. Then N is also a Euclidean norm on $F[x]$ so that $F[x]$ is also a Euclidean domain. In any Euclidean domain we can mimic the proofs of unique factorization in both $\mathbb{Z}$ and $F[x]$ to obtain the following:

Theorem 3.5.2. *Every Euclidean domain is a principal ideal domain and hence a unique factorization domain.*

Before proving this theorem we must develop some results on the *number theory* of general Euclidean domains. First some properties of the norm.

Lemma 3.5.3. *If R is a Euclidean domain then*

(a) $N(1)$ *is minimal among* $\{N(r) : r \in R^\star\}$.

(b) $N(u) = N(1)$ *if and only if u is a unit.*

(c) $N(a) = N(b)$ *for* $a, b \in R^\star$ *if a, b are associates.*

(d) $N(a) < N(ab)$ *unless b is a unit.*

Proof. (a) From property (1) of Euclidean norms we have

$$N(1) \leq N(1 \cdot r) = N(r) \quad \text{for any } r \in R^\star.$$

(b) Suppose u is a unit. Then there exists u^{-1} with $u \cdot u^{-1} = 1$. Then

$$N(u) \leq N(u \cdot u^{-1}) = N(1).$$

From the minimality of $N(1)$ it follows that $N(u) = N(1)$.

Conversely suppose $N(u) = N(1)$. Apply the division algorithm to get

$$1 = qu + r.$$

If $r \neq 0$ then $N(r) < N(u) = N(1)$ contradicting the minimality of $N(1)$. Therefore $r = 0$ and $1 = qu$. Then u has a multiplicative inverse and hence is a unit.

(c) Suppose $a, b \in R^\star$ are associates. Then $a = ub$ with u a unit. Then

$$N(b) \leq N(ub) = N(a).$$

On the other hand $b = u^{-1}a$ so

$$N(a) \leq N(u^{-1}a) = N(b).$$

Since $N(a) \leq N(b)$ and $N(b) \leq N(a)$ it follows that $N(a) = N(b)$.

(d) Suppose $N(a) = N(ab)$. Apply the division algorithm

$$a = q(ab) + r$$

where $r = 0$ or $N(r) < N(ab)$. If $r \neq 0$ then

$$r = a - qab = a(1 - qb) \implies N(ab) = N(a) \leq N(a(1 - qb)) = N(r)$$

contradicting that $N(r) < N(ab)$. Hence $r = 0$ and $a = q(ab) = (qb)a$. Then

$$a = (qb)a = 1 \cdot a \implies qb = 1$$

since there are no zero divisors in an integral domain. Hence b is a unit. Since $N(a) \leq N(ab)$ it follows that if b is not a unit we must have $N(a) < N(ab)$. □

We can now prove Theorem 3.5.2.

Proof. Let D be a Euclidean domain. We show that each ideal $I \neq D$ in D is principal. Let $I \neq D$ be an ideal in D. If $I = \{0\}$ then $I = \langle 0 \rangle$ and I is principal. Therefore we may assume that there are nonzero elements in I. Hence there are elements $x \in I$ with strictly positive norm. Let a be an element of I of minimal norm. We claim that $I = \langle a \rangle$. Let $b \in I$. We must show that b is a multiple of a. Now by the division algorithm

$$b = qa + r$$

where either $r = 0$ or $N(r) < N(a)$. As in $\mathbb{Z}$ and $F[x]$ we have a contradiction if $r \neq 0$. In this case $N(r) < N(a)$ but $r = b - qa \in I$ since I is an ideal contradicting the minimality of $N(a)$. Therefore $r = 0$ and $b = qa$ and hence $I = \langle a \rangle$. □

As a final example of a Euclidean domain we consider the *Gaussian integers*

$$\mathbb{Z}[i] = \{a + bi : a, b \in \mathbb{Z}\}.$$

It was first observed by Gauss that this set permits unique factorization. To show this we need a Euclidean norm on $\mathbb{Z}[i]$.

Definition 3.5.4. If $z = a + bi \in \mathbb{Z}[i]$ then its *norm* $N(z)$ is defined by

$$N(a + bi) = a^2 + b^2.$$

The basic properties of this norm follow directly from the definition (see exercises).

Lemma 3.5.5. *If $\alpha, \beta \in \mathbb{Z}[i]$ then:*

(1) *$N(\alpha)$ is an integer for all $\alpha \in \mathbb{Z}[i]$.*

(2) *$N(\alpha) \geq 0$ for all $\alpha \in \mathbb{Z}[i]$.*

(3) *$N(\alpha) = 0$ if and only if $\alpha = 0$.*

(4) *$N(\alpha) \geq 1$ for all $\alpha \neq 0$.*

(5) *$N(\alpha\beta) = N(\alpha)N(\beta)$, that is the norm is multiplicative.*

From the multiplicativity of the norm we have the following concerning primes and units in $\mathbb{Z}[i]$.

Lemma 3.5.6. (1) *$u \in \mathbb{Z}[i]$ is a unit if and only if $N(u) = 1$.*

(2) *If $\pi \in \mathbb{Z}[i]$ and $N(\pi) = p$ where p is an ordinary prime in $\mathbb{Z}$ then π is a prime in $\mathbb{Z}[i]$.*

Proof. Certainly u is a unit if and only if $N(u) = N(1)$. But in $\mathbb{Z}[i]$ we have $N(1) = 1$ so the first part follows.

Suppose next that $\pi \in \mathbb{Z}[i]$ with $N(\pi) = p$ for some $p \in \mathbb{Z}$. Suppose that $\pi = \pi_1 \pi_2$. From the multiplicativity of the norm we have

$$N(\pi) = p = N(\pi_1)N(\pi_2).$$

Since each norm is a positive ordinary integer and p is a prime it follows that either $N(\pi_1) = 1$ or $N(\pi_2) = 1$. Hence either π_1 or π_2 is a unit. Therefore π is a prime in $\mathbb{Z}[i]$. □

Armed with this norm we can show that $\mathbb{Z}[i]$ is a Euclidean domain.

Theorem 3.5.7. *The Gaussian integers $\mathbb{Z}[i]$ form a Euclidean domain.*

Proof. That $\mathbb{Z}[i]$ forms a commutative ring with an identity can be verified directly and easily. If $\alpha\beta = 0$ then $N(\alpha)N(\beta) = 0$ and since there are no zero divisors in $\mathbb{Z}$ we must have $N(\alpha) = 0$ or $N(\beta) = 0$. But then either $\alpha = 0$ or $\beta = 0$ and hence $\mathbb{Z}[i]$ is an integral domain. To complete the proof we show that the norm N is a Euclidean norm.

From the multiplicativity of the norm we have if $\alpha, \beta \neq 0$

$$N(\alpha\beta) = N(\alpha)N(\beta) \geq N(\alpha) \quad \text{since } N(\beta) \geq 1.$$

Therefore property (1) of Euclidean norms is satisfied. We must now show that the division algorithm holds.

Let $\alpha = a + bi$ and $\beta = c + di$ be Gaussian integers. Recall that for a nonzero complex number $z = x + iy$ its inverse is

$$\frac{1}{z} = \frac{\overline{z}}{|z|^2} = \frac{x - iy}{x^2 + y^2}.$$

Therefore as a complex number

$$\begin{aligned}\frac{\alpha}{\beta} = \alpha\frac{\overline{\beta}}{|\beta|^2} &= (a + bi)\frac{c - di}{c^2 + d^2} \\ &= \frac{ac + bd}{c^2 + d^2} + \frac{ac - bd}{c^2 + d^2}i = u + iv.\end{aligned}$$

Now since a, b, c, d are integers u, v must be rationals. The set

$$\{u + iv : u, v \in \mathbb{Q}\}$$

is called the set of the *Gaussian rationals*.

If $u, v \in \mathbb{Z}$ then $u + iv \in \mathbb{Z}[i]$, $\alpha = q\beta$ with $q = u + iv$, and we are done. Otherwise choose ordinary integers m, n satisfying $|u - m| \le \frac{1}{2}$ and $|v - n| \le \frac{1}{2}$ and let $q = m + in$. Then $q \in \mathbb{Z}[i]$. Let $r = \alpha - q\beta$. We must show that $N(r) < N(\beta)$.

Working with complex absolute value we get

$$|r| = |\alpha - q\beta| = |\beta|\left|\frac{\alpha}{\beta} - q\right|.$$

Now

$$\left|\frac{\alpha}{\beta} - q\right| = |(u-m) + i(v-n)| = \sqrt{(u-m)^2 + (v-n)^2} \le \sqrt{\left(\frac{1}{2}\right)^2 + \left(\frac{1}{2}\right)^2} < 1.$$

Therefore

$$|r| < |\beta| \implies |r|^2 < |\beta|^2 \implies N(r) < N(\beta)$$

completing the proof. □

Since $\mathbb{Z}[i]$ forms a Euclidean domain it follows from our previous results that $\mathbb{Z}[i]$ must be a principal ideal domain and hence a unique factorization domain.

Corollary 3.5.8. *The Gaussian integers are a UFD.*

Since we will now be dealing with many kinds of *integers* we will refer to the ordinary integers $\mathbb{Z}$ as the *rational integers* and the ordinary primes p as the *rational primes*. It is clear that $\mathbb{Z}$ can be embedded into $\mathbb{Z}[i]$. However not every rational prime is also prime in $\mathbb{Z}[i]$. The primes in $\mathbb{Z}[i]$ are called the *Gaussian primes*. For example we can show that both $1 + i$ and $1 - i$ are Gaussian primes, that is primes in $\mathbb{Z}[i]$. However $(1 + i)(1 - i) = 2$ so that the rational prime 2 is not a prime in $\mathbb{Z}[i]$. Using the multiplicativity of the Euclidean norm in $\mathbb{Z}[i]$ we can describe all the units and primes in $\mathbb{Z}[i]$.

Theorem 3.5.9. (1) *The only units in* $\mathbb{Z}[i]$ *are* $\pm 1, \pm i$.

(2) *Suppose* π *is a Gaussian prime. Then* π *is either:*

(a) *a positive rational prime* $p \equiv 3 \bmod 4$ *or an associate of such a rational prime.*

(b) $1 + i$ *or an associate of* $1 + i$.

(c) $a + bi$ *or* $a - bi$ *where* $a > 0$, $b > 0$, a *is even and* $N(\pi) = a^2 + b^2 = p$ *with* p *a rational prime congruent to* $1 \bmod 4$ *or an associate of* $a + bi$ *or* $a - bi$.

Proof. (1) Suppose $u = x + iy \in \mathbb{Z}[i]$ is a unit. Then from Lemma 3.5.6 we have $N(u) = x^2 + y^2 = 1$ implying that $(x, y) = (0, \pm 1)$ or $(x, y) = (\pm 1, 0)$. Hence $u = \pm 1$ or $u = \pm i$.

(2) Now suppose that π is a Gaussian prime. Since $N(\pi) = \pi\overline{\pi}$ and $\overline{\pi} \in \mathbb{Z}[i]$ it follows that $\pi | N(\pi)$. $N(\pi)$ is a rational integer so $N(\pi) = p_1 \cdots p_k$ where the p_i's are rational primes. By Euclid's lemma $\pi | p_i$ for some p_i and hence a Gaussian prime must divide at least one rational prime. On the other hand suppose $\pi | p$ and $\pi | q$ where p, q are different primes. Then $(p, q) = 1$ and hence there exist $x, y \in \mathbb{Z}$ such that $1 = px + qy$. It follows that $\pi | 1$ a contradiction. Therefore a Gaussian prime divides one and only one rational prime.

Let p be the rational prime that π divides. Then $N(\pi) | N(p) = p^2$. Since $N(\pi)$ is a rational integer it follows that $N(\pi) = p$ or $N(\pi) = p^2$. If $\pi = a + bi$ then $a^2 + b^2 = p$ or $a^2 + b^2 = p^2$.

If $p = 2$ then $a^2 + b^2 = 2$ or $a^2 + b^2 = 4$. It follows that $\pi = \pm 2, \pm 2i$ or $\pi = 1 + i$ or an associate of $1 + i$. Since $(1 + i)(1 - i) = 2$ and neither $1 + i$ nor $1 - i$ are units it follows that neither 2 nor any of its associates are primes. Then $\pi = 1 + i$ or an associate of $1 + i$. To see that $1 + i$ is prime suppose $1 + i = \alpha\beta$. Then $N(1 + i) = 2 = N(\alpha)N(\beta)$. It follows that either $N(\alpha) = 1$ or $N(\beta) = 1$ and either α or β is a unit.

If $p \neq 2$ then either $p \equiv 3 \bmod 4$ or $p \equiv 1 \bmod 4$. Suppose first that $p \equiv 3 \bmod 4$. Then $a^2 + b^2 = p$ would imply from Fermat's two-square theorem (see [35]) that $p \equiv 1 \bmod 4$. Therefore from the remarks above $a^2 + b^2 = p^2$ and $N(\pi) = N(p)$. Since $\pi | p$ we have $\pi = \alpha p$ with $\alpha \in \mathbb{Z}[i]$. From $N(\pi) = N(p)$ we get that $N(\alpha) = 1$ and α is a unit. Therefore π and p are associates. Hence in this case π is an associate of a rational prime congruent to 3 mod 4.

Finally suppose $p \equiv 1 \bmod 4$. From the remarks above either $N(\pi) = p$ or $N(\pi) = p^2$. If $N(\pi) = p^2$ then $a^2 + b^2 = p^2$. Since $p \equiv 1 \bmod 4$ from Fermat's two square theorem there exist $m, n \in \mathbb{Z}$ with $m^2 + n^2 = p$. Let $u = m + in$ then the norm $N(u) = p$. Since p is a rational prime it follows that u is a Gaussian prime. Similarly its conjugate $\overline{u}$ is also a Gaussian prime. Now $u\overline{u} = p^2 = N(\pi)$. Since $\pi | N(\pi)$ it follows that $\pi | u\overline{u}$ and from Euclid's lemma either $\pi | u$ or $\pi | \overline{u}$. If $\pi | u$ they are associates since both are primes. But this is a contradiction since $N(\pi) \neq N(u)$. The same is true if $\pi | \overline{u}$.

It follows that if $p \equiv 1 \bmod 4$, then $N(\pi) \neq p^2$. Therefore in this case $N(\pi) = p = a^2 + b^2$. An associate of π has both $a, b > 0$ (see exercises). Further since $a^2 + b^2 = p$ one of a or b must be even. If a is odd then b is even then $i\pi$ is an associate of π with a even completing the proof. □

Finally we mention that the methods used in $\mathbb{Z}[i]$ cannot be applied to all quadratic integers. For example we have seen that there is not unique factorization in $\mathbb{Z}[\sqrt{-5}]$.

3.6 Overview of Integral Domains

Here we present some additional definitions for special types of integral domains.

Definition 3.6.1. (1) A *Dedekind domain* D is an integral domain such that each nonzero proper ideal A ($\{0\} \neq A \neq R$) can be written uniquely as a product of prime ideals

$$A = P_1 \cdots P_r$$

with each P_i a prime ideal and the factorization is unique up to ordering.

(2) A *Prüfer ring* is an integral domain such that

$$A \cdot (B \cap C) = AB \cap AC$$

for all ideals A, B, C in R.

Dedekind domains arise naturally in algebraic number theory. It can be proved that the rings of algebraic integers in any algebraic number field are Dedekind domains (see [35]).

If R is a Dedekind domain it is also a Prüfer Ring. If R is a Prüfer ring and a unique factorization domain then R is a principal ideal domain.

In the next chapter we will prove a theorem due to Gauss that if R is a UFD then the polynomial ring $R[x]$ is also a UFD. If K is a field we have already seen that $K[x]$ is a UFD. Hence the polynomial ring in several variables $K[x_1, \dots, x_n]$ is also a UFD. This fact plays an important role in algebraic geometry.

3.7 Exercises

1. Let R be an integral domain and let $\pi \in R \setminus (U(R) \cup \{0\})$. Show:
   (i) If for each $a \in R$ with $\pi \nmid a$ there exist $\lambda, \mu \in R$ with $\lambda\pi + \mu a = 1$ then π is a prime element of R.
   (ii) Give an example for a prime element π in an UFD R, which does not satisfy the conditions of (i).
2. Let R be an UFD and let $a_1, \dots, a_t$ be pairwise coprime elements of R. If $a_1 \cdots a_t$ is an m-th power ($m \in \mathbb{N}$), then all factors a_i are an associate of an m-th power. Is each a_i necessarily an m-th power?
3. Decide if the unit group of $\mathbb{Z}[\sqrt{3}]$, $\mathbb{Z}[\sqrt{5}]$ and $\mathbb{Z}[\sqrt{7}]$ is finite or infinite. For which $a \in \mathbb{Z}$ are $(1 - \sqrt{5})$ and $(a + \sqrt{5})$ associates in $\mathbb{Z}[\sqrt{5}]$?
4. Let $k \in \mathbb{Z}$ and $k \neq x^2$ for all $x \in \mathbb{Z}$. Let $\alpha = a + b\sqrt{k}$ and $\beta = c + d\sqrt{k}$ be elements of $\mathbb{Z}[\sqrt{k}]$ and $N(\alpha) = a^2 - kb^2$, $N(\beta) = c^2 - kd^2$. Show:

(i) The equality of the absolute values of $N(\alpha)$ and $N(\beta)$ is necessary for the association of α and β in $\mathbb{Z}[\sqrt{k}]$. Is this constraint also sufficient?

(ii) Sufficient for the irreducibility of α in $\mathbb{Z}[\sqrt{k}]$ is the irreducibility of $N(\alpha)$ in $\mathbb{Z}$. Is this also necessary?

5. In general irreducible elements are not prime. Consider the set of complex number given by

$$R = \mathbb{Z}[i\sqrt{5}] = \{x + i\sqrt{5}y : x, y \in \mathbb{Z}\}.$$

Show that they form a subring of $\mathbb{C}$.

6. For an element $x + iy\sqrt{5} \in R$ define its *norm* by

$$N(x + iy\sqrt{5}) = |x + iy\sqrt{5}| = x^2 + 5y^2.$$

Prove that the norm is multiplicative, that is $N(ab) = N(a)N(b)$.

7. Prove Lemma 3.4.4.

8. Prove that the set of polynomials $R[x]$ with coefficients in a ring R forms a ring.

9. Prove the basic properties of the norm of the Gaussian integers. If $\alpha, \beta \in \mathbb{Z}[i]$ then:

(i) $N(\alpha)$ is an integer for all $\alpha \in \mathbb{Z}[i]$.

(ii) $N(\alpha) \geq 0$ for all $\alpha \in \mathbb{Z}[i]$.

(iii) $N(\alpha) = 0$ if and only if $\alpha = 0$.

(iv) $N(\alpha) \geq 1$ for all $\alpha \neq 0$.

(v) $N(\alpha\beta) = N(\alpha)N(\beta)$, that is the norm is multiplicative.

Chapter 4
Polynomials and Polynomial Rings

4.1 Polynomials and Polynomial Rings

In the last chapter we saw that if K is a field then the set of polynomials with coefficients in K, which we denoted $K[x]$, forms a unique factorization domain. In this chapter we take a more detailed look at polynomials over a general ring R. We then prove that if R is a UFD then the polynomial ring $R[x]$ is also a UFD. We first take a formal look at polynomials.

Let R be a commutative ring with an identity. Consider the set $\tilde{R}$ of functions f from the nonnegative integers $\overline{N} = \mathbb{N} \cup \{0\}$ into R with only a finite number of values nonzero. That is

$$\tilde{R} = \{f : \overline{N} \to R : f(n) \neq 0 \text{ for only finitely many } n\}.$$

On $\tilde{R}$ we define the following addition and multiplication

$$(f+g)(m) = f(m) + g(m)$$
$$(f \cdot g)(m) = \sum_{i+j=m} f(i)g(j).$$

If we let $x = (0, 1, 0, \ldots)$ and identify $(r, 0, \ldots)$ with $r \in R$ then

$$x^0 = (1, 0, \ldots) = 1 \quad \text{and} \quad x^{i+1} = x \cdot x^i.$$

Now if $f = (r_0, r_1, r_2, \ldots)$ then f can be written as

$$f = \sum_{i=0}^{\infty} r_i x^i = \sum_{i=0}^{m} r_i x^i$$

for some $m \geq 0$ since $r_i \neq 0$ for only finitely many i. Further this presentation is unique.

We now call x an *indeterminate* over R and write each element of $\tilde{R}$ as $f(x) = \sum_{i=0}^{m} r_i x^i$ with $f(x) = 0$ or $r_m \neq 0$. We also now write $R[x]$ for $\tilde{R}$. Each element of $R[x]$ is called a *polynomial* over R. The elements $r_0, \ldots, r_m$ are called the *coefficients* of $f(x)$ with r_m the *leading coefficient*. If $r_m \neq 0$ the natural number m is called the *degree* of $f(x)$ which we denote by $\deg f(x)$. We say that $f(x) = 0$ has degree $-\infty$. The uniqueness of the representation of a polynomial implies that two nonzero polynomials are equal if and only if they have the same degree and exactly the same coefficients. A polynomial of degree 1 is called a *linear polynomial* while one of degree two is a *quadratic polynomial.* The set of polynomials of degree 0 together with 0 form a ring isomorphic to R and hence can be identified with R, the constant polynomials. Thus the ring R embeds in the set of polynomials $R[x]$. The following results are straightforward concerning degree.

Lemma 4.1.1. *Let* $f(x) \neq 0, g(x) \neq 0 \in R[x]$. *Then:*

(a) $\deg f(x)g(x) \leq \deg f(x) + \deg g(x)$.

(b) $\deg(f(x) \pm g(x)) \leq \operatorname{Max}(\deg f(x), \deg g(x))$.

If R is an integral domain then we have equality in (a).

Theorem 4.1.2. *Let R be a commutative ring with an identity. Then the set of polynomials* $R[x]$ *forms a ring called the ring of polynomials over R. The ring R identified with* 0 *and the polynomials of degree 0 naturally embeds into* $R[x]$. $R[x]$ *is commutative if and only if R is commutative. Further* $R[x]$ *is uniquely determined by R and* x.

Proof. Set $f(x) = \sum_{i=0}^{n} r_i x^i$ and $g(x) = \sum_{j=0}^{m} s_j x^j$. The ring properties follow directly by computation. The identification of $r \in R$ with the polynomial $r(x) = r$ provides the embedding of R into $R[x]$. From the definition of multiplication in $R[x]$ if R is commutative then $R[x]$ is commutative. Conversely if $R[x]$ is commutative then from the embedding of R into $R[x]$ it follows that R must be commutative. Note that if R has a multiplicative identity $1 \neq 0$ then this is also the multiplicative identity of $R[x]$.

Finally if S is a ring that contains R and $\alpha \in S$ then

$$R[\alpha] = \left\{ \sum_{i \geq 0} r_i \alpha^i : r_i \in R \text{ and } r_i \neq 0 \text{ for only a finite number of } i \right\}$$

is a homomorphic image of $R[x]$ via the map

$$\sum_{i \geq 0} r_i x^i \mapsto \sum_{i \geq 0} r_i \alpha^i.$$

Hence $R[x]$ is uniquely determined by R and x. □

If R is an integral domain then irreducible polynomials are defined as irreducibles in the ring $R[x]$. If R is a field then $f(x)$ is an *irreducible polynomial* if there is no factorization $f(x) = g(x)h(x)$ where $g(x)$ and $h(x)$ are polynomials of lower degree than $f(x)$. Otherwise $f(x)$ is called *reducible.* In elementary mathematics polynomials are considered as functions. We recover that idea via the concept of evaluation.

Definition 4.1.3. Let $f(x) = r_0 + r_1x + \cdots + r_mx^n$ be a polynomial over a commutative ring R with an identity and let $c \in R$. Then the element

$$f(c) = r_0 + r_1c + \cdots + r_nc^n \in R$$

is called the *evaluation* of $f(x)$ at c.

Definition 4.1.4. If $f(x) \in R[x]$ and $f(c) = 0$ for $c \in R$, then c is called a *zero* or a root of $f(x)$ in R.

4.2 Polynomial Rings over Fields

We now restate some of the result of the last chapter for $K[x]$ where K is a field. We then consider some consequences of these results to zeros of polynomials.

Theorem 4.2.1. *If F is a field, then $F[x]$ forms an integral domain. F can be naturally embedded into $F[x]$ by identifying each element of F with the corresponding constant polynomial. The only units in $F[x]$ are the nonzero elements of F.*

Proof. Verification of the basic ring properties is solely computational and is left to the exercises. Since $\deg P(x)Q(x) = \deg P(x) + \deg Q(x)$, it follows that if neither $P(x) \neq 0$ nor $Q(x) \neq 0$ then $P(x)Q(x) \neq 0$ and therefore $F[x]$ is an integral domain.

If $G(x)$ is a unit in $F[x]$, then there exists an $H(x) \in F[x]$ with $G(x)H(x) = 1$. From the degrees we have $\deg G(x) + \deg H(x) = 0$ and since $\deg G(x) \geq 0$, $\deg H(x) \geq 0$. This is possible only if $\deg G(x) = \deg H(x) = 0$. Therefore $G(x) \in F$. □

Now that we have $F[x]$ as an integral domain we proceed to show that $F[x]$ is a principal ideal domain and hence there is unique factorization into primes. We first repeat the definition of a prime in $F[x]$. If $0 \neq f(x)$ has no nontrivial, nonunit factors (it cannot be factorized into polynomials of lower degree) then $f(x)$ is a *prime* in $F[x]$ or a *prime polynomial*. A prime polynomial is also called an *irreducible polynomial* over F. Clearly, if $\deg g(x) = 1$ then $g(x)$ is irreducible.

The fact that $F[x]$ is a principal ideal domain follows from the division algorithm for polynomials, which is entirely analogous to the division algorithm for integers.

Theorem 4.2.2 (division algorithm in $F[x]$). *If $0 \neq f(x), 0 \neq g(x) \in F[x]$ then there exist unique polynomials $q(x), r(x) \in F[x]$ such that $f(x) = q(x)g(x) + r(x)$ where $r(x) = 0$ or $\deg r(x) < \deg g(x)$. (The polynomials $q(x)$ and $r(x)$ are called respectively the quotient and remainder.)*

This theorem is essentially long division of polynomials. A formal proof is based on induction on the degree of $g(x)$. We omit this but give some examples from $\mathbb{Q}[x]$.

Example 4.2.3. (a) Let $f(x) = 3x^4 - 6x^2 + 8x - 6$, $g(x) = 2x^2 + 4$. Then

$$\frac{3x^4 - 6x^2 + 8x - 6}{2x^2 + 4} = \frac{3}{2}x^2 - 6 \quad \text{with remainder } 8x + 18.$$

Thus here $q(x) = \frac{3}{2}x^2 - 6$ and $r(x) = 8x + 18$.

(b) Let $f(x) = 2x^5 + 2x^4 + 6x^3 + 10x^2 + 4x$, $g(x) = x^2 + x$. Then

$$\frac{2x^5 + 2x^4 + 6x^3 + 10x^2 + 4x}{x^2 + x} = 2x^3 + 6x + 4.$$

Thus here $q(x) = 2x^3 + 6x + 4$ and $r(x) = 0$.

Theorem 4.2.4. *Let F be a field. Then the polynomial ring $F[x]$ is a principal ideal domain and hence a unique factorization domain.*

We now give some consequences relative to zeros of polynomials in $F[x]$.

Theorem 4.2.5. *If $f(x) \in F[x]$ and $c \in F$ with $f(c) = 0$ then*

$$f(x) = (x - c)h(x),$$

where $\deg h(x) < \deg f(x)$.

Proof. Divide $f(x)$ by $x - c$. Then by the division algorithm we have

$$f(x) = (x - c)h(x) + r(x)$$

where $r(x) = 0$ or $\deg r(x) < \deg(x - c) = 1$. Hence if $r(x) \neq 0$ then $r(x)$ is a polynomial of degree 0, that is a constant polynomial, that is $r(x) = r$ for $r \in F$. Hence we have

$$f(x) = (x - c)h(x) + r.$$

This implies that

$$0 = f(x) = 0h(c) + r = r$$

and therefore $r = 0$ and $f(x) = (x - c)h(x)$. Since $\deg(x - c) = 1$ we must have that $\deg h(x) < \deg f(x)$. □

If $f(x) = (x - c)^k h(x)$ for some $k \geq 1$ with $h(c) \neq 0$ then c is called a zero of order k.

Theorem 4.2.6. *Let $f(x) \in F[x]$ with degree 2 or 3. Then f is irreducible if and only if $f(x)$ doesn't have a zero in F.*

Proof. Suppose that $f(x)$ is irreducible of degree 2 or 3. If $f(x)$ has a zero c then from Theorem 4.2.4 we have $f(x) = (x - c)h(x)$ with $h(x)$ of degree 1 or 2. Therefore $f(x)$ is reducible a contradiction and hence $f(x)$ cannot have a zero.

Conversely from Theorem 4.2.4 if $f(x)$ has a zero and if of degree greater than 1 then $f(x)$ is reducible. □

4.3 Polynomial Rings over Integral Domains

Here we consider $R[x]$ where R is an integral domain.

Definition 4.3.1. Let R be an integral domain. Then $a_1, a_2, \dots, a_n \in R$ are *coprime* if the set of all common divisors of $a_1, \dots, a_n$ consists only of units.

Notice for example that this concept depends on the ring R. For example 6 and 9 are not coprime over the integers $\mathbb{Z}$ since $3|6$ and $3|9$ and 3 is not a unit. However 6 and 9 are coprime over the rationals $\mathbb{Q}$. Here 3 is a unit.

Definition 4.3.2. Let $f(x) = \sum_{i=1}^{n} r_i x^i \in R[x]$ where R is an integral domain. Then $f(x)$ is a *primitive polynomial* or just *primitive* if $r_0, r_1, \dots, r_n$ are coprime in R.

Theorem 4.3.3. *Let R be an integral domain. Then*

(a) *The units of $R[x]$ are the units of R.*

(b) *If p is a prime element of R then p is a prime element of $R[x]$.*

Proof. If $r \in R$ is a unit then since R embeds into $R[x]$ it follows that r is also a unit in $R[x]$. Conversely suppose that $h(x) \in R[x]$ is a unit. Then there is a $g(x)$ such that $h(x)g(x) = 1$. Hence $\deg f(x) + \deg g(x) = \deg 1 = 0$. Since degrees are nonnegative integers it follows that $\deg f(x) = \deg g(x) = 0$ and hence $f(x) \in R$.

Now suppose that p is a prime element of R. Then $p \neq 0$ and pR is a prime ideal in R. We must show that $pR[x]$ is a prime ideal in $R[x]$. Consider the map

$$\tau : R[x] \to (R/pR)[x] \quad \text{given by}$$
$$\tau\Big(\sum_{i=0}^{n} r_i x^i\Big) = \sum_{i=0}^{n} (r_i + pR)x^i.$$

Then τ is an epimorphism with kernel $pR[x]$. Since pR is a prime ideal we know that R/pR is an integral domain. It follows that $(R/pR)[x]$ is also an integral domain. Hence $pR[x]$ must be a prime ideal in $R[x]$ and therefore p is also a prime element of $R[x]$. □

Recall that each integral domain R can be embedded into a unique field of fractions K. We can use results on $K[x]$ to deduce some results in $R[x]$.

Lemma 4.3.4. *If K is a field then each nonzero $f(x) \in K[x]$ is a primitive.*

Proof. Since K is a field each nonzero element of K is a unit. Therefore the only common divisors of the coefficients of $f(x)$ are units and hence $f(x) \in K[x]$ is primitive. □

Theorem 4.3.5. *Let R be an integral domain. Then each irreducible $f(x) \in R[x]$ of degree > 0 is primitive.*

Proof. Let $f(x)$ be an irreducible polynomial in $R[x]$ and let $r \in R$ be a common divisor of the coefficients of $f(x)$. Then $f(x) = rg(x)$ where $g(x) \in R[x]$. Then $\deg f(x) = \deg g(x) > 0$ so $g(x) \notin R$. Since the units of $R[x]$ are the units of R it follows that $g(x)$ is not a unit in $R[x]$. Since $f(x)$ is irreducible it follows that r must be a unit in $R[x]$ and hence r is a unit in R. Therefore $f(x)$ is primitive. □

Theorem 4.3.6. *Let R be an integral domain and K its field of fractions. If $f(x) \in R[x]$ is primitive and irreducible in $K[x]$ then $f(x)$ is irreducible in $R[x]$.*

Proof. Suppose that $f(x) \in R[x]$ is primitive and irreducible in $K[x]$ and suppose that $f(x) = g(x)h(x)$ where $g(x), h(x) \in R[x] \subset K[x]$. Since $f(x)$ is irreducible in $K[x]$ either $g(x)$ or $h(x)$ must be a unit in $K[x]$. Without loss of generality suppose that $g(x)$ is a unit in $K[x]$. Then $g(x) = g \in K$. But $g(x) \in R[x]$ and $K \cap R[x] = R$. Hence $g \in R$. Then g is a divisor of the coefficients of $f(x)$ and as $f(x)$ is primitive $g(x)$ must be a unit in R and therefore also a unit in $R[x]$. Therefore $f(x)$ is irreducible in $R[x]$. □

4.4 Polynomial Rings over Unique Factorization Domains

In this section we prove that if R is a UFD then the polynomial ring $R[x]$ is also a UFD. We first need the following due to Gauss.

Theorem 4.4.1 (Gauss' lemma). *Let R be a UFD and $f(x), g(x)$ primitive polynomials in $R[x]$. Then their product $f(x)g(x)$ is also primitive.*

Proof. Let R be a UFD and $f(x), g(x)$ primitive polynomials in $R[x]$. Suppose that $f(x)g(x)$ is not primitive. Then there is a prime element $p \in R$ that divides each of the coefficients of $f(x)g(x)$. Then $p | f(x)g(x)$. Since prime elements of R are also prime elements of $R[x]$ it follows that p is also a prime element of $R[x]$ and hence $p | f(x)$ or $p | g(x)$. Therefore either $f(x)$ or $g(x)$ is not primitive giving a contradiction. □

Theorem 4.4.2. *Let R be a UFD and K its field of fractions.*

(a) *If $g(x) \in K[x]$ is nonzero then there is a nonzero $a \in K$ such that $ag(x) \in R[x]$ is primitive.*

(b) *Let $f(x), g(x) \in R[x]$ with $g(x)$ primitive and $f(x) = ag(x)$ for some $a \in K$. Then $a \in R$.*

(c) *If $f(x) \in R[x]$ is nonzero then there is a $b \in R$ and a primitive $g(x) \in R[x]$ such that $f(x) = bg(x)$.*

Proof. (a) Suppose that $g(x) = \sum_{i=0}^{n} a_i x^i$ with $a_i = \frac{r_i}{s_i}$, $r_i, s_i \in R$. Set $s = s_0 s_1 \cdots s_n$. Then $sg(x)$ is a nonzero element of $R[x]$. Let d be a greatest common divisor of the coefficients of $sg(x)$. If we set $a = \frac{s}{d}$ then $ag(x)$ is primitive.

(b) For $a \in K$ there are coprime $r, s \in R$ satisfying $a = \frac{r}{s}$. Suppose that $a \notin R$. Then there is a prime element $p \in R$ dividing s. Since $g(x)$ is primitive p does not divide all the coefficients of $g(x)$. However we also have $f(x) = ag(x) = \frac{r}{s}g(x)$. Hence $sf(x) = rg(x)$ where $p|s$ and p doesn't divide r. Therefore p divides all the coefficients of $g(x)$ and hence $a \in R$.

(c) From part (a) there is a nonzero $a \in K$ such that $af(x)$ is primitive in $R[x]$. Then $f(x) = a^{-1}(af(x))$. From part (b) we must have $a^{-1} \in R$. Set $g(x) = af(x)$ and $b = a^{-1}$. □

Theorem 4.4.3. *Let R be a UFD and K its field of fractions. Let $f(x) \in R[x]$ be a polynomial of degree ≥ 1.*

(a) *If $f(x)$ is primitive and $f(x)|g(x)$ in $K[x]$ then $f(x)$ divides $g(x)$ also in $R[x]$.*

(b) *If $f(x)$ is irreducible in $R[x]$ then it is also irreducible in $K[x]$.*

(c) *If $f(x)$ is primitive and a prime element of $K[x]$ then $f(x)$ is also a prime element of $R[x]$.*

Proof. (a) Suppose that $g(x) = f(x)h(x)$ with $h(x) \in K[x]$. From Theorem 4.4.2 part (a) there is a nonzero $a \in K$ such that $h_1(x) = ah(x)$ is primitive in $R[x]$. Hence $g(x) = \frac{1}{a}(f(x)h_1(x)$. From Gauss' lemma $f(x)h_1(x)$ is primitive in $R[x]$ and therefore from Theorem 4.4.2 part (b) we have $\frac{1}{a} \in R$. It follows that $f(x)|g(x)$ in $R[x]$.

(b) Suppose that $g(x) \in K[x]$ is a factor of $f(x)$. From Theorem 4.4.2 part (a) there is a nonzero $a \in K$ with $g_1(x) = ag(x)$ primitive in $R[x]$. Since a is a unit in K it follows that

$$g(x)|f(x) \quad \text{in } K[x] \qquad \text{implies} \quad g_1(x)|f(x) \quad \text{in } K[x]$$

and hence since $g_1(x)$ is primitive

$$g_1(x)|f(x) \quad \text{in } R[x].$$

However by assumption $f(x)$ is irreducible in $R[x]$. This implies that either $g_1(x)$ is a unit in R or $g_1(x)$ is an associate of $f(x)$.

If $g_1(x)$ is a unit then $g_1 \in K$ and $g_1 = ga$ and hence $g \in K$, that is $g = g(x)$ is a unit.

If $g_1(x)$ is an associate of $f(x)$ then $f(x) = bg(x)$ where $b \in K$ since $g_1(x) = ag(x)$ with $a \in K$. Combining these it follows that $f(x)$ has only trivial factors in $K[x]$ and since by assumption $f(x)$ is nonconstant it follows that $f(x)$ is irreducible in $K[x]$.

(c) Suppose that $f(x)|g(x)h(x)$ with $g(x), h(x) \in R[x]$. Since $f(x)$ is a prime element in $K[x]$ we have that $f(x)|g(x)$ or $f(x)|h(x)$ in $K[x]$. From part (a) we have $f(x)|g(x)$ or $f(x)|h(x)$ in $R[x]$ implying that $f(x)$ is a prime element in $R[x]$. □

We can now state and prove our main result.

Theorem 4.4.4 (Gauss). *Let R be a UFD. Then the polynomial ring $R[x]$ is also a UFD*

Proof. By induction on degree we show that each nonunit $f(x) \in R[x]$, $f(x) \neq 0$, is a product of prime elements. Since R is an integral domain so is $R[x]$, and so the fact that $R[x]$ is a UFD then follows from Theorem 3.3.3.

If $\deg f(x) = 0$ then $f(x) = f$ is a nonunit in R. Since R is a UFD f is a product of prime elements in R. However from Theorem 4.3.3 each prime factor is then also prime in $R[x]$. Therefore $f(x)$ is a product of prime elements.

Now suppose $n > 0$ and that the claim is true for all polynomials $f(x)$ of degree $< n$. Let $f(x)$ be a polynomial of degree $n > 0$. From Theorem 4.4.2 (c) there is an $a \in R$ and a primitive $h(x) \in R[x]$ satisfying $f(x) = ah(x)$. Since R is a UFD the element a is a product of prime elements in R or a is a unit in R. Since the units in $R[x]$ are the units in R and a prime element in R is also a prime element in $R[x]$ it follows that a is a product of prime elements in $R[x]$ or a is a unit in $R[x]$. Let K be the field of fractions of R. Then $K[x]$ is a UFD. Hence $h(x)$ is a product of prime elements of $K[x]$. Let $p(x) \in K[x]$ be a prime divisor of $h(x)$. From Theorem 4.4.2 we can assume by multiplication of field elements that $p(x) \in R[x]$ and $p(x)$ is primitive. From Theorem 4.4.2 (c) it follows that $p(x)$ is a prime element of $R[x]$ and further from Theorem 4.4.3 (a) that $p(x)$ is a divisor of $h(x)$ in $R[x]$. Therefore

$$f(x) = ah(x) = ap(x)g(x) \in R[x]$$

where

(1) a is a product of prime elements of $R[x]$ or a is a unit in $R[x]$,

(2) $\deg p(x) > 0$, since $p(x)$ is a prime element in $K[x]$,

(3) $p(x)$ is a prime element in $R[x]$, and

(4) $\deg g(x) < \deg f(x)$ since $\deg p(x) > 0$.

By our inductive hypothesis we have then that $g(x)$ is a product of prime elements in $R[x]$ or $g(x)$ is a unit in $R[x]$. Therefore the claim holds for $f(x)$ and therefore holds for all $f(x)$ by induction. □

If $R[x]$ is a polynomial ring over R we can form a polynomial ring in a new indeterminate y over this ring to form $(R[x])[y]$. It is straightforward that $(R[x])[y]$ is isomorphic to $(R[y])[x]$. We denote both of these rings by $R[x, y]$ and consider this as the ring of polynomials in two commuting variables x, y with coefficients in R.

If R is a UFD then from Theorem 4.4.4 $R[x]$ is also a UFD and hence $R[x, y]$ is also a UFD. Inductively then the ring of polynomials in n commuting variables $R[x_1, x_2, \dots, x_n]$ is also a UFD.

Corollary 4.4.5. *If R is a UFD then the polynomial ring in n commuting variables $R[x_1, \dots, x_n]$ is also a UFD.*

We now give a condition for a polynomial in $R[x]$ to have a zero in $K[x]$ where K is the field of fractions of R.

Theorem 4.4.6. *Let R be a UFD and K its field of fractions. Let $f(x) = x^n + r_{n-1}x^{n-1} + \dots + r_0 \in R[x]$. Suppose that $\beta \in K$ is a zero of $f(x)$. Then β is in R and is a divisor of r_0.*

Proof. Let $\beta = \frac{r}{s}$ where $s \neq 0$ and $r, s \in R$ and r, s are coprime. Now

$$f\left(\frac{r}{s}\right) = 0 = \frac{r^n}{s^n} + r_{n-1}\frac{r^{n-1}}{s^{n-1}} + \dots + r_0.$$

Hence it follows that s must divide r^n. Since r and s are coprime s must be a unit and then without loss of generality we may assume that $s = 1$. Then $\beta \in R$ and

$$r(r^{n-1} + \dots + a_1) = -a_0$$

and so $r|a_0$. □

Note that since $\mathbb{Z}$ is a UFD, Gauss' theorem implies that $\mathbb{Z}[x]$ is also a UFD. However $\mathbb{Z}[x]$ is not a principal ideal domain. For example the set of integral polynomials with even constant term is an ideal but not principal. We leave the verification to the exercises. On the other hand we saw that if K is a field $K[x]$ is a PID. The question arises as to when $R[x]$ actually is a principal ideal domain. It turns out to be precisely when R is a field.

Theorem 4.4.7. *Let R be a commutative ring with an identity. Then the following are equivalent:*

(1) *R is a field.*

(2) *$R[x]$ is Euclidean.*

(3) *$R[x]$ is a principal ideal domain.*

Proof. From Section 4.2 we know that (a) implies (b) which in turn implies (c). Therefore we must show that (c) implies (a). Assume then that $R[x]$ is a principal ideal domain. Define the map

$$\tau : R[x] \to R$$

by

$$\tau(f(x)) = f(0).$$

It is easy to see that τ is a ring homomorphism with $R[x]/\ker(\tau) \cong R$. Therefore $\ker(\tau) \neq R[x]$. Since $R[x]$ is a principal ideal domain it is an integral domain. It follows that $\ker(\tau)$ must be a prime ideal since the quotient ring is an integral domain. However since $R[x]$ is a principal ideal domain prime ideals are maximal ideals and hence $\ker(\tau)$ is a maximal ideal. Therefore $R \cong R[x]/\ker(\tau)$ is a field. □

We now consider the relationship between irreducibles in $R[x]$ for a general integral domain and irreducibles in $K[x]$ where K is its field of fractions. This is handled by the next result called Eisenstein's criterion.

Theorem 4.4.8 (Eisenstein's criterion). *Let R be an integral domain and K its field of fractions. Let $f(x) = \sum_{i=0}^{n} a_i x^i \in R[x]$ of degree $n > 0$. Let p be a prime element of R satisfying*

(1) *$p|a_i$ for $i = 0, \ldots, n-1$.*

(2) *p does not divide a_n.*

(3) *p^2 does not divide a_0.*

Then:

(a) *If $f(x)$ is primitive then $f(x)$ is irreducible in $R[x]$.*

(b) *Suppose that R is a UFD. Then $f(x)$ is also irreducible in $K[x]$.*

Proof. (a) Suppose that $f(x) = g(x)h(x)$ with $g(x), h(x) \in R[x]$. Suppose that

$$g(x) = \sum_{i=1}^{k} b_i x^i, \quad b_k \neq 0 \quad \text{and} \quad h(x) = \sum_{i=1}^{l} c_j x^j, \quad c_l \neq 0.$$

Then $a_0 = b_0 c_0$. Now $p|a_0$ but p^2 does not divide a_0. This implies that either p doesn't divide b_0 or p doesn't divide c_0. Without loss of generality assume that $p|b_0$ and p doesn't divide c_0.

Since $a_n = b_k c_l$ and p does not divide a_n it follows that p does not divide b_k. Let b_j be the first coefficient of $g(x)$ which is not divisible by p. Consider

$$a_j = b_j c_0 + \cdots + b_0 c_j$$

where everything after the first term is divisible by p. Since p does not divide both b_j and c_0 it follows that p does not divide $b_j c_0$ and therefore p does not divide a_j which implies that $j = n$. Then from $j \leq k \leq n$ it follows that $k = n$. Therefore $\deg g(x) = \deg f(x)$ and hence $\deg h(x) = 0$. Thus $h \in R$. Then from $f(x) = hg(x)$ with f primitive it follows that h is a unit and therefore $f(x)$ is irreducible.

(b) Suppose that $f(x) = g(x)h(x)$ with $g(x), h(x) \in R[x]$. The fact that $f(x)$ was primitive was only used in the final part of part (a) so by the same arguments as in part (a) we may assume without loss of generality that $h \in R \subset K$. Therefore $f(x)$ is irreducible in $K[x]$. □

We give some examples.

Example 4.4.9. Let $R = \mathbb{Z}$ and p a prime number. Suppose that n, m are integers such that $n \geq 1$ and p does not divide m. Then $x^n \pm pm$ is irreducible in $\mathbb{Z}[x]$ and $\mathbb{Q}[x]$. In particular $(pm)^{\frac{1}{n}}$ is irrational.

Example 4.4.10. Let $R = \mathbb{Z}$ and p a prime number. Consider the polynomial

$$\Phi_p(x) = \frac{x^p - 1}{x - 1} = x^{p-1} = x^{p-2} + \cdots + 1.$$

Since all the coefficients of $\Phi_p(x)$ are equal to 1, Eisenstein's criterion is not directly applicable. However the fact that $\Phi_p(x)$ is irreducible implies that for any integer a the polynomial $\Phi_p(x + a)$ is also irreducible in $\mathbb{Z}[x]$. It follows that

$$\begin{aligned}\Phi_p(x+1) &= \frac{(x+1)^p - 1}{(x+1) - 1} = \frac{x^p + \binom{p}{1}x^{p-1} + \cdots + \binom{p}{p-1}x + 1^p - 1}{x} \\ &= x^{p-1} + \binom{p}{1}x^{p-1} + \cdots + \binom{p}{p-1}.\end{aligned}$$

Now $p | \binom{p}{i}$ for $1 \leq i \leq p-1$ (see exercises) and moreover $\binom{p}{p-1} = p$ is not divisible by p^2. Therefore we can apply the Eisenstein criterion to conclude that $\Phi_p(x)$ is irreducible in $\mathbb{Z}[x]$ and $\mathbb{Q}[x]$.

Theorem 4.4.11. *Let R be a UFD and K its field of fractions. Let $f(x) = \sum_{i=0}^{n} a_i x^i \in R[x]$ be a polynomial of degree ≥ 1. Let P be a prime ideal in R with $a_n \notin P$. Let $\overline{R} = R/P$ and let $\alpha : R[x] \to \overline{R}[x]$ be defined by*

$$\alpha\left(\sum_{i=0}^{m} r_i x^i\right) = \sum_{i=0}^{m} (r_i + P)x^i.$$

α is an epimorphism. Then if $\alpha(f(x))$ is irreducible in $\overline{R}[x]$ then $f(x)$ is irreducible in $K[x]$.

Proof. By Theorem 4.4.3 there is an $a \in R$ and a primitive $g(x) \in R[x]$ satisfying $f(x) = ag(x)$. Since $a_n \notin P$ we have that $\alpha(a) \neq 0$ and further the highest coefficient of $g(x)$ is also not an element of P. If $\alpha(g(x))$ is reducible then $\alpha(f(x))$ is also reducible. Thus $\alpha(g(x))$ is irreducible. However from Theorem 4.4.4 $g(x)$ is irreducible in $K[x]$ so $f(x) = ag(x)$ is also irreducible in $K[x]$. Therefore to prove the theorem it suffices to consider the case where $f(x)$ is primitive in $R[x]$.

Now suppose that $f(x)$ is primitive. We show that $f(x)$ is irreducible in $R[x]$.

Suppose that $f(x) = g(x)h(x)$, $g(x), h(x) \in R[x]$ with $f(x), g(x)$ nonunits in $R[x]$. Since $f(x)$ is primitive $g, h \notin R$ and so $\deg g(x) < \deg f(x)$ and $\deg h(x) < \deg f(x)$.

Now we have $\alpha(f(x)) = \alpha(g(x))\alpha(h(x))$. Since P is a prime ideal R/P is an integral domain and so in $\overline{R}[x]$ we have

$$\deg \alpha(g(x)) + \deg \alpha(g(x)) = \deg \alpha(f(x)) = \deg f(x)$$

since $a_n \notin P$. Since R is a UFD it has no zero divisors and so

$$\deg f(x) = \deg g(x) + \deg h(x).$$

Now

$$\deg \alpha(g(x)) \leq \deg g(x)$$
$$\deg \alpha(h(x)) \leq \deg h(x).$$

Therefore $\deg \alpha(g(x)) = \deg g(x)$ and $\deg \alpha(h(x)) = \deg h(x)$. Therefore $\alpha(f(x))$ is reducible and we have a contradiction. □

It is important to note that $\alpha(f(x))$ being reducible does not imply that $f(x)$ is reducible. For example $f(x) = x^2 + 1$ is irreducible in $\mathbb{Z}[x]$. However in $\mathbb{Z}_2[x]$ we have

$$x^2 + 1 = (x + 1)^2$$

and hence $f(x)$ is reducible in $\mathbb{Z}_2[x]$.

Example 4.4.12. Let $f(x) = x^5 - x^2 + 1 \in \mathbb{Z}[x]$. Choose $P = 2\mathbb{Z}$ so that

$$\alpha(f(x)) = x^5 + x^2 + 1 \in \mathbb{Z}_2[x].$$

Suppose that in $\mathbb{Z}_2[x]$ we have $\alpha(f(x)) = g(x)h(x)$. Without loss of generality we may assume that $g(x)$ is of degree 1 or 2.

If $\deg g(x) = 1$ then $\alpha(f(x))$ has a zero c in $\mathbb{Z}_2[x]$. The two possibilities for c are $c = 0$ or $c = 1$. Then

$$\text{If } c = 0 \quad \text{then } 0 + 0 + 1 = 1 \neq 0.$$
$$\text{If } c = 1 \quad \text{then } 1 + 1 + 1 = 1 \neq 0.$$

Hence the degree of $g(x)$ cannot be 1.

Suppose $\deg g(x) = 2$. The polynomials of degree 2 over $\mathbb{Z}_2[x]$ have the form

$$x^2 + x + 1, \quad x^2 + x, \quad x^2 + 1, \quad x^2.$$

The last three, x^2+x, x^2+1, x^2 all have zeros in $\mathbb{Z}_2[x]$ so they can't divide $\alpha(f(x))$. Therefore $g(x)$ must be $x^2 + x + 1$. Applying the division algorithm we obtain

$$\alpha(f(x)) = (x^3 + x^2)(x^2 + x + 1) + 1$$

and therefore x^2+x+1 does not divide $\alpha(f(x))$. It follows that $\alpha(f(x))$ is irreducible and from the previous theorem $f(x)$ must be irreducible in $\mathbb{Q}[x]$.

4.5 Exercises

1. For which $a, b \in \mathbb{Z}$ does the polynomial $x^2 + 3x + 1$ divide the polynomial $x^3 + x^2 + ax + b$?
2. Let $a + bi \in \mathbb{C}$ be a zero of $f(x) \in \mathbb{R}[x]$. Show that also $a - ib$ is a zero of $f(x)$.
3. Determine all irreducible polynomials over $\mathbb{R}$. Factorize $f(x) \in \mathbb{R}[x]$ in irreducible polynomials.
4. Let R be an integral domain, $\mathcal{A} \triangleleft R$ an ideal and $f \in R[x]$ a monic polynomial. Define $\bar{f} \in R[x]/\mathcal{A}[x]$ by the mapping $R[x] \to R/\mathcal{A}[x]$, $f = \sum a_i x^i \mapsto \bar{f} = \sum \bar{a}_i x^i$, where $\bar{a} := a + \mathcal{A}$. Show: If $\bar{f} \in R/\mathcal{A}[x]$ is irreducible then also $f \in R[x]$ is irreducible.
5. Decide if the following polynomials $f \in R[x]$ are irreducible:
   (i) $f(x) = x^3 + 2x^2 + 3$, $R = \mathbb{Z}$.
   (ii) $f(x) = x^5 - 2x + 1$, $R = \mathbb{Q}$.
   (iii) $f(x) = 3x^4 + 7x^2 + 14x + 7$, $R = \mathbb{Q}$.
   (iv) $f(x) = x^7 + (3 - i)x^2 + (3 + 4i)x + 4 + 2i$, $R = \mathbb{Z}[i]$.
   (v) $f(x) = x^4 + 3x^3 + 2x^2 + 3x + 4$, $R = \mathbb{Q}$.
   (vi) $f(x) = 8x^3 - 4x^2 + 2x - 1$, $R = \mathbb{Z}$.
6. Let R be an integral domain with characteristic 0, let $k \geq 1$ and $\alpha \in R$. In $R[x]$ define the derivatives $f^{(k)}(x)$, $k = 0, 1, 2, \ldots$, of a polynomial $f(x) \in R[x]$ by

$$f^0(x) := f(x),$$
$$f^{(k)}(x) := f^{(k-1)'}(x).$$

   Show that α is a zero of order k of the polynomial $f(x) \in R[x]$, if $f^{(k-1)}(\alpha) = 0$, but $f^{(k)}(\alpha) \neq 0$.
7. Prove that the set of integral polynomials with even constant term is an ideal but not principal.
8. Prove that $p \mid \binom{p}{i}$ for $1 \leq i \leq p - 1$.

Chapter 5
Field Extensions

5.1 Extension Fields and Finite Extensions

Much of algebra in general arose from the theory of equations, specifically polynomial equations. As discovered by Galois and Abel the solutions of polynomial equations over fields is intimately tied to the theory of field extensions. This theory eventually blossoms into Galois Theory. In this chapter we discuss the basic material concerning field extensions.

Recall that if L is a field and $K \subset L$ is also a field under the same operations as L then K is called a *subfield* of L. If we view this situation from the viewpoint of K we say that L is an *extension field* or *field extension* of K. If K, L are fields with $K \subset L$ we always assume that K is a subfield of L.

Definition 5.1.1. If K, L are fields with $K \subset L$ then we say that L is a *field extension* or *extension field* of K. We denote this by $L|K$.

Note that this is equivalent to having a field monomorphism

$$i : K \to L$$

and then identifying K and $i(K)$.

As examples we have that $\mathbb{R}$ is an extension field of $\mathbb{Q}$ and $\mathbb{C}$ is an extension field of both $\mathbb{C}$ and $\mathbb{Q}$. If K is any field then the ring of polynomials $K[x]$ over K is an integral domain. Let $K(x)$ be the field of fractions of $K[x]$. This is called the *field of rational functions* over K. Since K can be considered as part of $K[x]$ it follows that $K \subset K(x)$ and hence $K(x)$ is an extension field of K.

A crucial concept is that of the degree of a field extension. Recall that a *vector space* V over a field F consists of an abelian group V together with scalar multiplication from F satisfying:

(1) $fv \in V$ if $f \in F, v \in V$.

(2) $f(u + v) = fu + fv$ for $f \in F, u, v \in V$.

(3) $(f + g)v = fv + gv$ for $f, g \in F, v \in V$.

(4) $(fg)v = f(gv)$ for $f, g \in F, v \in V$.

(5) $1v = v$ for $v \in V$.

Notice that if K is a subfield of L then multiplication of elements of L by elements of K are still in L. Since L is an abelian group under addition, L can be considered as a vector space over K. Thus any extension field is a vector space over any of its subfields. Using this we define the degree $|L : K|$ of an extension $K \subset L$ as the dimension $\dim_K(L)$ of L as a vector space over K. We call L a finite extension of K if $|L : K| < \infty$.

Definition 5.1.2. If L is an extension field of K then the *degree of the extension* $L|K$ is defined as the dimension, $\dim_K(L)$, of L, as a vector space over K. We denote the degree by $|L : K|$. The field extension $L|K$ is a *finite extension* if the degree $|L : K|$ is finite.

Lemma 5.1.3. $|\mathbb{C} : \mathbb{R}| = 2$ *but* $|\mathbb{R} : \mathbb{Q}| = \infty$.

Proof. Every complex number can be written uniquely as $a + ib$ where $a, b \in \mathbb{R}$. Hence the elements $1, i$ constitute a basis for $\mathbb{C}$ over $\mathbb{R}$ and therefore the dimension is 2, that is $|\mathbb{C} : \mathbb{R}| = 2$.

The fact that $|\mathbb{R} : \mathbb{Q}| = \infty$ depends on the existence of *transcendental numbers*. An element $r \in \mathbb{R}$ is *algebraic* (over $\mathbb{Q}$) if it satisfies some nonzero polynomial with coefficients from $\mathbb{Q}$. That is, $P(r) = 0$, where

$$0 \neq P(x) = a_0 + a_1x + \cdots + a_nx^n \quad \text{with } a_i \in \mathbb{Q}.$$

Any $q \in \mathbb{Q}$ is algebraic since if $P(x) = x - q$ then $P(q) = 0$. However, many irrationals are also algebraic. For example $\sqrt{2}$ is algebraic since $x^2 - 2 = 0$ has $\sqrt{2}$ as a root. An element $r \in \mathbb{R}$ is *transcendental* if it is not algebraic.

In general it is very difficult to show that a particular element is transcendental. However there are uncountably many transcendental elements (see exercises). Specific examples are e and π. We will give a proof of their transcendence later in this book.

Since e is transcendental, for any natural number n the set of vectors $\{1, e, e^2, \dots, e^n\}$ must be independent over $\mathbb{Q}$, for otherwise there would be a polynomial that e would satisfy. Therefore, we have infinitely many independent vectors in $\mathbb{R}$ over $\mathbb{Q}$ which would be impossible if $\mathbb{R}$ had finite degree over $\mathbb{Q}$. □

Lemma 5.1.4. *If K is any field then* $|K(x) : K| = \infty$.

Proof. For any n the elements $1, x, x^2, \dots, x^n$ are independent over K. Therefore as in the proof of Lemma 5.1.3 $K(x)$ must be infinite dimensional over K. □

If $L|K$ and $L_1|K_1$ are field extensions then they are *isomorphic* field extensions if there exists a field isomorphism $f : L \to L_1$ such that $f_{|K}$ is an isomorphism from K to K_1.

Suppose that $K \subset L \subset M$ are fields. Below we show that the degrees multiply. In this situation where $K \subset L \subset M$ we call L an *intermediate field*.

Theorem 5.1.5. *Let K, L, M be fields with $K \subset L \subset M$. Then*

$$|M : K| = |M : L||L : K|.$$

Note that $|M : K| = \infty$ if and only if either $|M : L| = \infty$ or $|L : K| = \infty$.

Proof. Let $\{x_i : i \in I\}$ be a basis for L as a vector space over K and let $\{y_j : j \in J\}$ be a basis for M as a vector space over L. To prove the result it is sufficient to show that the set

$$B = \{x_i y_j : i \in I, j \in J\}$$

is a basis for M as a vector space over K. To show this we must show that B is a linearly independent set over K and that B spans M.

Suppose that

$$\sum_{i,j} k_{ij} x_i y_j = 0 \quad \text{where } k_{ij} \in K.$$

We can then write this sum as

$$\sum_j \Big(\sum_i k_{ij} x_i\Big) y_j = 0.$$

But $\sum_i k_{ij} x_i \in L$. Since $\{y_j : j \in J\}$ is a basis for M over L the y_j are independent over L and hence for each j we get, $\sum_i k_{ij} x_i = 0$. Now since $\{x_i : i \in I\}$ is a basis for L over K it follows that the x_i are linearly independent and since for each j we have $\sum_i k_{ij} x_i = 0$ it must be that $k_{ij} = 0$ for all i and for all j. Therefore the set B is linearly independent over K.

Now suppose that $m \in M$. Then since $\{y_j : j \in J\}$ spans M over L we have

$$m = \sum_j c_j y_j \quad \text{with } c_j \in L.$$

However $\{x_i : i \in I\}$ spans L over K and so for each c_j we have

$$c_j = \sum_i k_{ij} x_i \quad \text{with } k_{ij} \in K.$$

Combining these two sums we have

$$m = \sum_{ij} k_{ij} x_i y_j$$

and hence B spans M over K. Therefore B is a basis for M over K and the result is proved. □

Corollary 5.1.6. (a) *If* $|L : K|$ *is a prime number then there exists no proper intermediate field between* L *and* K.

(b) *If* $K \subset L$ *and* $|L : K| = 1$ *then* $L = K$.

Let $L|K$ be a field extension and suppose that $A \subset L$. Then certainly there are subrings of L containing both A and K, for example L. We denote by $K[A]$ the intersection of all subrings of L containing both K and A. Since the intersection of subrings is a subring it follows that $K[A]$ is a subring containing both K and A and the smallest such subring. We call $K[A]$ the *ring adjunction* of A to K.

In an analogous manner we let $K(A)$ be the intersection of all subfields of L containing both K and A. This is then a subfield of L and the smallest subfield of L containing both K and A. The subfield $K(A)$ is called the *field adjunction* of A to K.

Clearly $K[A] \subset K(A)$. If $A = \{a_1, \dots, a_n\}$ then we write

$$K[A] = K[a_1, \dots, a_n] \quad \text{and} \quad K(A) = K(a_1, \dots, a_n).$$

Definition 5.1.7. The field extension $L|K$ is *finitely generated* if there exist $a_1, \dots, a_n \in L$ such that $K = K(a_1, \dots, a_n)$. The extension $L|K$ is a *simple extension* if there is an $a \in L$ with $L = K(a)$. In this case a is called a *primitive* element of $L|K$.

Later we will look at an alternative way to view the adjunction constructions in terms of polynomials.

5.2 Finite and Algebraic Extensions

We now turn to the relationship between field extensions and the solution of polynomial equations.

Definition 5.2.1. Let $L|K$ be a field extension. An element $a \in L$ is *algebraic* over K if there exists a polynomial $p(x) \in K[x]$ with $p(a) = 0$. L is an *algebraic extension* of K if each element of L is algebraic over K. An element $a \in L$ that is not algebraic over K is called *transcendental*. L is a *transcendental extension* if there are transcendental elements, that is they are not algebraic over K.

For the remainder of this section we assume that $L|K$ is a field extension.

Lemma 5.2.2. *Each element of* K *is algebraic over* K.

Proof. Let $k \in K$. Then k is a root of the polynomial $p(x) = x - k \in K[x]$. □

We tie now algebraic extensions to finite extensions.

Theorem 5.2.3. *If $L|K$ is a finite extension then $L|K$ is an algebraic extension.*

Proof. Suppose that $L|K$ is a finite extension and $a \in L$. We must show that a is algebraic over K. Suppose that $|L : K| = n < \infty$ then $\dim_K(L) = n$. It follows that any $n + 1$ elements of L are linearly dependent over K.

Now consider the elements $1, a, a^2, \ldots, a^n$ in L. These are $n + 1$ distinct elements in L so they are dependent over K. Hence there exist $c_0, \ldots, c_n \in K$ not all zero such that

$$c_0 + c_1 a + \cdots + c_n a^n = 0.$$

Let $p(x) = c_0 + c_1 x + \cdots + c_n x^n$. Then $p(x) \in K[x]$ and $p(a) = 0$. Therefore a is algebraic over K. Since a was arbitrary it follows that L is an algebraic extension of K. □

From the previous theorem it follows that every finite extension is algebraic. The converse is not true, that is there are algebraic extensions that are not finite. We will give examples in Section 5.4.

The following lemma gives some examples of algebraic and transcendental extensions.

Lemma 5.2.4. *$\mathbb{C}|\mathbb{R}$ is algebraic but $\mathbb{R}|\mathbb{Q}$ and $\mathbb{C}|\mathbb{Q}$ are transcendental. If K is any field then $K(x)|K$ is transcendental.*

Proof. Since $1, i$ constitute a basis for $\mathbb{C}$ over $\mathbb{R}$ we have $|\mathbb{C} : \mathbb{R}| = 2$. Hence $\mathbb{C}$ is a finite extension of $\mathbb{R}$ and therefore from Theorem 5.2.3 an algebraic extension. More directly if $\alpha = a + ib \in \mathbb{C}$ then α is a zero of $x^2 - 2ax + (a^2 + b^2) \in \mathbb{R}[x]$.

The existence of transcendental numbers (we will discuss these more fully in Section 5.5) shows that both $\mathbb{R}|\mathbb{Q}$ and $\mathbb{C}|\mathbb{Q}$ are transcendental extensions.

Finally the element $x \in K(x)$ is not a zero of any polynomial in $K[x]$. Therefore x is a transcendental element so the extension $K(x)|K$ is transcendental. □

5.3 Minimal Polynomials and Simple Extensions

If $L|K$ is a field extension and $a \in L$ is algebraic over K then $p(a) = 0$ for some polynomial $p(x) \in K[x]$. In this section we consider the smallest such polynomial and tie it to a simple extension of K.

Definition 5.3.1. Suppose that $L|K$ is a field extension and $a \in L$ is algebraic over K. The polynomial $m_a(x) \in K[x]$ is the *minimal polynomial* of a over K if

(1) $m_a(x)$ has leading coefficient 1, that is, it is a monic polynomial.

(2) $m_a(a) = 0$.

(3) If $f(x) \in K[x]$ with $f(a) = 0$ then $m_a(x) | f(x)$.

Hence $m_a(x)$ is the monic polynomial of minimal degree that has a as a zero.

We prove next that every algebraic element has such a minimal polynomial.

Theorem 5.3.2. *Suppose that $L|K$ is a field extension and $a \in L$ is algebraic over K. Then*

(1) *The minimal polynomial $m_a(x) \in K[x]$ exists and is irreducible over K.*

(2) *$K[a] \cong K(a) \cong K[x]/(m_a(x))$ where $(m_a(x))$ is the principal ideal in $K[x]$ generated by $m_a(x)$.*

(3) *$|K(a) : K| = \deg(m_a(x))$. Therefore $K(a)|K$ is a finite extension.*

Proof. (1) Suppose that $a \in L$ is algebraic over K. Let

$$I = \{f(x) \in K[x] : f(a) = 0\}.$$

Since a is algebraic $I \neq \emptyset$. It is straightforward to show (see exercises) that I is an ideal in $K[x]$. Since K is a field we have that $K[x]$ is a principal ideal domain. Hence there exists $g(x) \in K[x]$ with $I = (g(x))$. Let b be the leading coefficient of $g(x)$. Then $m_a(x) = b^{-1}g(x)$ is a monic polynomial. We claim that $m_a(x)$ is the minimal polynomial of a and that $m_a(x)$ is irreducible.

First it is clear that $I = (g(x)) = (m_a(x))$. If $f(x) \in K[x]$ with $f(a) = 0$ then $f(x) = h(x)m_a(x)$ for some $h(x)$. Therefore $m_a(x)$ divides any polynomial that has a as a zero. It follows that $m_a(x)$ is the minimal polynomial.

Suppose that $m_a(x) = g_1(x)g_2(x)$. Then since $m_a(a) = 0$ it follows that either $g_1(a) = 0$ or $g_2(a) = 0$. Suppose $g_1(a) = 0$. Then from above $m_a(x)|g_1(x)$ and since $g_1(x)|m_a(x)$ we must then have that $g_2(x)$ is a unit. Therefore $m_a(x)$ is irreducible.

(2) Consider the map $\tau : K[x] \to K[a]$ given by

$$\tau\Big(\sum_i k_i x^i\Big) = \sum_i k_i a^i.$$

Then τ is a ring epimorphism (see exercises) and

$$\ker(\tau) = \{f(x) \in K[x] : f(a) = 0\} = (m_a(x))$$

from the argument in the proof of part (1). It follows that

$$K[x]/(m_a(x)) \cong K[a].$$

Since $m_a(x)$ is irreducible we have $K[x]/(m_a(x))$ is a field and therefore $K[a] = K(a)$.

(3) Let $n = \deg(m_a(x))$. We claim that the elements $1, a, \dots, a^{n-1}$ are a basis for $K[a] = K(a)$ over K. First suppose that

$$\sum_{i=1}^{n-1} c_i a^i = 0$$

with not all $c_i = 0$ and $c_i \in K$. Then $h(a) = 0$ where $h(x) = \sum_{i=0}^{n-1} c_i x^i$. But this contradicts the fact that $m_a(x)$ has minimal degree over all polynomials in $K[x]$ that have a as a zero. Therefore the set $1, a, \dots, a^{n-1}$ is linearly independent over K.

Now let $b \in K[a] \cong K[x]/(m_a(x))$. Then there is a $g(x) \in K[x]$ with $b = g(a)$. By the division algorithm

$$g(x) = h(x)m_a(x) + r(x)$$

where $r(x) = 0$ or $\deg(r(x)) < \deg(m_a(x))$. Now

$$r(a) = g(a) - h(a)m_a(a) = g(a) = b.$$

If $r(x) = 0$ then $b = 0$. If $r(x) \neq 0$ then since $\deg(r(x)) < n$ we have

$$r(x) = c_0 + c_1 x + \dots + c_{n-1}x^{n-1}$$

with $c_i \in K$ and some c_i but not all might be zero. This implies that

$$b = r(a) = c_0 + c_1 a + \dots + c_{n-1}a^{n-1}$$

and hence b is a linear combination over K of $1, a, \dots, a^{n-1}$. Hence $1, a, \dots, a^{n-1}$ spans $K[a]$ over K and hence forms a basis. □

Theorem 5.3.3. *Suppose that $L|K$ is a field extension and $a \in L$ is algebraic over K. Suppose that $f(x) \in K[x]$ is a monic polynomial with $f(a) = 0$. Then $f(x)$ is the minimal polynomial if and only if $f(x)$ is irreducible in $K[x]$.*

Proof. Suppose that $f(x)$ is the minimal polynomial of a. Then $f(x)$ is irreducible from the previous theorem.

Conversely suppose that $f(x)$ is monic, irreducible and $f(a) = 0$. From the previous theorem $m_a(x) | f(x)$. Since $f(x)$ is irreducible we have $f(x) = cm_a(x)$ with $c \in K$. However since both $f(x)$ and $m_a(x)$ are monic we must have $c = 1$ and $f(x) = m_a(x)$. □

We now show that a finite extension of K is actually finitely generated over K and further it is generated by finitely many algebraic elements.

Theorem 5.3.4. *Let $L|K$ be a field extension. Then the following are equivalent:*

(1) *$L|K$ is a finite extension.*

(2) *$L|K$ is an algebraic extension and there exist elements $a_1, \dots, a_n \in L$ such that $L = K(a_1, \dots, a_n)$.*

(3) *There exist algebraic elements $a_1, \dots, a_n \in L$ such that $L = K(a_1, \dots, a_n)$.*

Proof. (1) $\Rightarrow$ (2). We have seen in Theorem 5.2.3 that a finite extension is algebraic. Suppose that $a_1, \dots, a_n$ are a basis for L over K. Then clearly $L = K(a_1, \dots, a_n)$.

(2) $\Rightarrow$ (3). If $L|K$ is an algebraic extension and $L = K(a_1, \dots, a_n)$ then each a_i is algebraic over K.

(3) $\Rightarrow$ (1). Suppose that there exist algebraic elements $a_1, \dots, a_n \in L$ such that $L = K(a_1, \dots, a_n)$. We show that $L|K$ is a finite extension. We do this by induction on n. If $n = 1$ then $L = K(a)$ for some algebraic element a and the result follows from Theorem 5.3.2. Suppose now that $n \geq 2$. We assume then that an extension $K(a_1, \dots, a_{n-1})$ with $a_1, \dots, a_{n-1}$ algebraic elements is a finite extension. Now suppose that we have $L = K(a_1, \dots, a_n)$ with $a_1, \dots, a_n$ algebraic elements.

Then

$$\begin{aligned}&|K(a_1, \dots, a_n) : K| \\ &= |K(a_1, \dots, a_{n-1})(a_n) : K(a_1, \dots, a_{n-1})||K(a_1, \dots, a_{n-1}) : K|.\end{aligned}$$

The second term $|K(a_1, \dots, a_{n-1}) : K|$ is finite from the inductive hypothesis. The first term $|K(a_1, \dots, a_{n-1})(a_n) : K(a_1, \dots, a_{n-1})|$ is also finite from Theorem 5.3.2 since it is a simple extension of the field $K(a_1, \dots, a_{n-1})$ by the algebraic element a_n. Therefore $|K(a_1, \dots, a_n) : K|$ is finite. □

Theorem 5.3.5. *Suppose that K is a field and R is an integral domain with $K \subset R$. Then R can be viewed as a vector space over K. If $\dim_K(R) < \infty$ then R is a field.*

Proof. Let $r_0 \in R$ with $r_0 \neq 0$. Define the map from R to R given by

$$\tau(r) = rr_0.$$

It is easy to show (see exercises) that this is a linear transformation from R to R considered as a vector space over K.

Suppose that $\tau(r) = 0$. Then $rr_0 = 0$ and hence $r = 0$ since $r_0 \neq 0$ and R is an integral domain. It follows that τ is an injective map. Since R is a finite dimensional vector space over K and τ is an injective linear transformation it follows that τ must also be surjective. This implies that there exists and r_1 with $\tau(r_1) = 1$. Then $r_1 r_0 = 1$ and hence r_0 has an inverse within R. Since r_0 was an arbitrary nonzero element of R it follows that R is a field. □

Theorem 5.3.6. *Suppose that $K \subset L \subset M$ is a chain of field extensions. Then $M|K$ is algebraic if and only if $M|L$ is algebraic and $L|K$ is algebraic.*

Proof. If $M|K$ is algebraic then certainly $M|L$ and $L|K$ are algebraic.

Now suppose that $M|L$ and $L|K$ are algebraic. We show that $M|K$ is algebraic. Let $a \in M$. Then since a is algebraic over L there exist $b_0, b_1, \ldots, b_n \in L$ with

$$b_0 + b_1 a + \cdots + b_n a^n = 0.$$

Each b_i is algebraic over K and hence $K(b_0, \ldots, b_n)$ is finite dimensional over K. Therefore $K(b_0, \ldots, b_n)(a) = K(b_0, \ldots, b_n, a)$ is also finite dimensional over K. Therefore $K(b_0, \ldots, b_n, a)$ is a finite extension of K and hence an algebraic extension K. Since $a \in K(b_0, \ldots, b_n, a)$ it follows that a is algebraic over K and therefore M is algebraic over K. □

5.4 Algebraic Closures

As before suppose that $L|K$ is a field extension. Since each element of K is algebraic over K there are certainly algebraic elements over K within L. Let $\mathcal{A}_K$ denote the set of all elements of L that are algebraic over K. We prove that $\mathcal{A}_K$ is actually a subfield of L. It is called the *algebraic closure* of K within L.

Theorem 5.4.1. *Suppose that $L|K$ is a field extension and let $\mathcal{A}_K$ denote the set of all elements of L that are algebraic over K. Then $\mathcal{A}_K$ is a subfield of L. $\mathcal{A}_K$ is called the algebraic closure of K in L.*

Proof. Since $K \subset \mathcal{A}_K$ we have that $\mathcal{A}_K \neq \emptyset$. Let $a, b \in \mathcal{A}_K$. Since a, b are both algebraic over K from Theorem 5.3.4 we have that $K(a, b)$ is a finite extension of K. Therefore $K(a, b)$ is an algebraic extension of K and hence each element of $K(a, b)$ is algebraic over K. Now $a, b \in K(a, b)$, if $b \neq 0$, and $K(a, b)$ is a field so $a \pm b, ab$ and a/b are all in $K(a, b)$ and hence all algebraic over K. Therefore $a \pm b, ab, a/b$, if $b \neq 0$, are all in $\mathcal{A}_K$. It follows that $\mathcal{A}_K$ is a subfield of L. □

In Section 5.2 we showed that every finite extension is an algebraic extension. We mentioned that the converse is not necessarily true, that is there are algebraic extensions that are not finite. Here we give an example.

Theorem 5.4.2. *Let $\mathcal{A}$ be the algebraic closure of the rational numbers $\mathbb{Q}$ within the complex numbers $\mathbb{C}$. Then $\mathcal{A}$ is an algebraic extension of $\mathbb{Q}$ but $|\mathcal{A} : \mathbb{Q}| = \infty$.*

Proof. From the previous theorem $\mathcal{A}$ is an algebraic extension of $\mathbb{Q}$. We show that it cannot be a finite extension. By Eisenstein's criterion the rational polynomial $f(x) = x^p + p$ is irreducible over $\mathbb{Q}$ for any prime p. Let a be a zero in $\mathbb{C}$ of $f(x)$. Then

$a \in \mathcal{A}$ and $|\mathbb{Q}(a) : \mathbb{Q}| = p$. Therefore $|\mathcal{A} : \mathbb{Q}| \geq p$ for all primes p. Since there are infinitely many primes this implies that $|\mathcal{A} : \mathbb{Q}| = \infty$ □

5.5 Algebraic and Transcendental Numbers

In this section we consider the string of field extensions $\mathbb{Q} \subset \mathbb{R} \subset \mathbb{C}$.

Definition 5.5.1. An *algebraic number* α is an element of $\mathbb{C}$ which is algebraic over $\mathbb{Q}$. Hence an algebraic number is an $\alpha \in \mathbb{C}$ such that $f(\alpha) = 0$ for some $f(x) \in \mathbb{Q}[x]$. If $\alpha \in \mathbb{C}$ is not algebraic it is *transcendental*.

We will let $\mathcal{A}$ denote the totality of algebraic numbers within the complex numbers $\mathbb{C}$, and $\mathcal{T}$ the set of transcendentals so that $\mathbb{C} = \mathcal{A} \cup \mathcal{T}$. In the language of the last subsection, $\mathcal{A}$ is the algebraic closure of $\mathbb{Q}$ within $\mathbb{C}$. As in the general case, if $\alpha \in \mathbb{C}$ is algebraic we will let $m_\alpha(x)$ denote the minimal polynomial of α over $\mathbb{Q}$.

We now examine the sets $\mathcal{A}$ and $\mathcal{T}$ more closely. Since $\mathcal{A}$ is precisely the algebraic closure of $\mathbb{Q}$ in $\mathbb{C}$ we have from our general result that $\mathcal{A}$ actually forms a subfield of $\mathbb{C}$. Further since the intersection of subfields is again a subfield it follows that $\mathcal{A}' = \mathcal{A} \cap \mathbb{R}$ the real algebraic numbers form a subfield of the reals.

Theorem 5.5.2. *The set $\mathcal{A}$ of algebraic numbers forms a subfield of $\mathbb{C}$. The subset $\mathcal{A}' = \mathcal{A} \cap \mathbb{R}$ of real algebraic numbers forms a subfield of $\mathbb{R}$.*

Since each rational is algebraic it is clear that there are algebraic numbers. Further there are irrational algebraic numbers, $\sqrt{2}$ for example, since it satisfies the irreducible polynomial $x^2 - 2 = 0$ over $\mathbb{Q}$. On the other hand we haven't examined the question of whether transcendental numbers really exist. To show that any particular complex number is transcendental is in general quite difficult. However it is relatively easy to show that there are uncountably infinitely many transcendentals.

Theorem 5.5.3. *The set $\mathcal{A}$ of algebraic numbers is countably infinite. Therefore $\mathcal{T}$ the set of transcendental numbers and $\mathcal{T}' = \mathcal{T} \cap \mathbb{R}$, the real transcendental numbers, are uncountably infinite.*

Proof. Let

$$\mathcal{P}_n = \{f(x) \in \mathbb{Q}[x] : \deg(f(x)) \leq n\}.$$

Since if $f(x) \in \mathcal{P}_n$, $f(x) = q_o + q_1 x + \cdots + q_n x^n$ with $q_i \in \mathbb{Q}$ we can identify a polynomial of degree $\leq n$ with an $(n+1)$-tuple $(q_0, q_1, \ldots, q_n)$ of rational numbers. Therefore the set $\mathcal{P}_n$ has the same size as the $(n+1)$-fold Cartesian product of $\mathbb{Q}$:

$$\mathbb{Q}^{n+1} = \mathbb{Q} \times \mathbb{Q} \times \cdots \times \mathbb{Q}.$$

Since a finite Cartesian product of countable sets is still countable it follows that $\mathcal{P}_n$ is a countable set.

Now let

$$\mathcal{B}_n = \bigcup_{p(x)\in\mathcal{P}_n} \{\text{roots of} p(x)\},$$

that is $\mathcal{B}_n$ is the union of all roots in $\mathbb{C}$ of all rational polynomials of degree $\le n$. Since each such $p(x)$ has a maximum of n roots and since $\mathcal{P}_n$ is countable it follows that $\mathcal{B}_n$ is a countable union of finite sets and hence is still countable. Now

$$\mathcal{A} = \bigcup_{n=1}^{\infty} \mathcal{B}_n$$

so $\mathcal{A}$ is a countable union of countable sets and is therefore countable.

Since both $\mathbb{R}$ and $\mathbb{C}$ are uncountably infinite the second assertions follow directly from the countability of $\mathcal{A}$. If say $\mathcal{T}$ were countable then $\mathbb{C} = \mathcal{A} \cup \mathcal{T}$ would also be countable which is a contradiction. □

From Theorem 5.5.3 we know that there exist infinitely many transcendental numbers. Liouville in 1851 gave the first proof of the existence of transcendentals by exhibiting a few. He gave as one the following example.

Theorem 5.5.4. *The real number*

$$c = \sum_{j=1}^{\infty} \frac{1}{10^{j!}}$$

is transcendental.

Proof. First of all since $\frac{1}{10^{j!}} < \frac{1}{10^j}$, and $\sum_{j=1}^{\infty} \frac{1}{10^j}$ is a convergent geometric series, it follows from the comparison test that the infinite series defining c converges and defines a real number. Further since $\sum_{j=1}^{\infty} \frac{1}{10^j} = \frac{1}{9}$, it follows that $c < \frac{1}{9} < 1$.

Suppose that c is algebraic so that $g(c) = 0$ for some rational nonzero polynomial $g(x)$. Multiplying through by the least common multiple of all the denominators in $g(x)$ we may suppose that $f(c) = 0$ for some integral polynomial $f(x) = \sum_{j=0}^{n} m_j x^j$. Then c satisfies

$$\sum_{j=0}^{n} m_j c^j = 0$$

for some integers $m_0, \dots, m_n$.

If $0 < x < 1$ then by the triangle inequality

$$|f'(x)| = \left|\sum_{j=1}^{n} jm_j x^{j-1}\right| \le \sum_{j=1}^{n} |jm_j| = B$$

where B is a real constant depending only on the coefficients of $f(x)$.

Now let

$$c_k = \sum_{j=1}^{k} \frac{1}{10^{j!}}$$

be the k-th partial sum for c. Then

$$|c - c_k| = \sum_{j=k+1}^{\infty} \frac{1}{10^{j!}} < 2 \cdot \frac{1}{10^{(k+1)!}}.$$

Apply the mean value theorem to $f(x)$ at c and c_k to obtain

$$|f(c) - f(c_k)| = |c - c_k||f'(\zeta)|$$

for some ζ with $c_k < \zeta < c < 1$. Now since $0 < \zeta < 1$ we have

$$|c - c_k||f'(\zeta)| < 2B \frac{1}{10^{(k+1)!}}.$$

On the other hand, since $f(x)$ can have at most n roots, it follows that for all k large enough we would have $f(c_k) \neq 0$. Since $f(c) = 0$ we have

$$|f(c) - f(c_k)| = |f(c_k)| = \left|\sum_{j=1}^{n} m_j c_k^j\right| > \frac{1}{10^{nk!}}$$

since for each j, $m_j c_k^j$ is a rational number with denominator $10^{jk!}$. However if k is chosen sufficiently large and n is fixed we have

$$\frac{1}{10^{nk!}} > \frac{2B}{10^{(k+1)!}}$$

contradicting the equality from the mean value theorem. Therefore c is transcendental. □

In 1873 Hermite proved that e is transcendental while Lindemann in 1882 showed that π is transcendental. Schneider in 1934 showed that a^b is transcendental if $a \neq 0$, a and b are algebraic and b is irrational. Later in the book we will prove that both e and π are transcendental. An interesting open question is the following:

Is π transcendental over $\mathbb{Q}(e)$?

To close this section we show that in general if $a \in L$ is transcendental over K then $K(a)|K$ is isomorphic to the field of rational functions over K.

Theorem 5.5.5. *Suppose that $L|K$ is a field extension and $a \in L$ is transcendental over K. Then $K(a)|K$ is isomorphic to $K(x)|K$. Here the isomorphism $\mu : K(x) \to K(a)$ can be chosen such that $\mu(x) = a$.*

Proof. Define the map $\mu : K(x) \to K(a)$ by

$$\mu\left(\frac{f(x)}{g(x)}\right) = \frac{f(a)}{g(a)}$$

for $f(x), g(x) \in K[x]$ with $g(x) \neq 0$. Then μ is a homomorphism and $\mu(x) = a$. Since $\mu \neq 0$ it follows that μ is an isomorphism. □

5.6 Exercises

1. Let $a \in \mathbb{C}$ with $a^3 - 2a + 2 = 0$ and $b = a^2 - a$. Compute the minimal polynomial $m_b(x)$ of b over $\mathbb{Q}$ and compute the inverse of b in $\mathbb{Q}(a)$.
2. Determine the algebraic closure of $\mathbb{R}$ in $\mathbb{C}(x)$.
3. Let $a_n := \sqrt[2^n]{2} \in \mathbb{R}$, $n = 1, 2, 3, \dots$ and $A := \{a_n : n \in \mathbb{N}\}$ and $E := \mathbb{Q}(A)$. Show:
   (i) $|\mathbb{Q}(a_n) : \mathbb{Q}| = 2^n$.
   (ii) $|E : \mathbb{Q}| = \infty$.
   (iii) $E = \bigcup_{n=1}^{\infty} \mathbb{Q}(a_n)$.
   (iv) E is algebraic over $\mathbb{Q}$.
4. Determine $|E : \mathbb{Q}|$ for
   (i) $E = \mathbb{Q}(\sqrt{2}, \sqrt{-2})$.
   (ii) $E = \mathbb{Q}(\sqrt{3}, \sqrt{3} + \sqrt[3]{3})$.
   (iii) $E = \mathbb{Q}(\frac{1+i}{\sqrt{2}}, \frac{-1+i}{\sqrt{2}})$.
5. Show that $\mathbb{Q}(\sqrt{2}, \sqrt{3}) = \{a + b\sqrt{2} + c\sqrt{3} + d\sqrt{6} : a, b, c, d \in \mathbb{Q}\}$. Determine the degree of $\mathbb{Q}(\sqrt{2}, \sqrt{3})$ over $\mathbb{Q}$. Further show that $\mathbb{Q}(\sqrt{2}, \sqrt{3}) = \mathbb{Q}(\sqrt{2} + \sqrt{3})$.
6. Let K, E be fields and $a \in E$ be transcendental over K. Show:
   (i) Each element of $K(a)|K$ is transcendental over K.
   (ii) a^n is transcendental over K for each $n > 1$.
   (iii) If $L := K(\frac{a^3}{a+1})$ then a is algebraic over L. Determine the minimal polynomial $m_a(x)$ of a over L.

7. Let K be a field and $a \in K(x)|K$. Show:
   (i) x is algebraic over $K(a)$.
   (ii) If L is a field with $K \subset L \subseteq K(x)$, then $|K(x) : L| < \infty$.
   (iii) a is transcendental over K.

8. Suppose that $a \in L$ is algebraic over K. Let

$$I = \{f(x) \in K[x] : f(a) = 0\}.$$

   Since a is algebraic $I \neq \emptyset$. Prove that I is an ideal in $K[x]$.

9. Prove that there are uncountably many transcendental numbers. To do this show that the set $\mathcal{A}$ of algebraic numbers is countable. To do this:
   (i) Show that $\mathbb{Q}_n[x]$ the set of rational polynomials of degree $\leq n$ is countable (finite Cartesian product of countable sets).
   (ii) Let $\mathcal{B}_n = \{\text{Zeros of polynomials in } \mathbb{Q}_n\}$. Show that $\mathcal{B}$ is countable.
   (iii) Show that $\mathcal{A} = \bigcup_{n=1}^{\infty} \mathcal{B}_n$ and conclude that $\mathcal{A}$ is countable.
   (iv) Show that the transcendental numbers are uncountable.

10. Consider the map $\tau : K[x] \to K[a]$ given by

$$\tau\Big(\sum_i k_i x^i\Big) = \sum_i k_i a^i.$$

   Show that τ is a ring epimorphism.

11. Suppose that K is a field and R is an integral domain with $K \subset R$. Then R can be viewed as a vector space over K. Let $r_0 \in R$ with $r_0 \neq 0$. Define the map from R to R given by

$$\tau(r) = r r_0.$$

   Show that this is a linear transformation from R to R considered as a vector space over K.

Chapter 6
Field Extensions and Compass and Straightedge Constructions

6.1 Geometric Constructions

Greek mathematicians in the classical period posed the problem of constructing certain geometric figures in the Euclidean plane using only a straightedge and a compass. These are known as *geometric construction problems*.

Recall from elementary geometry that using a straightedge and compass it is possible to draw a line parallel to a given line segment through a given point, to extend a given line segment, and to erect a perpendicular to a given line at a given point on that line. There were other geometric construction problems that the Greeks could not determine straightedge and compass solutions but on the other hand were never able to prove that such constructions were impossible. In particular there were four famous insolvable (to the Greeks) construction problems. The first is the *squaring of the circle*. This problem is, given a circle, to construct using straightedge and compass a square having area equal to that of the given circle. The second is the *doubling of the cube*. This problem is given a cube of given side length, to construct using a straightedge and compass, a side of a cube having double the volume of the original cube. The third problem is the *trisection of an angle*. This problem is to trisect a given angle using only a straightedge and compass. The final problem is the *construction of a regular n-gon*. This problems asks which regular n-gons could be constructed using only straightedge and compass.

By translating each of these problems into the language of field extensions we can show that each of the first three problems are insolvable in general and we can give the complete solution to the construction of the regular n-gons.

6.2 Constructible Numbers and Field Extensions

We now translate the geometric construction problems into the language of field extensions. As a first step we define a *constructible number*.

Definition 6.2.1. Suppose we are given a line segment of unit length. An $\alpha \in \mathbb{R}$ is *constructible* if we can construct a line segment of length $|\alpha|$ in a finite number of steps from the unit segment using a straightedge and compass.

Our first result is that the set of all constructible numbers forms a subfield of $\mathbb{R}$.

Theorem 6.2.2. *The set $\mathcal{C}$ of all constructible numbers forms a subfield of $\mathbb{R}$. Further, $\mathbb{Q} \subset \mathcal{C}$.*

Proof. Let $\mathcal{C}$ be the set of all constructible numbers. Since the given unit length segment is constructible, we have $1 \in \mathcal{C}$. Therefore, $\mathcal{C} \neq \emptyset$, and thus to show that it is a field we must show that it is closed under the field operations.

Suppose α, β are constructible. We must show then that $\alpha \pm \beta, \alpha\beta$, and α/β for $\beta \neq 0$ are constructible. If $\alpha, \beta > 0$, construct a line segment of length $|\alpha|$. At one end of this line segment extend it by a segment of length $|\beta|$. This will construct a segment of length $\alpha + \beta$. Similarly, if $\alpha > \beta$, lay off a segment of length $|\beta|$ at the beginning of a segment of length $|\alpha|$. The remaining piece will be $\alpha - \beta$. By considering cases we can do this in the same manner if either α or β or both are negative. These constructions are pictured in Figure 6.1. Therefore, $\alpha \pm \beta$ are constructible.

Figure 6.1

In Figure 6.2 we show how to construct $\alpha\beta$. Let the line segment $\overline{OA}$ have length $|\alpha|$. Consider a line L through O not coincident with $\overline{OA}$. Let $\overline{OB}$ have length $|\beta|$ as in the diagram. Let P be on ray $\overline{OB}$ so that $\overline{OP}$ has length 1. Draw $\overline{AP}$ and then find Q on ray $\overline{OA}$ such that $\overline{BQ}$ is parallel to $\overline{AP}$. From similar triangles we then have

$$\frac{|\overline{OP}|}{|\overline{OB}|} = \frac{|\overline{OA}|}{|\overline{OQ}|} \Rightarrow \frac{1}{|\beta|} = \frac{|\alpha|}{|\overline{OQ}|}.$$

Then $|\overline{OQ}| = |\alpha||\beta|$, and so $\alpha\beta$ is constructible.

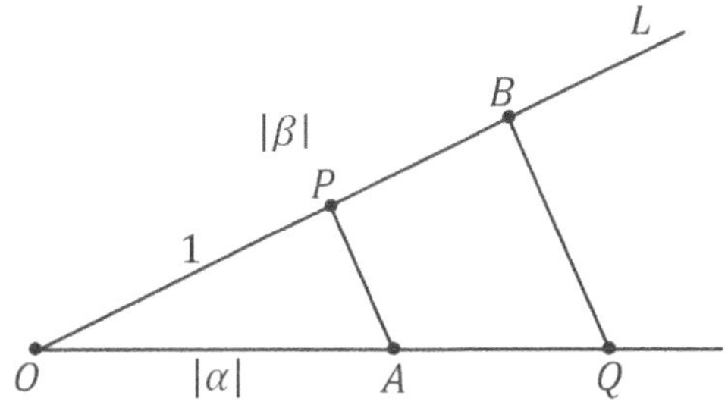

Figure 6.2

A similar construction, pictured in Figure 6.3, shows that α/β for $\beta \neq 0$ is constructible. Find $\overline{OA}, \overline{OB}, \overline{OP}$ as above. Now, connect A to B and let $\overline{PQ}$ be parallel to $\overline{AB}$. From similar triangles again we have

$$\frac{1}{|\beta|} = \frac{|\overline{OQ}|}{|\alpha|} \implies \frac{|\alpha|}{|\beta|} = |\overline{OQ}|.$$

Hence α/β is constructible.

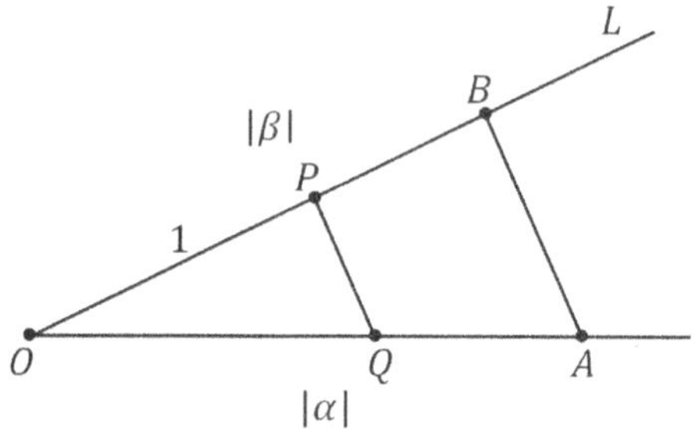

Figure 6.3

Therefore, $\mathcal{C}$ is a subfield of $\mathbb{R}$. Since char $\mathcal{C} = 0$, it follows that $\mathbb{Q} \subset \mathcal{C}$. □

Let us now consider analytically how a constructible number is found in the plane. Starting at the origin and using the unit length and the constructions above, we can locate any point in the plane with rational coordinates. That is, we can construct the point $P = (q_1, q_2)$ with $q_1, q_2 \in \mathbb{Q}$. Using only straightedge and compass, any further point in the plane can be determined in one of the following three ways.

1. The intersection point of two lines each of which passes through two known points each having rational coordinates.

2. The intersection point of a line passing through two known points having rational coordinates and a circle whose center has rational coordinates and whose radius squared is rational.

3. The intersection point of two circles each of whose centers has rational coordinates and each of whose radii is the square root of a rational number.

Analytically, the first case involves the solution of a pair of linear equations each with rational coefficients and thus only leads to other rational numbers. In cases two and three we must solve equations of the form $x^2 + y^2 + ax + by + c = 0$, with $a, b, c \in \mathbb{Q}$. These will then be quadratic equations over $\mathbb{Q}$, and thus the solutions will either be in $\mathbb{Q}$ or in a quadratic extension $\mathbb{Q}(\sqrt{\alpha})$ of $\mathbb{Q}$. Once a real quadratic extension of $\mathbb{Q}$ is found, the process can be iterated. Conversely it can be shown that if α is constructible, so is $\sqrt{\alpha}$. We thus can prove the following theorem.

Theorem 6.2.3. *If γ is constructible with $\gamma \notin \mathbb{Q}$, then there exists a finite number of elements $\alpha_1, \dots, \alpha_r \in \mathbb{R}$ with $\alpha_r = \gamma$ such that for $i = 1, \dots, r$, $\mathbb{Q}(\alpha_1, \dots, \alpha_i)$ is a quadratic extension of $\mathbb{Q}(\alpha_1, \dots, \alpha_{i-1})$. In particular, $|\mathbb{Q}(\gamma) : \mathbb{Q}| = 2^n$ for some $n \geq 1$.*

Therefore, the constructible numbers are precisely those real numbers that are contained in repeated quadratic extensions of $\mathbb{Q}$. In the next section we use this idea to show the impossibility of the first three mentioned construction problems.

6.3 Four Classical Construction Problems

We now consider the aforementioned construction problems. Our main technique will be to use Theorem 6.2.3. From this result we have that if γ is constructible with $\gamma \notin \mathbb{Q}$, then $|\mathbb{Q}(\gamma) : \mathbb{Q}| = 2^n$ for some $n \geq 1$.

6.3.1 Squaring the Circle

Theorem 6.3.1. *It is impossible to square the circle. That is, it is impossible in general, given a circle, to construct using straightedge and compass a square having area equal to that of the given circle.*

Proof. Suppose the given circle has radius 1. It is then constructible and would have an area of π. A corresponding square would then have to have a side of length $\sqrt{\pi}$. To be constructible a number α must have $|\mathbb{Q}(\alpha) : \mathbb{Q}| = 2^m < \infty$ and hence α must be algebraic. However π is transcendental, so $\sqrt{\pi}$ is also transcendental and therefore not constructible. □

6.3.2 The Doubling of the Cube

Theorem 6.3.2. *It is impossible to double the cube. This means that it is impossible in general, given a cube of given side length, to construct using a straightedge and compass, a side of a cube having double the volume of the original cube.*

Proof. Let the given side length be 1, so that the original volume is also 1. To double this we would have to construct a side of length $2^{1/3}$. However $|\mathbb{Q}(2^{1/3}) : \mathbb{Q}| = 3$ since the minimal polynomial over $\mathbb{Q}$ is $m_{2^{1/3}}(x) = x^3 - 2$. This is not a power of 2 so $2^{1/3}$ is not constructible. □

6.3.3 The Trisection of an Angle

Theorem 6.3.3. *It is impossible to trisect an angle. This means that it is impossible in general to trisect a given angle using only a straightedge and compass.*

Proof. An angle θ is constructible if and only if a segment of length $|\cos\theta|$ is constructible. Since $\cos(\pi/3) = 1/2$, therefore $\pi/3$ is constructible. We show that it cannot be trisected by straightedge and compass.

The following trigonometric identity holds

$$\cos(3\theta) = 4\cos^3(\theta) - 3\cos(\theta).$$

Let $\alpha = \cos(\pi/9)$. From the above identity we have $4\alpha^3 - 3\alpha - \frac{1}{2} = 0$. The polynomial $4x^3 - 3x - \frac{1}{2}$ is irreducible over $\mathbb{Q}$, and hence the minimal polynomial over $\mathbb{Q}$ is $m_\alpha(x) = x^3 - \frac{3}{4}x - \frac{1}{8}$. It follows that $|\mathbb{Q}(\alpha) : \mathbb{Q}| = 3$, and hence α is not constructible. Therefore, the corresponding angle $\pi/9$ is not constructible. Therefore, $\pi/3$ is constructible, but it cannot be trisected. □

6.3.4 Construction of a Regular n-Gon

The final construction problem we consider is the construction of regular n-gons. The algebraic study of the constructibility of regular n-gons was initiated by Gauss in the early part of the nineteenth century.

Notice first that a regular n-gon will be constructible for $n \geq 3$ if and only if the angle $\frac{2\pi}{n}$ is constructible, which is the case if and only if the length $\cos\frac{2\pi}{n}$ is a constructible number. From our techniques if $\cos\frac{2\pi}{n}$ is a constructible number then necessarily $|\mathbb{Q}(\cos(\frac{2\pi}{n})) : \mathbb{Q}| = 2^m$ for some m. After we discuss Galois theory we see that this condition is also sufficient. Therefore $\cos\frac{2\pi}{n}$ is a constructible number if and only if $|\mathbb{Q}(\cos(\frac{2\pi}{n})) : \mathbb{Q}| = 2^m$ for some m.

The solution of this problem, that is the determination of when $|\mathbb{Q}(\cos(\frac{2\pi}{n}) : \mathbb{Q}| = 2^m$ involves two concepts from number theory; the *Euler phi-function* and *Fermat primes*.

Definition 6.3.4. For any natural number n, the *Euler phi-function* is defined by

$$\phi(n) = \text{number of integers less than or equal to } n \text{ and relatively prime to } n.$$

Example 6.3.5. $\phi(6) = 2$ since among $1, 2, 3, 4, 5, 6$ only $1, 5$ are relatively prime to 6.

It is fairly straightforward to develop a formula for $\phi(n)$. A formula is first determined for primes and for prime powers and then pasted back together via the fundamental theorem of arithmetic.

Lemma 6.3.6. *For any prime p and $m > 0$,*

$$\phi(p^m) = p^m - p^{m-1} = p^m\left(1 - \frac{1}{p}\right).$$

Proof. If $1 \leq a \leq p$ then either $a = p$ or $(a, p) = 1$. It follows that the positive integers less than or equal to p^m which are not relatively prime to p^m are precisely the multiples of p that is $p, 2p, 3p, \ldots, p^{m-1} \cdot p$. All other positive $a < p^m$ are relatively prime to p^m. Hence the number relatively prime to p^m is

$$p^m - p^{m-1}. \qquad \square$$

Lemma 6.3.7. *If* $(a, b) = 1$ *then* $\phi(ab) = \phi(a)\phi(b)$.

Proof. Given a natural number n a *reduced residue system* modulo n is a set of integers $x_1, \ldots, x_k$ such that each x_i is relatively prime to n, $x_i \not\equiv x_j \bmod n$ unless $i = j$ and if $(x, n) = 1$ for some integer x then $x \equiv x_i \bmod n$ for some i. Clearly $\phi(n)$ is the size of a reduced residue system modulo n.

Let $R_a = \{x_1, \ldots, x_{\phi(a)}\}$ be a reduced residue system modulo a, $R_b = \{y_1, \ldots, y_{\phi(b)}\}$ be a reduced residue system modulo b, and let

$$S = \{ay_i + bx_j : i = 1, \ldots, \phi(b), j = 1, \ldots, \phi(a)\}.$$

We claim that S is a reduced residue system modulo ab. Since S has $\phi(a)\phi(b)$ elements it will follow that $\phi(ab) = \phi(a)\phi(b)$.

To show that S is a reduced residue system modulo ab we must show three things: first that each $x \in S$ is relatively prime to ab; second that the elements of S are distinct; and finally that given any integer n with $(n, ab) = 1$ then $n \equiv s \bmod ab$ for some $s \in S$.

Let $x = ay_i + bx_j$. Then since $(x_j, a) = 1$ and $(a, b) = 1$ it follows that $(x, a) = 1$. Analogously $(x, b) = 1$. Since x is relatively prime to both a and b we have $(x, ab) = 1$. This shows that each element of S is relatively prime to ab.

Next suppose that

$$ay_i + bx_j \equiv ay_k + bx_l \mod ab.$$

Then

$$ab|(ay_i + bx_j) - (ay_k + bx_l) \implies ay_i \equiv ay_k \mod b.$$

Since $(a, b) = 1$ it follows that $y_i \equiv y_k \bmod b$. But then $y_i = y_k$ since R_b is a reduced residue system. Similarly $x_j = x_l$. This shows that the elements of S are distinct modulo ab.

Finally suppose $(n, ab) = 1$. Since $(a, b) = 1$ there exist x, y with $ax + by = 1$. Then

$$anx + bny = n.$$

Since $(x, b) = 1$ and $(n, b) = 1$ it follows that $(nx, b) = 1$. Therefore there is an s_i with $nx = s_i + tb$. In the same manner $(ny, a) = 1$ and so there is an r_j with

$ny = r_j + ua$. Then

$$a(s_i + tb) + b(r_j + ua) = n \implies n = as_i + br_j + (t+u)ab$$
$$\implies n \equiv ar_i + bs_j \mod ab$$

and we are done. □

We now give the general formula for $\phi(n)$.

Theorem 6.3.8. *Suppose* $n = p_1^{e_1} \cdots p_k^{e_k}$ *then*

$$\phi(n) = (p_1^{e_1} - p_1^{e_1 - 1})(p_2^{e_2} - p_2^{e_2 - 1}) \cdots (p_k^{e_k} - p_k^{e_k - 1}).$$

Proof. From the previous lemma we have

$$\begin{aligned}\phi(n) &= \phi(p_1^{e_1})\phi(p_2^{e2}) \cdots \phi(p_k^{e_k}) \\ &= (p_1^{e_1} - p_1^{e_1-1})(p_2^{e_2} - p_2^{e_2-1}) \cdots (p_k^{e_k} - p_k^{e_k-1}) \\ &= p_1^{e_1}(1 - 1/p_1) \cdots p_k^{e_k}(1 - 1/p_k) = p_1^{e_1} \cdots p_k^{e_k} \cdot (1 - 1/p_1) \cdots (1 - 1/p_k) \\ &= n \prod_i (1 - 1/p_i).\end{aligned}$$

□

Example 6.3.9. Determine $\phi(126)$. Now

$$126 = 2 \cdot 3^2 \cdot 7 \implies \phi(126) = \phi(2)\phi(3^2)\phi(7) = (1)(3^2 - 3)(6) = 36.$$

Hence there are 36 units in $\mathbb{Z}_{126}$.

An interesting result with many generalizations in number theory is the following.

Theorem 6.3.10. *For* $n > 1$ *and for* $d \geq 1$

$$\sum_{d|n} \phi(d) = n.$$

Proof. We first prove the theorem for prime powers and then paste together via the fundamental theorem of arithmetic.

Suppose that $n = p^e$ for p a prime. Then the divisors of n are $1, p, p^2, \ldots, p^e$, so

$$\begin{aligned}\sum_{d|n} \phi(d) &= \phi(1) + \phi(p) + \phi(p^2) + \cdots + \phi(p^e) \\ &= 1 + (p-1) + (p^2 - p) + \cdots + (p^e - p^{e-1}).\end{aligned}$$

Notice that this sum telescopes, that is $1 + (p-1) = p$, $p + (p^2 - p) = p^2$ and so on. Hence the sum is just p^e and the result is proved for n a prime power.

We now do an induction on the number of distinct prime factors of n. The above argument shows that the result is true if n has only one distinct prime factor. Assume that the result is true whenever an integer has less than k distinct prime factors and suppose $n = p_1^{e_1} \cdots p_k^{e_k}$ has k distinct prime factors. Then $n = p^e c$ where $p = p_1$, $e = e_1$ and c has fewer than k distinct prime factors. By the inductive hypothesis

$$\sum_{d|c} \phi(d) = c.$$

Since $(c, p) = 1$ the divisors of n are all of the form $p^\alpha d_1$ where $d_1|c$ and $\alpha = 0, 1, \ldots, e$. It follows that

$$\sum_{d|n} \phi(d) = \sum_{d_1|c} \phi(d_1) + \sum_{d_1|c} \phi(pd_1) + \cdots + \sum_{d_1|c} \phi(p^e d_1).$$

Since $(d_1, p^\alpha) = 1$ for any divisor of c this sum equals

$$\begin{aligned}
&\sum_{d_1|c} \phi(d_1) + \sum_{d_1|c} \phi(p)\phi(d_1) + \cdots + \sum_{d_1|c} \phi(p^e)\phi(d_1) \\
&= \sum_{d_1|c} \phi(d_1) + (p-1)\sum_{d_1|c} \phi(d_1) + \cdots + (p^e - p^{e-1})\sum_{d_1|c} \phi(d_1) \\
&= c + (p-1)c + (p^2 - p)c + \cdots + (p^e - p^{e-1})c.
\end{aligned}$$

As in the case of prime powers this sum telescopes giving a final result

$$\sum_{d|n} \phi(d) = p^e c = n. \qquad \square$$

Example 6.3.11. Consider $n = 10$. The divisors are $1, 2, 5, 10$. Then $\phi(1) = 1$, $\phi(2) = 1$, $\phi(5) = 4$, $\phi(10) = 4$. Then

$$\phi(1) + \phi(2) + \phi(5) + \phi(10) = 1 + 1 + 4 + 4 = 10.$$

We will see later in the book that the Euler phi-function plays an important role in the structure theory of abelian groups.

We now turn to *Fermat primes.*

Definition 6.3.12. The *Fermat numbers* are the sequence (F_n) of positive integers defined by

$$F_n = 2^{2^n} + 1, \quad n = 0, 1, 2, 3, \ldots.$$

If a particular F_n is prime it is called a *Fermat prime.*

Fermat believed that all the numbers in this sequence were primes. In fact F_0, F_1, F_2, F_3, F_4 are all prime but F_5 is composite and divisible by 641 (see exercises). It is still an open question whether or not there are infinitely many Fermat primes. It has been conjectured that there are only finitely many. On the other hand if a number of the form $2^n + 1$ is a prime for some integer n then it must be a Fermat prime.

Theorem 6.3.13. *If $a \geq 2$ and $a^n + 1$, $n \geq 1$, is a prime then a is even and $n = 2^m$ for some nonnegative integer m. In particular if $p = 2^k + 1$, $k \geq 1$, is a prime then $k = 2^n$ for some n and p is a Fermat prime.*

Proof. If a is odd then $a^n + 1$ is even and hence not a prime. Suppose then that a is even and $n = kl$ with k odd and $k \geq 3$. Then

$$\frac{a^{kl} + 1}{a^l + 1} = a^{(k-1)l} - a^{(k-2)l} + \cdots + 1.$$

Therefore $a^l + 1$ divides $a^{kl} + 1$ if $k \geq 3$. Hence if $a^n + 1$ is a prime we must have $n = 2^m$. □

We can now state the solution to the constructibility of regular n-gons.

Theorem 6.3.14. *A regular n-gon is constructible with a straightedge and compass if and only if $n = 2^m p_1 \cdots p_k$ where $p_1, \ldots, p_k$ are distinct Fermat primes.*

Before proving the theorem notice for example that a regular 20-gon is constructible since $20 = 2^2 \cdot 5$ and 5 is a Fermat prime. On the other hand a regular 11-gon is not constructible.

Proof. Let $\mu = e^{\frac{2\pi i}{n}}$ be a primitive n-th root of unity. Since

$$e^{\frac{2\pi i}{n}} = \cos\left(\frac{2\pi}{n}\right) + i \sin\left(\frac{2\pi}{n}\right)$$

is easy to compute that (see exercises)

$$\mu + \frac{1}{\mu} = 2\cos\left(\frac{2\pi}{n}\right).$$

Therefore $\mathbb{Q}(\mu + \frac{1}{\mu}) = \mathbb{Q}(\cos(\frac{2\pi}{n}))$. After we discuss Galois theory in more detail we will prove that

$$\left|\mathbb{Q}\left(\mu + \frac{1}{\mu}\right) : \mathbb{Q}\right| = \frac{\phi(n)}{2}$$

where $\phi(n)$ is the Euler phi-function. Therefore $\cos(\frac{2\pi}{n})$ is constructible if and only if $\frac{\phi(n)}{2}$ and hence $\phi(n)$ is a power of 2.

Suppose that $n = 2^m p_1^{e_1} \cdots p_k^{e_k}$, all p_i odd primes. Then from Theorem 6.3.8

$$\phi(n) = 2^{m-1} \cdot (p_1^{e_1} - p_1^{e_1-1})(p_2^{e_2} - p_2^{e_2-1}) \cdots (p_k^{e_k} - p_k^{e_k-1}).$$

If this was a power of 2 each factor must also be a power of 2. Now

$$p_i^{e_i} - p_i^{e_i-1} = p_i^{e_i-1}(p_i - 1).$$

If this is to be a power of 2 we must have $e_i = 1$ and $p_i - 1 = 2^{k_i}$ for some k_i. Therefore each prime is distinct to the first power and $p_i = 2^{k_i} + 1$ is a Fermat prime proving the theorem. □

6.4 Exercises

1. Let ϕ be a given angle. In which of the following cases is the angle ψ constructible from the angle ϕ by compass and straightedge?
   (a) $\phi = \frac{\pi}{13}$, $\psi = \frac{\pi}{26}$.
   (b) $\phi = \frac{\pi}{33}$, $\psi = \frac{\pi}{11}$.
   (c) $\phi = \frac{\pi}{7}$, $\psi = \frac{\pi}{12}$.
2. (The golden section) In the plane let $\overline{AB}$ be a given segment from A to B with length a. The segment $\overline{AB}$ should be divided such that the proportion of $\overline{AB}$ to the length of the bigger subsegment is equal to the proportion of the length of the bigger subsegment to the length of the smaller subsegment:

   $$\frac{a}{b} = \frac{b}{a-b},$$

   where b is the length of the bigger subsegment. Such a division is called division by the golden section. If we write $b = ax$, $0 < x < 1$, then $\frac{1}{x} = \frac{x}{1-x}$, that is $x^2 = 1 - x$. Show:
   (a) $\frac{1}{x} = \frac{1+\sqrt{5}}{2} = \alpha$.
   (b) Construct the division of $\overline{AB}$ by the golden section with compass and straightedge.
   (c) If we divide the radius $r > 0$ of a circle by the golden section, then the bigger part of the so divided radius is the side of the regular 10-gon with its 10 vertices on the circle.
3. Given a regular 10-gon such that the 10 vertices are on the circle with radius $R > 0$. Show that the length of each side is equal to the bigger part of the, by the golden section divided, radius. Describe the procedure of the construction of the regular 10-gon and 5-gon.

4. Construct the regular 17-gon with compass and straightedge. Hint: We have to construct the number $\frac{1}{2}(\omega + \omega^{-1}) = \cos\frac{2\pi}{17}$, where $\omega = e^{\frac{2\pi i}{17}}$. First, construct the positive zero ω_1 of the polynomial $x^2 + x - 4$; we get

$$\omega_1 = \frac{1}{2}(\sqrt{17} - 1) = \omega + \omega^{-1} + \omega^2 + \omega^{-2} + \omega^4 + \omega^{-4} + \omega^8 + \omega^{-8}.$$

Then, construct the positive zero ω_2 of the polynomial $x^2 - \omega_1 x - 1$; we get

$$\omega_2 = \frac{1}{4}\left(\sqrt{17} - 1 + \sqrt{34 - 2\sqrt{17}}\right) = \omega + \omega^{-1} + \omega^4 + \omega^{-4}.$$

From ω_1 and ω_2 construct $\beta = \frac{1}{2}(\omega_2^2 - \omega_1 + \omega_2 - 4)$. Then $\omega_3 = 2\cos\frac{2\pi}{17}$ is the biggest of the two positive zeros of the polynomial $x^2 - \omega_2 x + \beta$.

5. The Fibonacci-numbers f_n, $n \in \mathbb{N} \cup \{0\}$, are defined by $f_0 = 0$, $f_1 = 1$ and $f_{n+2} = f_{n+1} + f_n$ for $n \in \mathbb{N} \cup \{0\}$. Show:
   (a) $f_n = \frac{\alpha^n - \beta^n}{\alpha - \beta}$ with $\alpha = \frac{1+\sqrt{5}}{2}$, $\beta = \frac{1-\sqrt{5}}{2}$.
   (b) $(\frac{f_{n+1}}{f_n})_{n\in\mathbb{N}}$ converges and $\lim_{n\to\infty} \frac{f_{n+1}}{f_n} = \frac{1+\sqrt{5}}{2} = \alpha$.
   (c) $\left(\begin{smallmatrix} 0 & 1 \\ 1 & 1 \end{smallmatrix}\right)^n = \left(\begin{smallmatrix} f_{n-1} & f_n \\ f_n & f_{n+1} \end{smallmatrix}\right)$, $n \in \mathbb{N}$.
   (d) $f_1 + f_2 + \cdots + f_n = f_{n+2} - 1$, $n \geq 1$.
   (e) $f_{n-1} f_{n+1} - f_n^2 = (-1)^n$, $n \in \mathbb{N}$.
   (f) $f_1^2 + f_2^2 + \cdots + f_n^2 = f_n f_{n+1}$, $n \in \mathbb{N}$.
   (g) The Fermat numbers F_0, F_1, F_2, F_3, F_4. are all prime but F_5 is composite and divisible by 641.

6. Let $\mu = e^{\frac{2\pi i}{n}}$ be a primitive n-th root of unity. Using

$$e^{\frac{2\pi i}{n}} = \cos\left(\frac{2\pi}{n}\right) + i\sin\left(\frac{2\pi}{n}\right)$$

show that

$$\mu + \frac{1}{\mu} = 2\cos\left(\frac{2\pi}{n}\right).$$

Chapter 7
Kronecker's Theorem and Algebraic Closures

7.1 Kronecker's Theorem

In the last chapter we proved that if $L|K$ is a field extension then there exists an intermediate field $K \subset \mathcal{A} \subset L$ such that $\mathcal{A}$ is algebraic over K and contains all the elements of L that are algebraic over K. We call $\mathcal{A}$ the *algebraic closure* of K within L. In this chapter we prove that starting with any field K we can construct an extension field $\overline{K}$ that is algebraic over K and is *algebraically closed*. By this we mean that there are no algebraic extensions of $\overline{K}$ or equivalently that there are no irreducible nonlinear polynomials in $\overline{K}[x]$. In the final section of this chapter we will give a proof of the famous fundamental theorem of algebra which in the language of this chapter says that the field $\mathbb{C}$ of complex numbers is algebraically closed. We will present another proof of this important result later in the book after we discuss Galois theory.

First we need the following crucial result of Kronecker which says that given a polynomial $f(x)$ in $K[x]$ where K is a field we can construct an extension field L of K in which $f(x)$ has a root α. We say that L has been constructed by *adjoining* α to K. Recall that if $f(x) \in K[x]$ is irreducible then $f(x)$ can have no roots in K. We first need the following concept.

Definition 7.1.1. Let $L|K$ and $L'|K$ be field extensions. Then a K-isomorphism is an isomorphism $\tau : L \to L'$ that is the identity map on K, that is fixes each element of K.

Theorem 7.1.2 (Kronecker's theorem). *Let K be a field and $f(x) \in K[x]$. Then there exists a finite extension K' of K where $f(x)$ has a root.*

Proof. Suppose that $f(x) \in K[x]$. We know that $f(x)$ factors into irreducible polynomials. Let $p(x)$ be an irreducible factor of $f(x)$. From the material in Chapter 4 we know that since $p(x)$ is irreducible the principal ideal $\langle p(x)\rangle$ in $K[x]$ is a maximal ideal. To see this suppose that $g(x) \notin \langle p(x)\rangle$, so that $g(x)$ is not a multiple of $p(x)$. Since $p(x)$ is irreducible, it follows that $(p(x), g(x)) = 1$. Thus there exist $h(x), k(x) \in K[x]$ with

$$h(x)p(x) + k(x)g(x) = 1.$$

The element on the left is in the ideal $(g(x), (p(x))$, so the identity, 1, is in this ideal. Therefore, the whole ring $K[x]$ is in this ideal. Since $g(x)$ was arbitrary, this implies that the principal ideal $\langle p(x)\rangle$ is maximal.

Now let $K' = K[x]/\langle p(x)\rangle$. Since $\langle p(x)\rangle$ is a maximal ideal it follows that K' is a field. We show that K can be embedded in K' and that $p(x)$ has a zero in K'.

First consider the map $\alpha : K[x] \to K'$ by $\alpha(f(x)) = f(x) + \langle p(x)\rangle$. This is a homomorphism. Since the identity element $1 \in K$ is not in $\langle p(x)\rangle$ it follows that α restricted to K is nontrivial. Therefore α restricted to K is a monomorphism since if $\ker(\alpha_{|K}) \neq K$ then $\ker(\alpha_{|K}) = \{0\}$. Therefore K can be embedded into $\alpha(K)$ which is contained in K'. Therefore K' can be considered as an extension field of K. Consider the element $a = x + \langle p(x)\rangle \in K'$. Then $p(a) = p(x) + \langle p(x)\rangle = 0 + \langle p(x)\rangle$ since $p(x) \in \langle p(x)\rangle$. But $0 + \langle p(x)\rangle$ is the zero element 0 of the factor ring $K[x]/\langle p(x)\rangle$. Therefore in K' we have $p(a) = 0$ and hence $p(x)$ has a zero in K'. Since $p(x)$ divides $f(x)$ we must have $f(a) = 0$ in K' also. Therefore we have constructed an extension field of K in which $f(x)$ has a zero. □

We now outline a slightly more constructive proof of Kronecker's theorem. From this construction we say that the field K' constructed by *adjoining the root* α to K.

Proof of Kronecker's theorem. We can assume that $f(x)$ is irreducible. Suppose that $f(x) = a_0 + a_1x + \cdots + a_nx^n$ with $a_n \neq 0$. Define α to satisfy

$$a_0 + a_1\alpha + \cdots + a_n\alpha^n = 0.$$

Now define $K' = K(\alpha)$ in the following manner. We let

$$K(\alpha) = \{c_0 + c_1\alpha + \cdots + c_{n-1}\alpha^{n-1} : c_i \in K\}.$$

Then on $K(\alpha)$ define addition and subtraction componentwise and define multiplication by algebraic manipulation, replacing powers of α higher than α^n by using

$$\alpha^n = \frac{-a_0 - a_1\alpha - \cdots - a_{n-1}\alpha^{n-1}}{a_n}.$$

We claim that $K' = K(\alpha)$ then forms a field of finite degree over K. The basic ring properties follow easily by computation (see exercises) using the definitions. We must show then that every nonzero element of $K(\alpha)$ has a multiplicative inverse. Let $g(\alpha) \in K(\alpha)$. Then the corresponding polynomial $g(x) \in K[x]$ is a polynomial of degree $\leq n - 1$. Since $f(x)$ is irreducible of degree n it follows that $f(x)$ and $g(x)$ must be relatively prime, that is $(f(x), g(x)) = 1$. Hence there exist $a(x), b(x) \in K[x]$ with

$$a(x)f(x) + b(x)g(x) = 1.$$

Evaluate these polynomials at α to get

$$a(\alpha)f(\alpha) + b(\alpha)g(\alpha) = 1.$$

Since by definition we have $f(\alpha) = 0$ this becomes

$$h(\alpha)g(\alpha) = 1.$$

Now $h(\alpha)$ might have degree higher than $n - 1$ in α. However using the relation that $f(\alpha) = 0$ we can rewrite $h(\alpha)$ as $\overline{h}(\alpha)$ where $\overline{h}(\alpha)$ now has degree $\leq n - 1$ in α and hence is in $K(\alpha)$. Therefore

$$\overline{h}(\alpha)g(\alpha) = 1$$

and hence $g(\alpha)$ has a multiplicative inverse. It follows that $K(\alpha)$ is a field and by definition $f(\alpha) = 0$. The elements $1, \alpha, \dots, \alpha^{n-1}$ form a basis for $K(\alpha)$ over K and hence

$$|K(\alpha) : K| = n. \qquad \square$$

Example 7.1.3. Let $f(x) = x^2 + 1 \in \mathbb{R}[x]$. This is irreducible over $\mathbb{R}$. We construct the field in which this has a root.

Let α be an indeterminate with $\alpha^2 + 1 = 0$ or $\alpha^2 = -1$. The extension field $\mathbb{R}(\alpha)$ then has the form

$$\mathbb{R}(\alpha) = \{x + \alpha y : x, y \in \mathbb{R}, \alpha^2 = -1\}.$$

It is clear that this field is $\mathbb{R}$-isomorphic to the complex numbers $\mathbb{C}$, that is, $\mathbb{R}(\alpha) \cong \mathbb{R}(i) \cong \mathbb{C}$.

In Chapter 5 we showed that if $L|K$ is a field extension and $a \in L$ is algebraic over K then there is a smallest algebraic extension $K(a)$ of K within L. Further $K(a)$ is determined by the minimal polynomial $m_a(x)$. The difference between this construction and the construction in Kronecker's theorem is that in the proof of Kronecker's theorem α is defined to be the root and we constructed the field around it, whereas in the previous construction α was assumed to satisfy the polynomial and $K(\alpha)$ was an already existing field that contained α. However the next theorem says that these constructions are the same up to K-isomorphism.

Theorem 7.1.4. *Let $p(x) \in K[x]$ be an irreducible polynomial and let $K' = K(\alpha)$ be the extension field of K constructed in Kronecker's theorem in which $p(x)$ has a zero α. Let L be an extension field of K and suppose that $a \in L$ is algebraic with minimal polynomial $m_\alpha(x) = p(x)$. Then $K(\alpha)$ is K-isomorphic to $K(a)$.*

Proof. If $L|K$ is a field extension and $a \in L$ is algebraic then the construction of the subfield $K(a)$ within L is identical to the construction outlined above for $K(\alpha)$. If $\deg(p(x)) = n$ then the elements $1, a, \dots, a^{n-1}$ constitute a basis for $K(a)$ over K and the elements $1, \alpha, \dots, \alpha^{n-1}$ constitute a basis for $K(\alpha)$ over K. The mapping

$$\tau : K(a) \to K(\alpha)$$

defined by $\tau(k) = k$ if $k \in K$ and $\tau(a) = \alpha$ and then extended by linearity is easily shown to be a K-isomorphism (see exercises). $\square$

Theorem 7.1.5. *Let K be a field. Then the following are equivalent.*

(1) *Each nonconstant polynomial in $K[x]$ has a zero in K.*

(2) *Each nonconstant polynomial in $K[x]$ factors into linear factors over K. That is, for each $f(x) \in K[x]$ there exist elements $a_1, \dots, a_n, b \in K$ with*

$$f(x) = b(x - a_1) \cdots (x - a_n).$$

(3) *An element of $K[x]$ is irreducible if and only if it is of degree one.*

(4) *If $L|K$ is an algebraic extension then $L = K$.*

Proof. Suppose that each nonconstant polynomial in $K[x]$ has a zero in K. Let $f(x) \in K[x]$ with $\deg(f(x)) = n$. Suppose that a_1 is a zero of $f(x)$ then

$$f(x) = (x - a_1)h(x)$$

where the degree of $h(x)$ is $n - 1$. Now $h(x)$ has a zero a_2 in K so that

$$f(x) = (x - a_1)(x - a_2)g(x)$$

with $\deg(g(x)) = n - 2$. Continue in this manner and $f(x)$ factors completely into linear factors. Hence (1) implies (2).

Now suppose (2), that is that each nonconstant polynomial in $K[x]$ factors into linear factors over K. Suppose that $f(x)$ is irreducible. If $\deg(f(x)) > 1$ then $f(x)$ factors into linear factors and hence is not irreducible. Therefore $f(x)$ must be of degree 1 and (2) implies (3).

Now suppose that an element of $K[x]$ is irreducible if and only if it is of degree one and suppose that $L|K$ is an algebraic extension. Let $a \in L$. Then a is algebraic over K. Its minimal polynomial $m_a(x)$ is monic and irreducible over K and hence from (3) is linear. Therefore $m_a(x) = x - a \in K[x]$. It follows that $a \in K$ and hence $K = L$. Therefore (3) implies (4).

Finally suppose that whenever $L|K$ is an algebraic extension then $L = K$. Suppose that $f(x)$ is a nonconstant polynomial in $K[x]$. From Kronecker's theorem there exists a field extension L and $a \in L$ with $f(a) = 0$. However L is an algebraic extension so by supposition $K = L$. Therefore $a \in K$ and $f(x)$ has a zero in K. Therefore (4) implies (1) completing the proof. □

In the next section we will prove that given a field K we can always find an extension field $\overline{K}$ with the properties of the last theorem.

7.2 Algebraic Closures and Algebraically Closed Fields

A field K is termed *algebraically closed* if K has no algebraic extensions other than K itself. This is equivalent to any one of the conditions of Theorem 7.1.5.

Definition 7.2.1. A field K is *algebraically closed* if every nonconstant polynomial $f(x) \in K[x]$ has a zero in K.

The following theorem is just a restatement of Theorem 7.1.5.

Theorem 7.2.2. *A field K is algebraically closed if and only it satisfies any one of the following conditions.*

(1) *Each nonconstant polynomial in $K[x]$ has a zero in K.*

(2) *Each nonconstant polynomial in $K[x]$ factors into linear factors over K. That is, for each $f(x) \in K[x]$ there exist elements $a_1, \dots, a_n, b \in K$ with*

$$f(x) = b(x - a_1) \cdots (x - a_n).$$

(3) *An element of $K[x]$ is irreducible if and only if it is of degree one.*

(4) *If $L|K$ is an algebraic extension then $L = K$.*

The prime example of an algebraically closed field is the field $\mathbb{C}$ of complex numbers. The fundamental theorem of algebra says that any nonconstant complex polynomial has a complex root.

We now show that the algebraic closure of one field within an algebraically closed field is algebraically closed. First we define a general algebraic closure.

Definition 7.2.3. An extension field $\overline{K}$ of a field K is an *algebraic closure* of K if $\overline{K}$ is algebraically closed and $\overline{K}|K$ is algebraic.

Theorem 7.2.4. *Let K be a field and $L|K$ an extension of K with L algebraically closed. Let $\overline{K} = \mathcal{A}_K$ be the algebraic closure of K within L. Then $\overline{K}$ is an algebraic closure of K.*

Proof. Let $\overline{K} = \mathcal{A}_K$ be the algebraic closure of K within L. We know that $\overline{K}|K$ is algebraic therefore we must show that $\overline{K}$ is algebraically closed.

Let $f(x)$ be a nonconstant polynomial in $\overline{K}[x]$. Then $f(x) \in L[x]$. Since L is algebraically closed $f(x)$ has a zero a in L. Since $f(a) = 0$ and $f(x) \in \overline{K}[x]$ it follows that a is algebraic over $\overline{K}$. However $\overline{K}$ is algebraic over K and therefore a is also algebraic over K. Hence $a \in \overline{K}$ and $f(x)$ has a zero in $\overline{K}$. Therefore $\overline{K}$ is algebraically closed. □

We want to note the distinction between being algebraically closed and being an algebraic closure.

Lemma 7.2.5. *The complex numbers $\mathbb{C}$ is an algebraic closure of $\mathbb{R}$ but not an algebraic closure of $\mathbb{Q}$. An algebraic closure of $\mathbb{Q}$ is $\mathcal{A}$ the field of algebraic numbers within $\mathbb{C}$.*

Proof. $\mathbb{C}$ is algebraically closed (the fundamental theorem of algebra) and since $|\mathbb{C} : \mathbb{R}| = 2$ it is algebraic over $\mathbb{R}$. Therefore $\mathbb{C}$ is an algebraic closure of $\mathbb{R}$. Although $\mathbb{C}$ is algebraically closed and contains the rational numbers $\mathbb{Q}$ it is not an algebraic closure of $\mathbb{Q}$ since it is not algebraic over $\mathbb{Q}$ since there exist transcendental elements.

On the other hand, $\mathcal{A}$, the field of algebraic numbers within $\mathbb{Q}$, is an algebraic closure of $\mathbb{Q}$ from Theorem 7.2.4. □

We now show that every field has an algebraic closure. To do this we first show that any field can be embedded into an algebraically closed field.

Theorem 7.2.6. *Let K be a field. Then K can be embedded into an algebraically closed field.*

Proof. We show first that there is an extension field L of K in which each nonconstant polynomial $f(x) \in K[x]$ has a zero in L.

Assign to each nonconstant $f(x) \in K[x]$ the symbol y_f and consider

$$R = K[y_f : f(x) \in K[x]]$$

the polynomial ring over K in the variables y_f. Set

$$I = \left\{ \sum_{j=1}^{n} f_j(y_{f_j}) r_j : r_j \in R, f_j(x) \in K[x] \right\}.$$

It is straightforward that I is an ideal in R. Suppose that $I = R$. Then $1 \in I$. Hence there is a linear combination

$$1 = g_1 f_1(y_{f_1}) + \cdots + g_n f_n(y_{f_n})$$

where $g_i \in I = R$.

In the n polynomials $g_1, \ldots, g_n$ there are only a finite number of variables, say for example

$$y_{f_1}, \ldots, y_{f_n}, \ldots, y_{f_m}.$$

Hence

$$1 = \sum_{i=1}^{n} g_i(y_{f_1}, \ldots, y_{f_m}) f_i(y_{f_i})(*).$$

Successive applications of Kronecker's theorem lead us to construct an extension field P of K in which each f_i has a zero a_i. Substituting a_i for y_{f_i} in (8) above we get that $1 = 0$ a contradiction. Therefore $I \neq R$.

Since I is a ideal not equal to the whole ring R it follows that I is contained in a maximal ideal M of R. Set $L = R/M$. Since M is maximal K is a field. Now $K \cap M = \{0\}$. If not suppose that $a \in K \cap M$ with $a \neq 0$. Then $a^{-1}a = 1 \in M$

and then $M = R$. Now define $\tau : K \to L$ by $\tau(k) = k + M$. Since $K \cap M = \{0\}$ it follows that $\ker(\tau) = \{0\}$ so τ is a monomorphism. This allows us to identify K and $\tau(K)$ and shows that K embeds into L.

Now suppose that $f(x)$ is a nonconstant polynomial in $K[x]$. Then

$$f(y_f + M) = f(y_f) + M.$$

However by the construction $f(y_f) \in M$ so that

$$f(y_f + M) = M + M = \text{the zero element of } L.$$

Therefore $y_f + M$ is a zero of $f(x)$.

Therefore we have constructed a field L in which every nonconstant polynomial in $K[x]$ has a zero in L.

We now iterate this procedure to form a chain of fields

$$K \subset K_1(= L) \subset K_2 \subset \cdots$$

such that each nonconstant polynomial of $K_i[x]$ has a zero in K_{i+1}.

Now let $\hat{K} = \bigcup_I K_i$. It is easy to show (see exercises) that $\hat{K}$ is a field. If $f(x)$ is a nonconstant polynomial in $\hat{K}[x]$ then there is some i with $f(x) \in K_i[x]$. Therefore $f(x)$ has a zero in $K_{i+1}[x] \subset \hat{K}$. Hence $f(x)$ has a zero in $\hat{K}$ and $\hat{K}$ is algebraically closed. □

Theorem 7.2.7. *Let K be a field. Then K has an algebraic closure.*

Proof. Let $\hat{K}$ be an algebraically closed field containing K which exists from Theorem 7.2.6.

Now let $\overline{K} = \mathcal{A}_{\hat{K}}$ be the set of elements of $\hat{K}$ that are algebraic over K. From Theorem 7.2.4 $\hat{K}$ is an algebraic closure of K. □

The following lemma is straightforward. We leave the proof to the exercises.

Lemma 7.2.8. *Let K, K' be fields and $\phi : K \to K'$ a homomorphism. Then*

$$\tilde{\phi} : K[x] \to K'[x] \quad \textit{given by}$$

$$\tilde{\phi}\left(\sum_{i=1}^{n} k_i x^i\right) = \sum_{i=0}^{n} (\phi(k_i))x^i$$

is also a homomorphism. By convention we identify ϕ and $\tilde{\phi}$ and write $\phi = \tilde{\phi}$. If ϕ is an isomorphism then so is $\tilde{\phi}$.

Lemma 7.2.9. *Let K, K' be fields and $\phi : K \to K'$ an isomorphism. Let $f(x) \in K[x]$ be irreducible. Let $K \subset K(a)$ and $K' \subset K'(a')$ where a is a zero of $f(x)$ and a' is a zero of $\phi(f(x))$. Then there is an isomorphism $\psi : K(a) \to K'(a')$ with $\psi_{|K} = \phi$ and $\psi(a) = a'$. Further ψ is uniquely determined.*

Proof. This is a generalized version of Theorem 7.1.4. If $b \in K(a)$ then from the construction of $K(a)$ there is a polynomial $g(x) \in K[x]$ with $b = g(a)$. Define a map

$$\psi : K(a) \to K'(a')$$

by

$$\psi(b) = \phi(g(x))(a').$$

We show that ψ is an isomorphism.

First ψ is well-defined. Suppose that $b = g(a) = h(a)$ with $h(x) \in K[x]$. Then $(g-h)(a) = 0$. Since $f(x)$ is irreducible this implies that $f(x) = cm_a(x)$ and since a is a zero of $(g-h)(x)$ then $f(x)|(g-h)(x)$. Then

$$\phi(f(x))|(\phi(g(x)) - \phi(h(x))).$$

Since $\phi(f(x))(a') = 0$ this implies that $\phi(g(x))(a') = \phi(h(x))(a')$ and hence the map ψ is well-defined.

It is easy to show that ψ is a homomorphism. Let $b_1 = g_1(a)$, $b_2 = g_2(a)$. Then $b_1 b_1 = g_1 g_2(a)$. Hence

$$\psi(b_1 b_2) = (\phi(g_1 g_2))(a') = \phi(g_1)(a')\phi(g_2)(a') = \psi(b_1)\psi(b_2).$$

In the same manner we have $\psi(b_1 + b_2) = \psi(b_1) + \psi(b_2)$.

Now suppose that $k \in K$ so that $k \in K[x]$ is a constant polynomial. Then $\psi(k) = (\phi(k))(a') = \phi(k)$. Therefore ψ restricted to K is precisely ϕ.

As ψ is not the zero mapping it follows that ψ is a monomorphism.

Finally since $K(a)$ is generated from K and a, and ψ restricted to K is ϕ it follows that ψ is uniquely determined by ϕ and $\psi(a) = a'$. Hence ψ is unique. □

Theorem 7.2.10. *Let $L|K$ be an algebraic extension. Suppose that L_1 is an algebraically closed field and ϕ is an isomorphism from K to $K_1 \subset L_1$. Then there exists a monomorphism ψ from L to L_1 with $\psi_{|K} = \phi$.*

Before we give the proof we note that the theorem gives the following diagram:

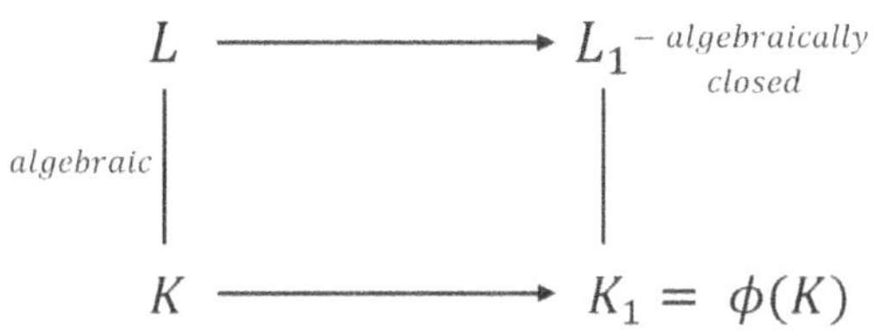

In particular the theorem can be applied to monomorphisms of a field K within an algebraic closure $\overline{K}$ of K. Specifically suppose that $K \subset \overline{K}$ where $\overline{K}$ is an algebraic closure of K and let $\alpha : K \to \overline{K}$ be a monomorphism with $\alpha(K) = K$. Then there exists an automorphism α^* of $\overline{K}$ with $\alpha^*_{|K} = \alpha$.

Proof of Theorem 7.2.10. Consider the set

$$\mathcal{M} = \{(M, \tau) : M \text{ is a field with } K \subset M \subset L, \\ \text{where there exists a monomorphism } \tau : M \to L_1 \text{ with } \tau_{|K} = \phi\}.$$

Now the set $\mathcal{M}$ is nonempty since $(K, \phi) \in \mathcal{M}$. Order $\mathcal{M}$ by $(M_1, \tau_1) < (M_2, \tau_2)$ if $M_1 \subset M_2$ and $(\tau_2)_{|M_1} = \tau_1$. Let

$$\mathcal{K} = \{(M_i, \tau_i) : i \in I\}$$

be a chain in $\mathcal{M}$. Let (M, τ) be defined by

$$M = \bigcup_{i \in I} M_i \quad \text{with } \tau(a) = \tau_i(a) \text{ for all } a \in M_i.$$

It is clear that M is an upper bound for the chain $\mathcal{K}$. Since each chain has an upper bound it follows from Zorn's lemma that $\mathcal{M}$ has a maximal element (N, ρ). We show that $N = L$.

Suppose that $N \subsetneq L$. Let $a \in L \setminus N$. Then a is algebraic over N and further algebraic over K since $L|K$ is algebraic. Let $m_a(x) \in N[x]$ be the minimal polynomial of a relative to N. Since L_1 is algebraically closed $\rho(m_a(x))$ has a zero $a' \in L_1$. Therefore there is a monomorphism $\rho' : N(a) \to L_1$ with ρ' restricted to N the same as ρ. It follows that $(N, \rho) < (N(a), \rho')$ since $a \notin N$. This contradicts the maximality of N. Therefore $N = L$ completing the proof. □

Combining the previous two theorems we can now prove that any two algebraic closures of a field K are unique up to K-isomorphism, that is up to an isomorphism that is the identity on K.

Theorem 7.2.11. *Let L_1 and L_2 be algebraic closures of the field K. Then there is a K-isomorphism $\tau : L \to L_1$. Again by K-isomorphism we mean that τ is the identity on K.*

Proof. From Theorem 7.2.7 there is a monomorphism $\tau : L_1 \to L_2$ with τ the identity on K. However since L_1 is algebraically closed so is $\tau(L_1)$. Then $L_2|\tau(L_1)$ is an algebraic extension and since L_2 is algebraically closed we must have $L_2 = \tau(L_1)$. Therefore τ is also surjective and hence an isomorphism. □

The following corollary is immediate.

Corollary 7.2.12. *Let $L|K$ and $L'|K$ be field extensions with $a \in L$ and $a' \in L'$ algebraic elements over K. Then $K(a)$ is K-isomorphic to $K(a')$ if and only if $|K(a) : K| = |K(a') : K|$ and there is an element $a'' \in K(a')$ with $m_a(x) = m_{a''}(x)$.*

7.3 The Fundamental Theorem of Algebra

In this section we give a proof of the fact that the complex numbers form an algebraically closed field. This is known as the fundamental theorem of algebra. First we need the concept of a *splitting field* for a polynomial. In the next chapter we will examine this concept more deeply.

7.3.1 Splitting Fields

We have just seen that given an irreducible polynomial over a field F we could always find a field extension in which this polynomial has a root. We now push this further to obtain field extensions where a given polynomial has all its roots.

Definition 7.3.1. If K is a field and $0 \neq f(x) \in K[x]$ and K' is an extension field of K, then $f(x)$ *splits* in K' (K' may be K), if $f(x)$ factors into linear factors in $K'[x]$. Equivalently, this means that all the roots of $f(x)$ are in K'.

K' is a *splitting field* for $f(x)$ over K if K' is the smallest extension field of K in which $f(x)$ splits. (A splitting field for $f(x)$ is the smallest extension field in which $f(x)$ has all its possible roots.)

K' is a *splitting field* over K if it is the splitting field for some finite set of polynomials over K.

Theorem 7.3.2. *If K is a field and $0 \neq f(x) \in K[x]$, then there exists a splitting field for $f(x)$ over K.*

Proof. The splitting field is constructed by repeated adjoining of roots. Suppose without loss of generality that $f(x)$ is irreducible of degree n over F. From Theorem 6.2.2 there exists a field F' containing α with $f(\alpha) = 0$. Then $f(x) = (x-\alpha)g(x) \in F'[x]$

with $\deg g(x) = n - 1$. By an inductive argument $g(x)$ has a splitting field and therefore so does $f(x)$. □

In the next chapter we will give a further characterization of splitting fields.

7.3.2 Permutations and Symmetric Polynomials

To obtain a proof of the fundamental theorem of algebra we need to go a bit outside of our main discussions of rings and fields and introduce symmetric polynomials. In order to introduce this concept we first review some basic ideas from elementary group theory which we will look at in detail later in the book.

Definition 7.3.3. A *group* G is a set with one binary operation which we will denote by multiplication, such that

(1) The operation is associative, that is, $(g_1g_2)g_3 = g_1(g_2g_3)$ for all $g_1, g_2, g_3 \in G$.

(2) There exists an identity for this operation, that is, an element 1 such that $1g = g$ for each $g \in G$.

(3) Each $g \in G$ has an inverse for this operation, that is, for each g there exists a g^{-1} with the property that $gg^{-1} = 1$.

If in addition the operation is commutative ($g_1g_2 = g_2g_1$ for all $g_1, g_2 \in G$), the group G is called an *abelian group*. The *order* of G is the number of elements in G, denoted $|G|$. If $|G| < \infty$, G is a *finite group*. $H \subset G$ is a *subgroup* if H is also a group under the same operation as G. Equivalently, H is a subgroup if $H \neq \emptyset$ and H is closed under the operation and inverses.

Groups most often arise from invertible mappings of a set onto itself. Such mappings are called *permutations*.

Definition 7.3.4. If T is a set, a *permutation* on T is a one-to-one mapping of T onto itself. We denote by S_T the set of all permutations on T.

Theorem 7.3.5. *For any set T, S_T forms a group under composition called the symmetric group on T. If T, T_1 have the same cardinality (size), then $S_T \cong S_{T_1}$. If T is a finite set with $|T| = n$, then S_T is a finite group and $|S_T| = n!$.*

Proof. If S_T is the set of all permutations on the set T, we must show that composition is an operation on S_T that is associative and has an identity and inverses.

Let $f, g \in S_T$. Then f, g are one-to-one mappings of T onto itself. Consider $f \circ g : T \to T$. If $f \circ g(t_1) = f \circ g(t_2)$, then $f(g(t_1)) = f(g(t_2))$ and $g(t_1) = g(t_2)$, since f is one-to-one. But then $t_1 = t_2$ since g is one-to-one.

If $t \in T$, there exists $t_1 \in T$ with $f(t_1) = t$ since f is onto. Then there exists $t_2 \in T$ with $g(t_2) = t_1$ since g is onto. Putting these together, $f(g(t_2)) = t$, and therefore $f \circ g$ is onto. Therefore, $f \circ g$ is also a permutation and composition gives a valid binary operation on S_T.

The identity function $1(t) = t$ for all $t \in T$ will serve as the identity for S_T, while the inverse function for each permutation will be the inverse. Such unique inverse functions exist since each permutation is a bijection.

Finally, composition of functions is always associative and therefore S_T forms a group.

If T, T_1 have the same cardinality, then there exists a bijection $\sigma : T \to T_1$. Define a map $F : S_T \to S_{T_1}$ in the following manner: if $f \in S_T$, let $F(f)$ be the permutation on T_1 given by $F(f)(t_1) = \sigma(f(\sigma^{-1}(t_1)))$. It is straightforward to verify that F is an isomorphism (see the exercises).

Finally, suppose $|T| = n < \infty$. Then $T = \{t_1, \dots, t_n\}$. Each $f \in S_T$ can be pictured as

$$f = \begin{pmatrix} t_1 & \dots & t_n \\ f(t_1) & \dots & f(t_n) \end{pmatrix}.$$

For t_1 there are n choices for $f(t_1)$. For t_2 there are only $n - 1$ choices since f is one-to-one. This continues down to only one choice for t_n. Using the multiplication principle, the number of choices for f and therefore the size of S_T is

$$n(n-1)\cdots 1 = n!.$$

For a set with n elements we denote S_T by S_n called the *symmetric group on n symbols*. □

Example 7.3.6. Write down the six elements of S_3 and give the multiplication table for the group.

Name the three elements $1, 2, 3$. The six elements of S_3 are then:

$$1 = \begin{pmatrix} 1 & 2 & 3 \\ 1 & 2 & 3 \end{pmatrix}, \quad a = \begin{pmatrix} 1 & 2 & 3 \\ 2 & 3 & 1 \end{pmatrix}, \quad b = \begin{pmatrix} 1 & 2 & 3 \\ 3 & 1 & 2 \end{pmatrix}$$

$$c = \begin{pmatrix} 1 & 2 & 3 \\ 2 & 1 & 3 \end{pmatrix}, \quad d = \begin{pmatrix} 1 & 2 & 3 \\ 3 & 2 & 1 \end{pmatrix}, \quad e = \begin{pmatrix} 1 & 2 & 3 \\ 1 & 3 & 2 \end{pmatrix}.$$

The multiplication table for S_3 can be written down directly by doing the required composition. For example,

$$ac = \begin{pmatrix} 1 & 2 & 3 \\ 2 & 3 & 1 \end{pmatrix}\begin{pmatrix} 1 & 2 & 3 \\ 2 & 1 & 3 \end{pmatrix} = \begin{pmatrix} 1 & 2 & 3 \\ 3 & 2 & 1 \end{pmatrix} = d.$$

To see this, note that $a : 1 \to 2, 2 \to 3, 3 \to 1$; $c : 1 \to 2, 2 \to 1, 3 \to 3$ and so $ac : 1 \to 3, 2 \to 2, 3 \to 1$.

It is somewhat easier to construct the multiplication table if we make some observations. First, $a^2 = b$ and $a^3 = 1$. Next, $c^2 = 1$, $d = ac$, $e = a^2c$ and finally $ac = ca^2$.

From these relations the following multiplication table can be constructed:

	1	a	a^2	c	ac	a^2c
1	1	a	a^2	c	ac	a^2c
a	a	a^2	1	ac	a^2c	c
a^2	a^2	1	a	a^2c	c	ac
c	c	a^2c	ac	1	a^2	a
ac	ac	c	a^2c	a	1	a^2
a^2c	a^2c	ac	c	a^2	a	1

To see this, consider, for example, $(ac)a^2 = a(ca^2) = a(ac) = a^2c$.

More generally, we can say that S_3 has a *presentation* given by

$$S_3 = \langle a, c; a^3 = c^2 = 1, ac = ca^2 \rangle.$$

By this we mean that S_3 is *generated by* a, c, or that S_3 has *generators* a, c and the whole group and its multiplication table can be generated by using the *relations* $a^3 = c^2 = 1, ac = ca^2$.

An important result, the form of which we will see later in our work on extension fields, is the following.

Lemma 7.3.7. *Let T be a set and $T_1 \subset T$ a subset. Let H be the subset of S_T that fixes each element of T_1, that is, $f \in H$ if $f(t) = t$ for all $t \in T_1$. Then H is a subgroup.*

Proof. $H \neq \emptyset$ since $1 \in H$. Now suppose $h_1, h_2 \in H$. Let $t_1 \in T_1$ and consider $h_1 \circ h_2(t_1) = h_1(h_2(t_1))$. Now $h_2(t_1) = t_1$ since $h_2 \in H$, but then $h_1(t_1) = t_1$ since $h_1 \in H$. Therefore, $h_1 \circ h_2 \in H$ and H is closed under composition. If h_1 fixes t_1 then h_1^{-1} also fixes t_1 so H is also closed under inverses and is therefore a subgroup. □

We now apply these ideas of permutations to certain polynomials in independent indeterminates over a field. We will look at these in detail later in this book.

Definition 7.3.8. Let $y_1, \dots, y_n$ be (independent) indeterminates over a field K. A polynomial $f(y_1, \dots, y_n) \in K[y_1, \dots, y_n]$ is a *symmetric polynomial* in y_1, $\dots, y_n$ if $f(y_1, \dots, y_n)$ is unchanged by any permutation σ of $\{y_1, \dots, y_n\}$, that is, $f(y_1, \dots, y_n) = f(\sigma(y_1), \dots, \sigma(y_n))$.

If $K \subset K'$ are fields and $\alpha_1, \dots, \alpha_n$ are in F', then we call a polynomial $f(\alpha_1, \dots, \alpha_n)$ with coefficients in K *symmetric* in $\alpha_1, \dots, \alpha_n$ if $f(\alpha_1, \dots, \alpha_n)$ is unchanged by any permutation σ of $\{\alpha_1, \dots, \alpha_n\}$.

Example 7.3.9. Let K be a field and $k_0, k_1 \in K$. Let $h(y_1, y_2) = k_0(y_1 + y_2) + k_1(y_1 y_2)$.

There are two permutations on $\{y_1, y_2\}$, namely $\sigma_1 : y_1 \to y_1, y_2 \to y_2$ and $\sigma_2 : y_1 \to y_2, y_2 \to y_1$. Applying either one of these two to $\{y_1, y_2\}$ leaves $h(y_1, y_2)$ invariant. Therefore, $h(y_1, y_2)$ is a symmetric polynomial.

Definition 7.3.10. Let $x, y_1, \dots, y_n$ be indeterminates over a field K (or elements of an extension field K' of K). Form the polynomial

$$p(x, y_1, \dots, y_n) = (x - y_1) \cdots (x - y_n).$$

The *i-th elementary symmetric polynomial* s_i in $y_1, \dots, y_n$ for $i = 1, \dots, n$, is $(-1)^i a_i$, where a_i is the coefficient of x^{n-i} in $p(x, y_1, \dots, y_n)$.

Example 7.3.11. Consider y_1, y_2, y_3. Then

$$\begin{aligned} p(x, y_1, y_2, y_3) &= (x - y_1)(x - y_2)(x - y_3) \\ &= x^3 - (y_1 + y_2 + y_3)x^2 + (y_1 y_2 + y_1 y_3 + y_2 y_3)x - y_1 y_2 y_3. \end{aligned}$$

Therefore, the three elementary symmetric polynomials in y_1, y_2, y_3 over any field are

(1) $s_1 = y_1 + y_2 + y_3$.

(2) $s_2 = y_1 y_2 + y_1 y_3 + y_2 y_3$.

(3) $s_3 = y_1 y_2 y_3$.

In general, the pattern of the last example holds for $y_1, \dots, y_n$. That is,

$$\begin{aligned} s_1 &= y_1 + y_2 + \cdots + y_n \\ s_2 &= y_1 y_2 + y_1 y_3 + \cdots + y_{n-1} y_n \\ s_3 &= y_1 y_2 y_3 + y_1 y_2 y_4 + \cdots + y_{n-2} y_{n-1} y_n \\ &\vdots \\ s_n &= y_1 \cdots y_n. \end{aligned}$$

The importance of the elementary symmetric polynomials is that any symmetric polynomial can be built up from the elementary symmetric polynomials. We make this precise in the next theorem called the *fundamental theorem of symmetric polynomials*. We will use this important result several times, and we will give a complete proof in Section 7.5.

Theorem 7.3.12 (fundamental theorem of symmetric polynomials). *If P is a symmetric polynomial in the indeterminates $y_1, \dots, y_n$ over a field K, that is, $P \in K[y_1, \dots, y_n]$ and P is symmetric, then there exists a unique $g \in K[y_1, \dots, y_n]$ such that $f(y_1, \dots, y_n) = g(s_1, \dots, s_n)$. That is, any symmetric polynomial in $y_1, \dots, y_n$ is a polynomial expression in the elementary symmetric polynomials in $y_1, \dots, y_n$.*

From this theorem we obtain the following two lemmas, which will be crucial in our proof of the fundamental theorem of algebra.

Lemma 7.3.13. *Let $p(x) \in K[x]$ and suppose $p(x)$ has the roots $\alpha_1, \dots, \alpha_n$ in the splitting field K'. Then the elementary symmetric polynomials in $\alpha_1, \dots, \alpha_n$ are in K.*

Proof. Suppose $p(x) = c_0 + c_1 x + \cdots + c_n x^n \in K[x]$. In $K'[x]$, $p(x)$ splits, with roots $\alpha_1, \dots, \alpha_n$, and thus in $K'[x]$,

$$p(x) = c_n(x - \alpha_1) \cdots (x - \alpha_n).$$

The coefficients are then $c_n(-1)^i s_i(\alpha_1, \dots, \alpha_n)$, where the $s_i(\alpha_1, \dots, \alpha_n)$ are the elementary symmetric polynomials in $\alpha_1, \dots, \alpha_n$. However, $p(x) \in K[x]$, so each coefficient is in K. It follows then that for each i, $c_n(-1)^i s_i(\alpha_1, \dots, \alpha_n) \in K$, and hence $s_i(\alpha_1, \dots, \alpha_n) \in K$ since $c_n \in K$. □

Lemma 7.3.14. *Let $p(x) \in K[x]$ and suppose $p(x)$ has the roots $\alpha_1, \dots, \alpha_n$ in the splitting field K'. Suppose further that $g(x) = g(x, \alpha_1, \dots, \alpha_n) \in K'[x]$. If $g(x)$ is a symmetric polynomial in $\alpha_1, \dots, \alpha_n$, then $g(x) \in K[x]$.*

Proof. If $g(x) = g(x, \alpha_1, \dots, \alpha_n)$ is symmetric in $\alpha_1, \dots, \alpha_n$, then from Theorem 7.3.12 it is a symmetric polynomial in the elementary symmetric polynomials in $\alpha_1, \dots, \alpha_n$. From Lemma 7.3.13 these are in the ground field K, so the coefficients of $g(x)$ are in K. Therefore, $g(x) \in K[x]$. □

7.4 The Fundamental Theorem of Algebra

We now present a proof of the fundamental theorem of algebra.

Theorem 7.4.1 (fundamental theorem of algebra). *Any nonconstant complex polynomial has a complex root. In other words, the complex number field $\mathbb{C}$ is algebraically closed.*

The proof depends on the following sequence lemmas. The crucial one now is the last, which says that any real polynomial must have a complex root.

Lemma 7.4.2. *Any odd-degree real polynomial must have a real root.*

Proof. This is a consequence of the intermediate value theorem from analysis.

Suppose $P(x) \in \mathbb{R}[x]$ with $\deg P(x) = n = 2k + 1$ and suppose the leading coefficient $a_n > 0$ (the proof is almost identical if $a_n < 0$). Then

$$P(x) = a_n x^n + (\text{lower terms})$$

and n is odd. Then,

(1) $\lim_{x \to \infty} P(x) = \lim_{x \to \infty} a_n x^n = \infty$ since $a_n > 0$.

(2) $\lim_{x \to -\infty} P(x) = \lim_{x \to -\infty} a_n x^n = -\infty$ since $a_n > 0$ and n is odd.

From (1), $P(x)$ gets arbitrarily large positively, so there exists an x_1 with $P(x_1) > 0$. Similarly, from (2) there exists an x_2 with $P(x_2) < 0$.

A real polynomial is a continuous real-valued function for all $x \in \mathbb{R}$. Since $P(x_1)P(x_2) < 0$, it follows from the intermediate value theorem that there exists an x_3, between x_1 and x_2, such that $P(x_3) = 0$. □

Lemma 7.4.3. *Any degree-two complex polynomial must have a complex root.*

Proof. This is a consequence of the quadratic formula and of the fact that any complex number has a square root.

If $P(x) = ax^2 + bx + c$, $a \neq 0$, then the roots formally are

$$x_1 = \frac{-b + \sqrt{b^2 - 4ac}}{2a}, \quad x_2 = \frac{-b - \sqrt{b^2 - 4ac}}{2a}.$$

From DeMoivre's theorem every complex number has a square root, hence x_1, x_2 exist in $\mathbb{C}$. They of course are the same if $b^2 - 4ac = 0$. □

To go further we need the concept of the *conjugate of a polynomial* and some straightforward consequences of this idea.

Definition 7.4.4. If $P(x) = a_0 + \cdots + a_n x^n$ is a complex polynomial then its *conjugate* is the polynomial $\overline{P}(x) = \overline{a_0} + \cdots + \overline{a_n} x^n$. That is, the conjugate is the polynomial whose coefficients are the complex conjugates of those of $P(x)$.

Lemma 7.4.5. *For any* $P(x) \in \mathbb{C}[x]$ *we have:*

(1) $\overline{P(z)} = \overline{P}(\overline{z})$ *if* $z \in \mathbb{C}$.

(2) $P(x)$ *is a real polynomial if and only if* $P(x) = \overline{P}(x)$.

(3) *If* $P(x)Q(x) = H(x)$ *then* $\overline{H}(x) = (\overline{P}(x))(\overline{Q}(x))$.

Proof. (1) Suppose $z \in \mathbb{C}$ and $P(z) = a_0 + \cdots + a_n z^n$. Then

$$\overline{P(z)} = \overline{a_0 + \cdots + a_n z^n} = \overline{a_0} + \overline{a_1 z} + \cdots + \overline{a_n z^n} = \overline{P}(\overline{z}).$$

(2) Suppose $P(x)$ is real then $a_i = \overline{a_i}$ for all its coefficients and hence $P(x) = \overline{P}(x)$. Conversely suppose $P(x) = \overline{P}(x)$. Then $a_i = \overline{a_i}$ for all its coefficients and hence $a_i \in \mathbb{R}$ for each a_i and so $P(x)$ is a real polynomial.

(3) The proof is a computation and left to the exercises. □

Lemma 7.4.6. *Suppose $G(x) \in \mathbb{C}[x]$. Then $H(x) = G(x)\overline{G}(x) \in \mathbb{R}[x]$.*

Proof. $\overline{H}(x) = \overline{G(x)\overline{G}(x)} = \overline{G}(x)\overline{\overline{G}}(x) = \overline{G}(x)G(x) = G(x)\overline{G}(x) = H(x)$. Therefore, $H(x)$ is a real polynomial. □

Lemma 7.4.7. *If every nonconstant real polynomial has a complex root, then every nonconstant complex polynomial has a complex root.*

Proof. Let $P(x) \in \mathbb{C}[x]$ and suppose that every nonconstant real polynomial has at least one complex root. Let $H(x) = P(x)\overline{P}(x)$. From Lemma 7.4.6, $H(x) \in \mathbb{R}[x]$. By supposition there exists a $z_0 \in \mathbb{C}$ with $H(z_0) = 0$. Then $P(z_0)\overline{P}(z_0) = 0$, and since $\mathbb{C}$ is a field it has no zero divisors. Hence either $P(z_0) = 0$ or $\overline{P}(z_0) = 0$. In the first case z_0 is a root of $P(x)$. In the second case $\overline{P}(z_0) = 0$. Then from Lemma 7.4.5 $\overline{P}(z_0) = \overline{\overline{P}(\overline{z_0})} = P(\overline{z_0}) = 0$. Therefore, $\overline{z_0}$ is a root of $P(x)$. □

Now we come to the crucial lemma.

Lemma 7.4.8. *Any nonconstant real polynomial has a complex root.*

Proof. Let $f(x) = a_0 + a_1 x + \cdots + a_n x^n \in \mathbb{R}[x]$ with $n \geq 1$, $a_n \neq 0$. The proof is an induction on the degree n of $f(x)$.

Suppose $n = 2^m q$ where q is odd. We do the induction on m. If $m = 0$ then $f(x)$ has odd degree and the theorem is true from Lemma 7.4.2. Assume then that the theorem is true for all degrees $d = 2^k q'$ where $k < m$ and q' is odd. Now assume that the degree of $f(x)$ is $n = 2^m q$.

Suppose K' is the splitting field for $f(x)$ over $\mathbb{R}$ in which the roots are $\alpha_1, \ldots, \alpha_n$. We show that at least one of these roots must be in $\mathbb{C}$. (In fact, all are in $\mathbb{C}$ but to prove the lemma we need only show at least one.)

Let $h \in \mathbb{Z}$ and form the polynomial

$$H(x) = \prod_{i<j} (x - (\alpha_i + \alpha_j + h\alpha_i\alpha_j)).$$

This is in $K'[x]$. In forming $H(x)$ we chose pairs of roots $\{\alpha_i, \alpha_j\}$, so the number of such pairs is the number of ways of choosing two elements out of $n = 2^m q$ elements. This is given by

$$\frac{(2^m q)(2^m q - 1)}{2} = 2^{m-1} q(2^m q - 1) = 2^{m-1} q'$$

with q' odd. Therefore, the degree of $H(x)$ is $2^{m-1}q'$.

$H(x)$ is a symmetric polynomial in the roots $\alpha_1, \ldots, \alpha_n$. Since $\alpha_1, \ldots, \alpha_n$ are the roots of a real polynomial, from Lemma 7.3.14 any polynomial in the splitting field symmetric in these roots must be a real polynomial.

Therefore, $H(x) \in \mathbb{R}[x]$ with degree $2^{m-1}q'$. By the inductive hypothesis, then, $H(x)$ must have a complex root. This implies that there exists a pair $\{\alpha_i, \alpha_j\}$ with

$$\alpha_i + \alpha_j + h\alpha_i\alpha_j \in \mathbb{C}.$$

Since h was an arbitrary integer, for any integer h_1 there must exist such a pair $\{\alpha_i, \alpha_j\}$ with

$$\alpha_i + \alpha_j + h_1\alpha_i\alpha_j \in \mathbb{C}.$$

Now let h_1 vary over the integers. Since there are only finitely many such pairs $\{\alpha_i, \alpha_j\}$, it follows that there must be at least two different integers h_1, h_2 such that

$$z_1 = \alpha_i + \alpha_j + h_1\alpha_i\alpha_j \in \mathbb{C} \quad \text{and} \quad z_2 = \alpha_i + \alpha_j + h_2\alpha_i\alpha_j \in \mathbb{C}.$$

Then $z_1 - z_2 = (h_1 - h_2)\alpha_i\alpha_j \in \mathbb{C}$ and since $h_1, h_2 \in \mathbb{Z} \subset \mathbb{C}$ it follows that $\alpha_i\alpha_j \in \mathbb{C}$. But then $h_1\alpha_i\alpha_j \in \mathbb{C}$, from which it follows that $\alpha_i + \alpha_j \in \mathbb{C}$. Then,

$$p(x) = (x - \alpha_i)(x - \alpha_j) = x^2 - (\alpha_i + \alpha_j)x + \alpha_i\alpha_j \in \mathbb{C}[x].$$

However, $p(x)$ is then a degree-two complex polynomial and so from Lemma 7.4.3 its roots are complex. Therefore, $\alpha_i, \alpha_j \in \mathbb{C}$, and therefore $f(x)$ has a complex root. □

It is now easy to give a proof of the fundamental theorem of algebra. From Lemma 7.4.8 every nonconstant real polynomial has a complex root. From Lemma 7.4.7 if every nonconstant real polynomial has a complex root, then every nonconstant complex polynomial has a complex root proving the fundamental theorem.

Theorem 7.4.9. *If E is a finite dimensional field extension of $\mathbb{C}$, then $E = \mathbb{C}$.*

Proof. Let $a \in E$. Regard the elements $1, a, a^2, \ldots$. These elements become linearly dependent over $\mathbb{C}$ and we get a nonconstant polynomial over $\mathbb{C}$ with root a. By the fundamental theorem of algebra we know that $a \in \mathbb{C}$. □

Corollary 7.4.10. *If E is a finite dimensional field extension of $\mathbb{R}$, then $E = \mathbb{R}$ or $E = \mathbb{C}$.*

7.5 The Fundamental Theorem of Symmetric Polynomials

In the proof of the fundamental theorem of algebra that was given in the previous section we used the fact that any symmetric polynomial in n indeterminates is a polynomial in the elementary symmetric polynomials in these indeterminates. In this section we give a proof of this theorem.

Let R be an integral domain with $x_1, \dots, x_n$ (independent) indeterminates over R and let $R[x_1, \dots, x_n]$ be the polynomial ring in these indeterminates. Any polynomial $f(x_1, \dots, x_n) \in R[x_1, \dots, x_n]$ is composed of a sum of *pieces* of the form $ax_1^{i_1} \dots x_n^{i_n}$ with $a \in R$. We first put an order on these pieces of a polynomial.

The piece $ax_1^{i_1} \dots x_n^{i_n}$ with $a \neq 0$ is called *higher* than the piece $bx_1^{j_1} \dots x_n^{j_n}$ with $b \neq 0$ if the first one of the differences

$$i_1 - j_1, i_2 - j_2, \dots, i_n - j_n$$

that differs from zero is in fact positive. The highest piece of a polynomial $f(x_1, \dots, x_n)$ is denoted by $\mathrm{HG}(f)$.

Lemma 7.5.1. *For $f(x_1, \dots, x_n), g(x_1, \dots, x_n) \in R[x_1, \dots, x_n]$ we have*

$$\mathrm{HG}(fg) = \mathrm{HG}(f)\,\mathrm{HG}(g).$$

Proof. We use an induction on n, the number of indeterminates. It is clearly true for $n = 1$, and now assume that the statement holds for all polynomials in k indeterminates with $k < n$ and $n \geq 2$. Order the polynomials via exponents on the first indeterminate x_1 so that

$$\begin{aligned} f(x_1, \dots, x_n) &= x_1^r \phi_r(x_2, \dots, x_n) + x_1^{r-1} \phi_{r-1}(x_2, \dots, x_n) \\ &\quad + \dots + \phi_0(x_2, \dots, x_n) \\ g(x_1, \dots, x_n) &= x_1^s \psi_s(x_2, \dots, x_n) + x_1^{s-1} \psi_{s-1}(x_2, \dots, x_n) \\ &\quad + \dots + \psi_0(x_2, \dots, x_n). \end{aligned}$$

Then $\mathrm{HG}(fg) = x_1^{r+s}\,\mathrm{HG}(\phi_r \psi_s)$. By the inductive hypothesis

$$\mathrm{HG}(\phi_r \psi_s) = \mathrm{HG}(\phi_r)\,\mathrm{HG}(\psi_s).$$

Hence

$$\begin{aligned} \mathrm{HG}(fg) &= x_1^{r+s}\,\mathrm{HG}(\phi_r)\,\mathrm{HG}(\psi_s) \\ &= (x_1^r\,\mathrm{HG}(\phi_r))(x_1^s\,\mathrm{HG}(\psi_s)) = \mathrm{HG}(f)\,\mathrm{HG}(g). \end{aligned}$$ □

The elementary symmetric polynomials in n indeterminates $x_1, \dots, x_n$ are

$$\begin{aligned}
s_1 &= x_1 + x_2 + \cdots + x_n \\
s_2 &= x_1x_2 + x_1x_3 + \cdots + x_{n-1}x_n \\
s_3 &= x_1x_2x_3 + x_1x_2x_4 + \cdots + x_{n-2}x_{n-1}x_n \\
&\ \vdots \\
s_n &= x_1 \cdots x_n.
\end{aligned}$$

These were found by forming the polynomial $p(x, x_1, \dots, x_n) = (x - x_1)\cdots(x - x_n)$. The i-th elementary symmetric polynomial s_i in $x_1, \dots, x_n$ is then $(-1)^i a_i$, where a_i is the coefficient of x^{n-i} in $p(x, x_1, \dots, x_n)$.

In general,

$$s_k = \sum_{i_1 < i_2 < \cdots < i_k, 1 \le k \le n} x_{i_1} x_{i_2} \cdots x_{i_k},$$

where the sum is taken over all the $\binom{n}{k}$ different systems of indices $i_1, \dots, i_k$ with $i_1 < i_2 < \cdots < i_k$.

Further, a polynomial $s(x_1, \dots, x_n)$ is a *symmetric polynomial* if $s(x_1, \dots, x_n)$ is unchanged by any permutation σ of $\{x_1, \dots, x_n\}$, that is, $s(x_1, \dots, x_n) = s(\sigma(x_1), \dots, \sigma(x_n))$.

Lemma 7.5.2. *In the highest piece $ax_1^{k_1} \cdots x_n^{k_n}$, $a \neq 0$, of a symmetric polynomial $s(x_1, \dots, x_n)$ we have $k_1 \ge k_2 \ge \cdots \ge k_n$.*

Proof. Assume that $k_i < k_j$ for some $i < j$. As a symmetric polynomial, $s(x_1, \dots, x_n)$ also must then contain the piece $ax_1^{k_1} \cdots x_i^{k_j} \cdots x_j^{k_i} \cdots x_n^{k_n}$, which is higher than $ax_1^{k_1} \cdots x_i^{k_i} \cdots x_j^{k_j} \cdots x_n^{k_n}$, giving a contradiction. □

Lemma 7.5.3. *The product $s_1^{k_1-k_2} s_2^{k_2-k_3} \cdots s_{n-1}^{k_{n-1}-k_n} s_n^{k_n}$ with $k_1 \ge k_2 \ge \cdots \ge k_n$ has the highest piece $x_1^{k_1} x_2^{k_2} \cdots x_n^{k_n}$.*

Proof. From the definition of the elementary symmetric polynomials we have that

$$\mathrm{HG}(s_k^t) = (x_1x_2\cdots x_k)^t, \quad 1 \le k \le n,\ t \ge 1.$$

From Lemma 7.4.2,

$$\begin{aligned}
&\mathrm{HG}(s_1^{k_1-k_2} s_2^{k_2-k_3} \cdots s_{n-1}^{k_{n-1}-k_n} s_n^{k_n}) \\
&\quad = x_1^{k_1-k_2}(x_1x_2)^{k_2-k_3} \cdots (x_1 \cdots x_{n-1}^{k_{n-1}-k_n})(x_1 \cdots x_n)^{k_n} \\
&\quad = x_1^{k_1} x_2^{k_2} \cdots x_n^{k_n}.
\end{aligned}$$

□

Theorem 7.5.4. *Let $s(x_1, \dots, x_n) \in R[x_1, \dots, x_n]$ be a symmetric polynomial. Then $s(x_1, \dots, x_n)$ can be uniquely expressed as a polynomial $f(s_1, \dots, s_n)$ in the elementary symmetric polynomials $s_1, \dots, s_n$ with coefficients from R.*

Proof. We prove the existence of the polynomial f by induction on the size of the highest pieces. If in the highest piece of a symmetric polynomial all exponents are zero, then it is constant, that is, an element of R and there is nothing to prove.

Now we assume that each symmetric polynomial with highest piece smaller than that of $s(x_1, \dots, x_n)$ can be written as a polynomial in the elementary symmetric polynomials. Let $ax_1^{k_1} \cdots x_n^{k_n}$, $a \neq 0$, be the highest piece of $s(x_1, \dots, x_n)$. Let

$$t(x_1, \dots, x_n) = s(x_1, \dots, x_n) - as_1^{k_1-k_2} \cdots s_{n-1}^{k_{n-1}-k_n} s_n^{k_n}.$$

Clearly, $t(x_1, \dots, x_n)$ is another symmetric polynomial, and from Lemma 7.4.5 the highest piece of $t(x_1, \dots, x_n)$ is smaller than that of $s(x_1, \dots, x_n)$. Therefore, $t(x_1, \dots, x_n)$ and hence $s(x_1, \dots, x_n) = t(x_1, \dots, x_n) + as_1^{k_1-k_2} \cdots s_{n-1}^{k_{n-1}-k_n} s_n^{k_n}$ can be written as a polynomial in $s_1, \dots, s_n$.

To prove the uniqueness of this expression assume that $s(x_1, \dots, x_n) = f(s_1, \dots, s_n) = g(s_1, \dots, s_n)$. Then $f(s_1, \dots, s_n) - g(s_1, \dots, s_n) = h(s_1, \dots, s_n) = \phi(x_1, \dots, x_n)$ is the zero polynomial in $x_1, \dots, x_n$. Hence, if we write $h(s_1, \dots, s_n)$ as a sum of products of powers of the $s_1, \dots, s_n$, all coefficients disappear because two different products of powers in the $s_1, \dots, s_n$ have different highest pieces. This follows from previous set of lemmas. Therefore, f and g are the same, proving the theorem. □

7.6 Exercises

1. Suppose that $f(x) = a_0 + a_1 x + \cdots + a_n x^n \in K[x]$, K a field, with $a_n \neq 0$ and $f(x)$ irreducible. Define α to satisfy

$$a_0 + a_1\alpha + \cdots + a_n\alpha^n = 0$$

and define $K' = K(\alpha)$ in the following manner. We let

$$K(\alpha) = \{c_0 + c_1\alpha + \cdots + c_{n-1}\alpha^{n-1} : c_i \in K\}.$$

Then on $K(\alpha)$ define addition and subtraction componentwise and define multiplication by algebraic manipulation, replacing powers of α higher than α^n by using

$$\alpha^n = \frac{-a_0 - a_1\alpha - \cdots - a_{n-1}\alpha^{n-1}}{a_n}.$$

Prove that $K' = K(\alpha)$ forms a ring.

2. Let $f, g \in K[x]$ be irreducible polynomials of degree 2 over the field K. Let α_1, α_2 respectively β_1, β_2 be zeros of f and g. For $1 \le i, j \le 2$ let $v_{ij} = \alpha_i + \beta_j$. Show:
   (a) $|K(v_{ij}) : K| \in \{1, 2, 3\}$.
   (b) For fixed f, g there are at most two different degrees in (a).
   (c) Decide, which combinations of degrees in (b) (with f, g variable) are possible, and give an example in each case.
3. Let $L|K$ be a field extension, let $v \in L$ and $f(x) \in L[x]$ a polynomial of degree ≥ 1. Let all coefficients of $f(x)$ be algebraic over K. If $f(v) = 0$, then v is algebraic over K.
4. Let $L|K$ be a field extension and let M be an intermediate field. The extension $M|K$ is algebraic. For $v \in L$ the following are equivalent:
   (a) v is algebraic over M.
   (b) v is algebraic over K.
5. Let $L|K$ be a field extension and $v_1, v_2 \in L$. Then the following are equivalent:
   (a) v_1 and v_2 are algebraic over K.
   (b) $v_1 + v_2$ and $v_1 v_2$ are algebraic over K.
6. Let $L|K$ be a simple field extension. Then there is an extension field L' of L of the form $L' = K(v_1, v_2)$ with:
   (a) v_1 and v_2 are transcendental over K.
   (b) The set of all over K algebraic elements of L' is L.
7. In the proof of Theorem 7.1.4 show that the mapping

$$\tau : K(a) \to K(\alpha)$$

   defined by $\tau(k) = k$ if $k \in K$ and $\tau(a) = \alpha$ and then extended by linearity is a K-isomorphism.
8. For each $i \in I$ let K_i be a field. Now let $\hat{K} = \bigcup_{i \in I} K_i$. Show that $\hat{K}$ is a field.
9. Prove Lemma 7.2.8.
10. If T, T_1 are sets with the same cardinality, then there exists a bijection $\sigma : T \to T_1$. Define a map $F : S_T \to S_{T_1}$ in the following manner: if $f \in S_T$, let $F(f)$ be the permutation on T_1 given by $F(f)(t_1) = \sigma(f(\sigma^{-1}(t_1)))$. Prove that F is an isomorphism.
11. Prove that if $P(X), Q(x), H(x) \in \mathbb{C}$ then if $P(x)Q(x) = H(x)$ then $\overline{H}(x) = (\overline{P}(x))(\overline{Q}(x))$.

Chapter 8
Splitting Fields and Normal Extensions

8.1 Splitting Fields

In the last chapter we introduced *splitting fields* and used this idea to present a proof of the fundamental theorem of algebra. The concept of a splitting field is essential to the Galois theory of equations so in this chapter we look more deeply at this idea.

Definition 8.1.1. Let K be a field and $f(x)$ a nonconstant polynomial in $K[x]$. An extension field L of K is a *splitting field* for $f(x)$ over K if

(a) $f(x)$ splits into linear factors in $L[x]$.

(b) If $K \subset M \subset L$ and $M \neq L$ then $f(x)$ does not split into linear factors in $M[x]$.

From part (b) in the definition the following is clear.

Lemma 8.1.2. *L is a splitting field for $f(x) \in K[x]$ if and only if $f(x)$ splits into linear factors in $L[x]$ and if $f(x) = b(x - a_1)\cdots(x - a_n)$ with $b \in K$ then $L = K(a_1, \ldots, a_n)$.*

Example 8.1.3. The field $\mathbb{C}$ of complex numbers is a splitting field for the polynomial $p(x) = x^2 + 1$ in $\mathbb{R}[x]$. In fact since $\mathbb{C}$ is algebraically closed it is a splitting field for any real polynomial $f(x) \in \mathbb{R}[x]$ which has at least one nonreal zero.

The field $\mathbb{Q}(i)$ adjoining i to $\mathbb{Q}$ is a splitting field for $x^2 + 1$ over $\mathbb{Q}[x]$.

The next result was used in the previous chapter. We restate and reprove it here.

Theorem 8.1.4. *Let K be a field. Then each nonconstant polynomial in $K[x]$ has a splitting field.*

Proof. Let $\overline{K}$ be an algebraic closure of K. Then $f(x)$ splits in $\overline{K}[x]$, that is $f(x) = b(x - a_1)\cdots(x - a_n)$ with $b \in K$ and $a_i \in \overline{K}$. Let $L = K(a_1, \ldots, a_n)$. Then L is the splitting field for $f(x)$ over K. □

We next show that the splitting field over K of a given polynomial is unique up to K-isomorphism.

Theorem 8.1.5. *Let K, K' be field and $\phi : K \to K'$ an isomorphism. Let $f(x)$ be a nonconstant polynomial in $K[x]$ and $f'(x) = \phi(f(x))$ its image in $K'[x]$. Suppose that L is a splitting field for $f(x)$ over K and L' is a splitting field for $f'(x)$ over K'.*

(a) *Suppose that $L' \subset L''$. Then if $\psi : L \to L''$ is a monomorphism with $\psi_{|K} = \phi$ then ψ is an isomorphism from L onto L'. Moreover ψ maps the set of zeros of $f(x)$ in L onto the set of zeros of $f'(x)$ in L'. The map ψ is uniquely determined by the values of the zeros of $f(x)$.*

(b) *If $g(x)$ is an irreducible factor of $f(x)$ in $K[x]$ and a is a zero of $g(x)$ in L and a' is a zero of $g'(x) = \phi(g(x))$ in L' then there is an isomorphism ψ from L to L' with $\psi_{|K} = \phi$ and $\psi(a) = \psi(a')$.*

Before giving the proof of this theorem we note that the following important result is a direct consequence of it.

Theorem 8.1.6. *A splitting field for $f(x) \in K[x]$ is unique up to K-isomorphism.*

Proof of Theorem 8.1.5. Suppose that $f(x) = b(x - a_1) \cdots (x - a_n) \in L[x]$ and that $f'(x) = b'(x - a'_1) \cdots (x - a'_n) \in L'[x]$. Then

$$f'(x) = \phi(f(x)) = \psi(f(x)) = (\psi(b))(x - \psi(a_1)) \cdots (x - \psi(a_n)).$$

We have proved that polynomials have unique factorization over fields. Since $L' \subset L''$ it follows that the set of zeros $(\psi(a_1), \dots, \psi(a_n))$ is a permutation of the set of zeros $(a'_1, \dots, a'_n)$. In particular this implies that $\psi(a_i) \in L'$, that is

$$\operatorname{im}(\psi) = L' = K'(a_1, \dots, a'_n).$$

Since the image of ψ is $K'(a_1, \dots, a'_n) = K'(\psi(a_i), \dots, \psi(a_n))$ it is clear that ψ is uniquely determined by the images $\psi(a_i)$. This proves part (a).

For part (b) embed L' in an algebraic closure L''. Hence there is a monomorphism

$$\phi' : K(a) \to L''$$

with $\phi'_{|K} = \phi$ and $\phi'(a) = a'$. Hence there is a monomorphism $\psi : L \to L''$ with $\psi_{|K(a)} = \phi'$. Then from part (a) it follows that $\psi : L \to L'$ is an isomorphism. □

Example 8.1.7. Let $f(x) = x^3 - 7 \in \mathbb{Q}[x]$. This has no zeros in $\mathbb{Q}$ and since it is of degree 3 it follows that it must be irreducible in $\mathbb{Q}[x]$.

Let $\omega = -\frac{1}{2} + \frac{\sqrt{3}}{2}i \in \mathbb{C}$. Then it is easy to show by computation that $\omega^2 = -\frac{1}{2} - \frac{\sqrt{3}}{2}i$ and $\omega^3 = 1$. Therefore the three zeros of $f(x)$ in $\mathbb{C}$ are

$$a_1 = 7^{1/3}$$
$$a_2 = \omega \cdot 7^{1/3}$$
$$a_3 = \omega^2 \cdot 7^{1/3}.$$

Hence $L = \mathbb{Q}(a_1, a_2, a_3)$ the splitting field of $f(x)$. Since the minimal polynomial of all three zeros over $\mathbb{Q}$ is the same ($f(x)$) it follows that

$$\mathbb{Q}(a_1) \cong \mathbb{Q}(a_2) \cong \mathbb{Q}(a_3).$$

Since $\mathbb{Q}(a_1) \subset \mathbb{R}$ and a_2, a_3 are nonreal it is clear that $a_2, a_3 \notin \mathbb{Q}(a_1)$.

Suppose that $\mathbb{Q}(a_2) = \mathbb{Q}(a_3)$. Then $\omega = a_3 a_2^{-1} \in \mathbb{Q}(a_2)$ and so $7^{1/3} = \omega^{-1} a_2 \in \mathbb{Q}(a_2)$. Hence $Q(a_1) \subset \mathbb{Q}(a_2)$ and therefore $\mathbb{Q}(a_1) = \mathbb{Q}(a_2)$ since they have the same degree over $\mathbb{Q}$. This contradiction shows that $\mathbb{Q}(a_2)$ and $\mathbb{Q}(a_3)$ are distinct.

By computation we have $a_3 = a_1^{-1} a_2^2$ and hence

$$L = \mathbb{Q}(a_1, a_2, a_3) = \mathbb{Q}(a_1, a_2) = \mathbb{Q}(7^{1/3}, \omega).$$

Now the degree of L over $\mathbb{Q}$ is

$$|L : \mathbb{Q}| = |Q(7^{1/3}, \omega) : \mathbb{Q}(\omega)||\mathbb{Q}(\omega) : \mathbb{Q}|.$$

Now $|\mathbb{Q}(\omega) : \mathbb{Q}| = 2$ since the minimal polynomial of ω over $\mathbb{Q}$ is $x^2 + x + 1$. Since no zero of $f(x)$ lies in $\mathbb{Q}(\omega)$ and the degree of $f(x)$ is 3 it follows that $f(x)$ is irreducible over $\mathbb{Q}(\omega)$. Therefore we have that the degree of L over $\mathbb{Q}(\omega)$ is 3. Hence $|L : \mathbb{Q}| = (2)(3) = 6$.

We now have the following lattice diagram of fields and subfields:

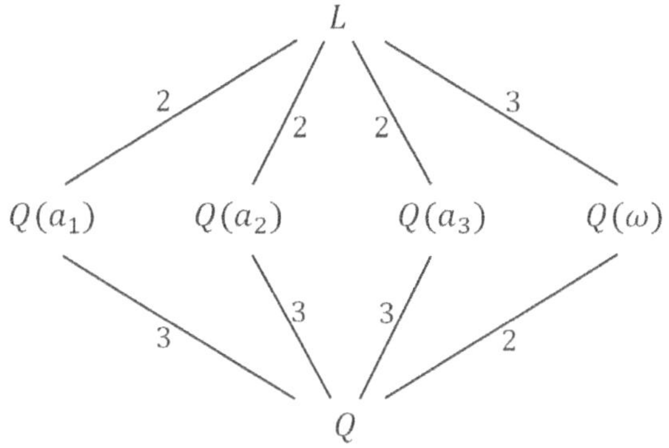

We do not know however if there are any more intermediate fields. There could for example be infinitely many. However as we will see when we do the Galois theory there are no others.

8.2 Normal Extensions

We now consider algebraic field extensions L of K which have the property that if $f(x) \in K[x]$ has a zero in L then $f(x)$ must split in L. In particular we show that if L is a splitting field of finite degree for some $g(x) \in K[x]$ then L has this property.

Definition 8.2.1. A field extension L of a field K is a *normal extension* if

(a) $L|K$ is algebraic.

(b) Each irreducible polynomial $f(x) \in K[x]$ that has a zero in L splits into linear factors in $L[x]$.

Note that in Example 8.1.7 the extension fields $Q(\alpha_i)|\mathbb{Q}$ are not normal extensions. Although $f(x)$ has a zero in $\mathbb{Q}(\alpha_i)$ the polynomial $f(x)$ does not split into linear factors in $\mathbb{Q}(\alpha_i)[x]$.

We now show that $L|K$ is a finite normal extension if and only if L is the splitting field for some $f(x) \in K[x]$.

Theorem 8.2.2. *Let $L|K$ be a finite extension. Then the following are equivalent.*

(a) *$L|K$ is a normal extension.*

(b) *$L|K$ is a splitting field for some $f(x) \in K[x]$.*

(c) *If $L \subset L'$ and $\psi : L \to L'$ is a monomorphism with $\psi_{|K}$ the identity map on K then ψ is an automorphism of L, that is $\psi(L) = L$.*

Proof. Suppose that $L|K$ is a finite normal extension. Since $L|K$ is a finite extension L is algebraic over K and since of finite degree we have $L = K(a_1, \dots, a_n)$ with a_i algebraic over K.

Let $f_i(x) \in K[x]$ be the minimal polynomial of a_i. Since $L|K$ is a normal extension $f_i(x)$ splits in $L[x]$. This is true for each $i = 1, \dots, n$. Let $f(x) = f_1(x) f_2(x) \cdots f_n(x)$. Then $f(x)$ splits into linear factors in $L[x]$. Since $K = K(a_1, \dots, a_n)$ the polynomial $f(x)$ cannot have all its zeros in any intermediate extension between K and L. Therefore L is the splitting field for $f(x)$. Hence (a) implies (b).

Now suppose that $L \subset L'$ and $\psi : L \to L'$ is a monomorphism with $\psi_{|K}$ the identity map on K. Then the extension field $\psi(L)$ of K is also a splitting field for $f(x)$ since $\psi_{|K}$ is the identity on K. Hence ψ maps the zeros of $f(x)$ in $L \subset L'$ onto the zeros of $f(x)$ in $\psi(L) \subset L'$ it follows that $\psi(L) = L$. Hence (b) implies (c).

Finally suppose (c). Hence we assume that if $L \subset L'$ and $\psi : L \to L'$ is a monomorphism with $\psi_{|K}$ the identity map on K then ψ is an automorphism of L, that is $\psi(L) = L$.

As before $L|K$ is algebraic since $L|K$ is finite. Suppose that $f(x) \in K[x]$ is irreducible and that $a \in L$ is a zero of $f(x)$. There are algebraic elements $a_1, \dots, a_n \in L$ with $L = K(a_1, \dots, a_n)$ since $L|K$ is finite. For $i = 1, \dots, n$ let $f_i(x) \in K[x]$ be the minimal polynomial of a_i and let $g(x) = f(x) f_1(x) \cdots f_n(x)$. Let L' be the splitting field of $g(X)$. Clearly $L \subset L'$. Let $b \in L'$ be a zero of $f(x)$. From Theorem 8.1.5 there is an automorphism ψ of L' with $\psi(a) = b$ and $\psi_{|K}$ the identity on K. Hence by our assumption $\psi_{|L}$ is an automorphism of L. It follows that $b \in L$ and hence $f(x)$ splits in $L[x]$. Therefore (c) implies (a) completing the proof. □

To give simple examples of normal extensions we have the following lemma.

Lemma 8.2.3. *If L is an extension of K with $|L : K| = 2$ then L is a normal extension of K.*

Proof. Suppose that $|L : K| = 2$. Then $L|K$ is algebraic since its finite. Let $f(x) \in K[x]$ be irreducible with leading coefficient 1 and which has a zero in L. Let a be one zero. Then $f(x)$ must be the minimal polynomial of a. However $\deg(m_a(x)) \leq |L : K| = 2$ and hence $f(x)$ is of degree 1 or 2. Since $f(x)$ has a zero in L it follows that it must split into linear factors in $L[x]$ and therefore L is a normal extension. □

Later we will tie this result to group theory when we prove that a subgroup of index 2 must be a normal subgroup.

Example 8.2.4. As a first example of the lemma consider the polynomial $f(x) = x^2 - 2$. In $\mathbb{R}$ this splits as $(x - \sqrt{2})(x + \sqrt{2})$ and hence the field $\mathbb{Q}(\sqrt{2})$ is the splitting field of $f(x) = x^2 - 2$ over $\mathbb{Q}$. Therefore $\mathbb{Q}(\sqrt{2})$ is a normal extension of $\mathbb{Q}$.

Example 8.2.5. As a second example consider the polynomial $x^4 - 2$ in $\mathbb{Q}[x]$. The zeros in $\mathbb{C}$ are

$$2^{1/4}, \quad 2^{1/4}i, \quad 2^{1/4}i^2, \quad 2^{1/4}i^3.$$

Hence

$$L = \mathbb{Q}(2^{1/4}, 2^{1/4}i, 2^{1/4}i^2, 2^{1/4}i^3)$$

is the splitting field of $x^4 - 2$ over $\mathbb{Q}$.

Now

$$L = \mathbb{Q}(2^{1/4}, 2^{1/4}i, 2^{1/4}i^2, 2^{1/4}i^3) = \mathbb{Q}(2^{1/4}, i).$$

Therefore we have

$$|L : \mathbb{Q}| = |L : \mathbb{Q}(2^{1/4})||\mathbb{Q}(2^{1/4}) : \mathbb{Q}|.$$

Since $x^4 - 2$ is irreducible over $\mathbb{Q}$ we have $|\mathbb{Q}(2^{1/4}) : \mathbb{Q}| = 4$. Since i has degree 2 over any real field we have $|L : \mathbb{Q}(2^{1/4})| = 2$. Therefore L is a normal extension of $\mathbb{Q}(2^{1/4})$ and $x^2 - \sqrt{2} \in \mathbb{Q}(\sqrt{2})[x]$ has the splitting field $\mathbb{Q}(2^{1/4})$.

Altogether we have $L|\mathbb{Q}(2^{1/4})$, $\mathbb{Q}(2^{1/4})|\mathbb{Q}(2^{1/2})$, $\mathbb{Q}(2^{1/2})|\mathbb{Q}$ and $L|\mathbb{Q}$ are normal extensions. However $\mathbb{Q}(2^{1/4})|\mathbb{Q}$ is not normal since $2^{1/4}$ is a zero of $x^4 - 2$ but $\mathbb{Q}(2^{1/4})$ does not contain all the zeros of $x^4 - 2$.

Hence, we have the following figure.

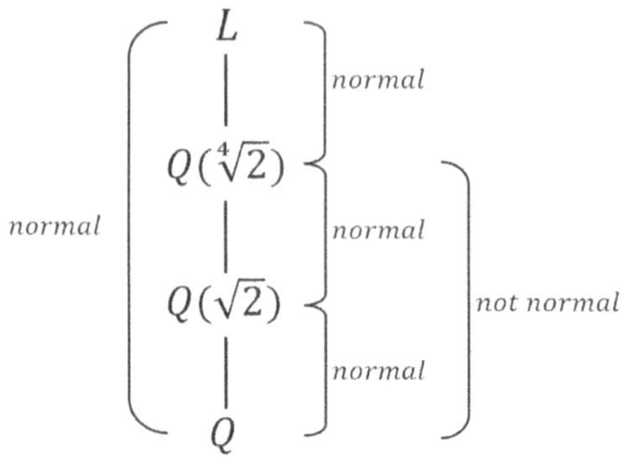

Figure 8.1

8.3 Exercises

1. Determine the splitting field of $f(x) \in \mathbb{Q}[x]$ and its degree over $\mathbb{Q}$ in the following cases:
   (a) $f(x) = x^4 - p$, where p is a prime.
   (b) $f(x) = x^p - 2$, where p is a prime.
2. Determine the degree of the splitting field of the polynomial $x^4 + 4$ over $\mathbb{Q}$. Determine the splitting field of $x^6 + 4x^4 + 4x^2 + 3$ over $\mathbb{Q}$.
3. For each $a \in \mathbb{Z}$ let $f_a(x) = x^3 - ax^2 + (a-3)x + 1 \in \mathbb{Q}[x]$ be given.
   (a) f_a is irreducible over $\mathbb{Q}$ for each $a \in \mathbb{Z}$.
   (b) If $b \in \mathbb{R}$ is a zero of f_a, then also $(1-b)^{-1}$ and $(b-1)b^{-1}$ are zeros of f_a.
   (c) Determine the splitting field L of $f_a(x)$ over $\mathbb{Q}$ and its degree $|L:\mathbb{Q}|$.
4. Let K be a field and $f(x) \in K[x]$ a polynomial of degree n. Let L be a splitting field of $f(x)$. Show:
   (a) If $a_1, \dots, a_n \in L$ are the zeros of f, then $|K(a_1, \dots, a_t) : K| \le n \cdot (n-1) \cdots (n-t+1)$ for each t with $1 \le t \le n$.
   (b) L over K is of degree at most $n!$.
   (c) If $f(x)$ is irreducible over K then n divides $|L:K|$.

Chapter 9
Groups, Subgroups and Examples

9.1 Groups, Subgroups and Isomorphisms

Recall from Chapter 1 that the three most commonly studied algebraic structures are groups, rings and fields. We have now looked rather extensively at rings and fields and in this chapter we consider the basic concepts of group theory. Groups arise in many different areas of mathematics. For example they arise in geometry as groups of congruence motions and in topology as groups of various types of continuous functions. Later in this book they will appear in Galois theory as groups of automorphisms of fields. First we recall the definition of a group given previously in Chapter 1.

Definition 9.1.1. A *group* G is a set with one binary operation which we will denote by multiplication, such that

(1) The operation is associative, that is, $(g_1g_2)g_3 = g_1(g_2g_3)$ for all $g_1, g_2, g_3 \in G$.

(2) There exists an identity for this operation, that is, an element 1 such that $1g = g$ and $g1 = g$ for each $g \in G$.

(3) Each $g \in G$ has an inverse for this operation, that is, for each g there exists a g^{-1} with the property that $gg^{-1} = 1$ and $g^{-1}g = 1$.

If in addition the operation is commutative, that is $g_1g_2 = g_2g_1$ for all $g_1, g_2 \in G$, the group G is called an *abelian group*.

The *order* of G, denoted $|G|$, is the number of elements in the group G. If $|G| < \infty$, G is a *finite group* otherwise it is an *infinite group*.

It follows easily from the definition that the identity is unique and that each element has a unique inverse.

Lemma 9.1.2. *If G is a group then there is a unique identity. Further if $g \in G$ its inverse is unique. Finally if $g_1, g_2 \in G$ then $(g_1g_2)^{-1} = g_2^{-1}g_1^{-1}$.*

Proof. Suppose that 1 and e are both identities for G. Then $1e = e$ since e is an identity and $1e = 1$ since 1 is an identity. Therefore $1 = e$ and there is only one identity.

Next suppose that $g \in G$ and g_1 and g_2 are inverses for g. Then

$$g_1gg_2 = (g_1g)g_2 = 1g_2 = g_2$$

since $g_1 g = 1$. On the other hand

$$g_1 g g_2 = g_1(g g_2) = g_1 1 = g_1$$

since $g g_2 = 1$. It follows that $g_1 = g_2$ and g has a unique inverse.

Finally consider

$$(g_1 g_2)(g_2^{-1} g_1^{-1}) = g_1(g_2 g_2^{-1}) g_1^{-1} = g_1 1 g_1^{-1} = g_1 g_1^{-1} = 1.$$

Therefore $g_2^{-1} g_1^{-1}$ is an inverse for $g_1 g_2$ and since inverses are unique it is the inverse of the product. □

Groups most often arise as permutations on a set. We will see this, as well as other specific examples of groups, in the next sections.

Finite groups can be completely described by their *group tables* or multiplication tables. These are sometimes called *Cayley tables*. In general, let $G = \{g_1, \dots, g_n\}$ be a group, then the *multiplication table* of G is:

$$\begin{array}{c|ccccc}
 & g_1 & g_2 & \cdots & g_j & \cdots\ g_n \\
\hline
g_1 & \cdots & & & & \\
g_2 & \cdots & & & & \\
\vdots & & & & & \\
g_i & \cdots & \cdots & \cdots & g_i g_j & \\
\vdots & & & & & \\
g_n & \cdots & & & &
\end{array}$$

The entry in the row of $g_i \in G$ and column of $g_j \in G$ is the product (in that order) $g_i g_j$ in G.

Groups satisfy the *cancellation law for multiplication*.

Lemma 9.1.3. *If G is a group and $a, b, c \in G$ with $ab = ac$ or $ba = ca$ then $b = c$.*

Proof. Suppose that $ab = ac$. Then a has an inverse a^{-1} so we have

$$a^{-1}(ab) = a^{-1}(ac).$$

From the associativity of the group operation we then have

$$(a^{-1}a)b = (a^{-1}a)c \implies 1 \cdot b = 1 \cdot c \implies b = c.$$ □

A consequence of Lemma 9.1.3 is that each row and each column in a group table is just a permutation of the group elements. That is each group element appears exactly once in each row and each column.

A subset $H \subset G$ is a *subgroup* of G if H is also a group under the same operation as G. As for rings and fields a subset of a group is a subgroup if it is nonempty and closed under both the group operation and inverses.

Lemma 9.1.4. *A subset $H \subset G$ is a subgroup if $H \neq \emptyset$ and H is closed under the operation and inverses. That is, if $a, b \in H$ then $ab \in H$ and $a^{-1}, b^{-1} \in H$.*

We leave the proof of this to the exercises.

Let G be a group and $g \in G$; we denote by g^n, $n \in \mathbb{N}$, as with numbers, the product of g taken n times. A negative exponent will indicate the inverse of the positive exponent. As usual, let $g^0 = 1$. Clearly group exponentiation will satisfy the standard laws of exponents. Now consider the set

$$H = \{1 = g^0, g, g^{-1}, g^2, g^{-2}, \dots\}$$

of all powers of g. We will denote this by $\langle g \rangle$.

Lemma 9.1.5. *If G is a group and $g \in G$ then $\langle g \rangle$ forms a subgroup of G called the cyclic subgroup generated by g. $\langle g \rangle$ is abelian even if G is not.*

Proof. If $g \in G$ then $g \in \langle g \rangle$ and hence $\langle g \rangle$ is nonempty. Suppose then that $a = g^n, b = g^m$ are elements of $\langle g \rangle$. Then $ab = g^n g^m = g^{n+m} \in \langle g \rangle$ so $\langle g \rangle$ is closed under the group operation. Further $a^{-1} = (g^n)^{-1} = g^{-n} \in \langle g \rangle$ so $\langle g \rangle$ is closed under inverses. Therefore $\langle g \rangle$ is a subgroup.

Finally $ab = g^n g^m = g^{n+m} = g^{m+n} = g^m g^n = ba$ and hence $\langle g \rangle$ is abelian. □

Suppose that $g \in G$ and $g^m = 1$ for some positive integer m. Then let n be the smallest positive integer such that $g^n = 1$. It follows that the set of elements $\{1, g, g^2, \dots, g^{n-1}\}$ are all distinct but for any other power g^k we have $g^k = g^t$ for some $k = 0, 1, \dots, n-1$ (see exercises). The cyclic subgroup generated by g then has order n and we say that g has *order* n which we denote by $o(g) = n$. If no such n exists we say that g has *infinite order*. We will look more deeply at cyclic groups and subgroups in Section 9.5.

We introduce one more concept before looking at examples.

Definition 9.1.6. If G and H are groups then a mapping $f : G \to H$ is a (group) *homomorphism* if $f(g_1 g_2) = f(g_1) f(g_2)$ for any $g_1, g_2 \in G$. If f is also a bijection then it is an *isomorphism*.

As with rings and fields we say that two groups G and H are *isomorphic*, denoted by $G \cong H$, if there exists an isomorphism $F : G \to H$. This means that abstractly G and H have exactly the same algebraic structure.

9.2 Examples of Groups

As already mentioned groups arise in many diverse areas of mathematics. In this section and the next we present specific examples of groups.

First of all any ring or field under addition forms an abelian group. Hence, for example $(\mathbb{Z}, +), (\mathbb{Q}, +), (\mathbb{R}, +), (\mathbb{C}, +)$ where $\mathbb{Z}, \mathbb{Q}, \mathbb{R}, \mathbb{C}$ are respectively the integers, the rationals, the reals and the complex numbers, all are infinite abelian groups. If $\mathbb{Z}_n$ is the modular ring $\mathbb{Z}/n\mathbb{Z}$ then for any natural number n, $(\mathbb{Z}_n, +)$ forms a finite abelian group. In abelian groups the group operation is often denoted by $+$ and the identity element by 0 (zero).

In a field F, the nonzero elements are all invertible and form a group under multiplication. This is called the *multiplicative group* of the field F and is usually denoted by F^*. Since multiplication in a field is commutative the multiplicative group of a field is an abelian group. Hence $\mathbb{Q}^*, \mathbb{R}^*, \mathbb{C}^*$ are all infinite abelian groups while if p is a prime $\mathbb{Z}_p^*$ forms a finite abelian group. Recall that if p is a prime then the modular ring $\mathbb{Z}_p$ is a field.

Within $\mathbb{Q}^*, \mathbb{R}^*, \mathbb{C}^*$ there are certain multiplicative subgroups. Since the positive rationals $\mathbb{Q}_+$ and the positive reals $\mathbb{R}_+$ are closed under multiplication and inverse they form subgroups of $\mathbb{Q}^*$ and $\mathbb{R}^*$ respectively. In $\mathbb{C}$ if we consider the set of all complex numbers z with $|z| = 1$ then these form a multiplicative subgroup. Further within this subgroup if we consider the set of n-th roots of unity z (that is $z^n = 1$) for a fixed n this forms a subgroup, this time of finite order.

The multiplicative group of a field is a special case of the unit group of a ring. If R is a ring with identity, recall that a *unit* is an element of R with a multiplicative inverse. Hence in $\mathbb{Z}$ the only units are ± 1 while in any field every nonzero element is a unit.

Lemma 9.2.1. *If R is a ring with identity then the set of units in R forms a group under multiplication called the unit group of R and is denoted by $U(R)$. If R is a field then $U(R) = R^*$.*

Proof. Let R be a ring with identity. Then the identity 1 itself is a unit so $1 \in U(R)$ and hence $U(R)$ is nonempty. If $e \in R$ is a unit then it has a multiplicative inverse e^{-1}. Clearly then the multiplicative inverse has an inverse, namely e so $e^{-1} \in U(R)$ if e is. Hence to show $U(R)$ is a group we must show that it is closed under product.

Let $e_1, e_2 \in U(R)$. Then there exist e_1^{-1}, e_2^{-1}. It follows that $e_2^{-1}e_1^{-1}$ is an inverse for e_1e_2. Hence e_1e_2 is also a unit and $U(R)$ is closed under product. Therefore for any ring R with identity $U(R)$ forms a multiplicative group. □

To present examples of nonabelian groups we turn to matrices. If F is a field we let

$$\mathrm{GL}(n, F) = \{n \times n \text{ matrices over } F \text{ with nonzero determinant}\}$$

and

$$\mathrm{SL}(n, F) = \{n \times n \text{ matrices over } F \text{ with determinant one}\}.$$

Lemma 9.2.2. *If F is a field then for $n \geq 2$, $\mathrm{GL}(n, F)$ forms a nonabelian group under matrix multiplication and $\mathrm{SL}(n, F)$ forms a subgroup.*

$\mathrm{GL}(n, F)$ is called the n-dimensional general linear group over F, while $\mathrm{SL}(n, F)$ is called the n-dimensional special linear group over F.

Proof. Recall that for two $n \times n$ matrices A, B with $n \geq 2$ over a field we have

$$\det(AB) = \det(A)\det(B)$$

where det is the determinant.

Now for any field the $n \times n$ identity matrix I has determinant 1 and hence $I \in \mathrm{GL}(n, F)$. Since determinants multiply the product of two matrices with nonzero determinant has nonzero determinant so $\mathrm{GL}(n, F)$ is closed under product. Further over a field F if A is an invertible matrix then

$$\det(A^{-1}) = \frac{1}{\det A}$$

and so if A has nonzero determinant so does its inverse. It follows that $\mathrm{GL}(n, F)$ has the inverse of any of its elements. Since matrix multiplication is associative it follows that $\mathrm{GL}(n, F)$ form a group. It is nonabelian since in general matrix multiplication is noncommutative.

We leave the fact that $\mathrm{SL}(n, F)$ forms a subgroup to the exercises. □

Groups play an important role in geometry. In any metric geometry an *isometry* is a mapping that preserves distance. To understand a geometry one must understand the group of isometries. We look briefly at the Euclidean geometry of the plane $\mathcal{E}^2$.

An *isometry* or *congruence motion* of $\mathcal{E}^2$ is a transformation or bijection T of $\mathcal{E}^2$ that preserves distance, that is $d(a, b) = d(T(a), T(b))$ for all points $a, b \in \mathcal{E}^2$.

Theorem 9.2.3. *The set of congruence motions of $\mathcal{E}^2$ forms a group called the Euclidean group. We denote the Euclidean group by $\mathcal{E}$.*

Proof. The identity map I is clearly an isometry and since composition of mappings is associative we need only show that the product of isometries is an isometry and that the inverse of an isometry is an isometry.

Let T, U be isometries. Then $d(a, b) = d(T(a), T(b))$ and $d(a, b) = d(U(a), U(b))$ for any points a, b. Now consider

$$d(TU(a), TU(b)) = d(T(U(a)), T(U(b))) = d(U(a), U(b))$$

since T is an isometry. However

$$d(U(a), U(b)) = d(a, b)$$

since U is an isometry. Combining these we have that TU is also an isometry.

Consider T^{-1} and points a, b. Then

$$d(T^{-1}(a), T^{-1}(b)) = d(TT^{-1}(a), TT^{-1}(b))$$

since T is an isometry. But $TT^{-1} = I$ and hence

$$d(T^{-1}(a), T^{-1}(b)) = d(TT^{-1}(a), TT^{-1}(b)) = d(a, b).$$

Therefore T^{-1} is also an isometry and hence $\mathcal{E}$ is a group. □

One of the major results concerning $\mathcal{E}$ is the following. We refer to [25], [26] and [22] for a more thorough treatment.

Theorem 9.2.4. *If $T \in \mathcal{E}$ then T is either a translation, rotation, reflection or glide reflection. The set of translations and rotations forms a subgroup.*

Proof. We outline a brief proof. If T is an isometry and T fixes the origin $(0, 0)$ then T is a linear mapping. It follows that T is a rotation or a reflection. If T does not fix the origin then there is a translation T_0 such that T_0T fixes the origin. This gives translations and glide reflections. In the exercises we expand out more of the proof. □

If D is a geometric figure in $\mathcal{E}^2$, such as a triangle or square, then a *symmetry* of D is a congruence motion $T : \mathcal{E}^2 \to \mathcal{E}^2$ that leaves D in place. It may move the individual elements of D however. So for example a rotation about the center of a circle is a symmetry of the circle.

Lemma 9.2.5. *If D is a geometric figure in $\mathcal{E}^2$ then the set of symmetries of D forms a subgroup of $\mathcal{E}$ called the symmetry group of D denoted by* $\text{Sym}(D)$.

Proof. We show that $\text{Sym}(D)$ is a subgroup of $\mathcal{E}$. The identity map I fixes D so $I \in \text{Sym}(D)$ and so $\text{Sym}(D)$ is nonempty. Let $T, U \in \text{Sym}(D)$. Then T maps D to D and so does U. It follows directly that so does the composition TU and hence $TU \in \text{Sym}(D)$. If T maps D to D then certainly the inverse does. Therefore $\text{Sym}(D)$ is a subgroup of $\mathcal{E}$. □

Example 9.2.6. Let T be an equilateral triangle. Then there are exactly six symmetries of T (see exercises). These are

$$\begin{aligned}
I &= \text{the identity}\\
r &= \text{a rotation of } 120^\circ \text{ around the center of } T\\
r^2 &= \text{a rotation of } 240^\circ \text{ around the center of } T\\
f &= \text{a reflection over the perpendicular bisector of one of the sides}\\
fr &= \text{the composition of } f \text{ and } r\\
fr^2 &= \text{the composition of } f \text{ and } r^2.
\end{aligned}$$

Sym(T) is called the *dihedral group* D_3. In the next section we will see that it is isomorphic to S_3, the symmetric group on 3 symbols.

9.3 Permutation Groups

Groups most often appear as groups of transformations or permutations on a set. In this section we will take a short look at permutation groups and then examine them more deeply in the next chapter. We recall some ideas first introduced in Chapter 7 in relation to the proof of the fundamental theorem of algebra.

Definition 9.3.1. If A is a set, a *permutation* on A is a one-to-one mapping of A onto itself. We denote by S_A the set of all permutations on A.

Theorem 9.3.2. *For any set A, S_A forms a group under composition called the symmetric group on A. If $|A| > 2$ then S_A is nonabelian. Further if A, B have the same cardinality, then $S_A \cong S_B$.*

Proof. If S_A is the set of all permutations on the set A, we must show that composition is an operation on S_A that is associative and has an identity and inverses.

Let $f, g \in S_A$. Then f, g are one-to-one mappings of A onto itself. Consider $f \circ g : A \to A$. If $f \circ g(a_1) = f \circ g(a_2)$, then $f(g(a_1)) = f(g(a_2))$ and $g(a_1) = g(a_2)$, since f is one-to-one. But then $a_1 = a_2$ since g is one-to-one.

If $a \in A$, there exists $a_1 \in A$ with $f(a_1) = a$ since f is onto. Then there exists $a_2 \in A$ with $g(a_2) = a_1$ since g is onto. Putting these together, $f(g(a_2)) = a$, and therefore $f \circ g$ is onto. Therefore, $f \circ g$ is also a permutation and composition gives a valid binary operation on S_A.

The identity function $1(a) = a$ for all $a \in A$ will serve as the identity for S_A, while the inverse function for each permutation will be the inverse. Such unique inverse functions exist since each permutation is a bijection.

Finally, composition of functions is always associative and therefore S_A forms a group.

Suppose that $|A| > 2$. Then A has at least 3 elements. Call them a_1, a_2, a_2. Consider the 2 permutations f and g which fix (leave unchanged) all of A except a_1, a_2, a_3 and on these three elements

$$f(a_1) = a_2, \qquad f(a_2) = a_3, \qquad f(a_3) = a_1$$
$$g(a_1) = a_2, \qquad g(a_2) = a_1, \qquad g(a_3) = a_3.$$

Then under composition

$$f(g(a_1)) = a_3, \quad f(g(a_2)) = a_2, \quad f(g(a_3)) = a_1$$

while

$$g(f(a_1)) = a_1, \quad g(f(a_2)) = a_3, \quad g(f(a_3)) = a_2.$$

Therefore $f \circ g \neq g \circ f$ and hence S_A is not abelian.

If A, B have the same cardinality, then there exists a bijection $\sigma : A \to B$. Define a map $F : S_A \to S_B$ in the following manner: if $f \in S_A$, let $F(f)$ be the permutation on B given by $F(f)(b) = \sigma(f(\sigma^{-1}(b)))$. It is straightforward to verify that F is an isomorphism (see the exercises). □

If $A_1 \subset A$ then those permutations on A that map A_1 to A_1 form a subgroup of S_A called the *stabilizer* of A_1 denoted stab(A_1). We leave the proof to the exercises.

Lemma 9.3.3. *If* $A_1 \subset A$ *then* $\operatorname{stab}(A_1) = \{f \in S_A : f : A_1 \to A_1\}$ *forms a subgroup of* S_A.

A *permutation group* is any subgroup of S_A for some set A.

We now look at finite permutation groups. Let A be a finite set, say $A = \{a_1, a_2, \dots, a_n\}$. Then each $f \in S_A$ can be pictured as

$$f = \begin{pmatrix} a_1 & \dots & a_n \\ f(a_1) & \dots & f(a_n) \end{pmatrix}.$$

For a_1 there are n choices for $f(a_1)$. For a_2 there are only $n-1$ choices since f is one-to-one. This continues down to only one choice for a_n. Using the multiplication principle, the number of choices for f and therefore the size of S_A is

$$n(n-1)\cdots 1 = n!.$$

We have thus proved the following theorem.

Theorem 9.3.4. *If* $|A| = n$ *then* $|S_A| = n!$.

For a set with n elements we denote S_A by S_n, called the *symmetric group on n symbols*.

Example 9.3.5. Write down the six elements of S_3 and give the multiplication table for the group.

Name the three elements $1, 2, 3$. The six elements of S_3 are then

$$1 = \begin{pmatrix} 1 & 2 & 3 \\ 1 & 2 & 3 \end{pmatrix}, \quad a = \begin{pmatrix} 1 & 2 & 3 \\ 2 & 3 & 1 \end{pmatrix}, \quad b = \begin{pmatrix} 1 & 2 & 3 \\ 3 & 1 & 2 \end{pmatrix}$$

$$c = \begin{pmatrix} 1 & 2 & 3 \\ 2 & 1 & 3 \end{pmatrix}, \quad d = \begin{pmatrix} 1 & 2 & 3 \\ 3 & 2 & 1 \end{pmatrix}, \quad e = \begin{pmatrix} 1 & 2 & 3 \\ 1 & 3 & 2 \end{pmatrix}.$$

The multiplication table for S_3 can be written down directly by doing the required composition. For example,

$$ac = \begin{pmatrix} 1 & 2 & 3 \\ 2 & 3 & 1 \end{pmatrix} \begin{pmatrix} 1 & 2 & 3 \\ 2 & 1 & 3 \end{pmatrix} = \begin{pmatrix} 1 & 2 & 3 \\ 3 & 2 & 1 \end{pmatrix} = d.$$

To see this, note that $a : 1 \to 2, 2 \to 3, 3 \to 1$; $c : 1 \to 2, 2 \to 1, 3 \to 3$ and so $ac : 1 \to 3, 2 \to 2, 3 \to 1$.

It is somewhat easier to construct the multiplication table if we make some observations. First, $a^2 = b$ and $a^3 = 1$. Next, $c^2 = 1$, $d = ac$, $e = a^2c$ and finally $ac = ca^2$.

From these relations the following multiplication table can be constructed:

	1	a	a^2	c	ac	a^2c
1	1	a	a^2	c	ac	a^2c
a	a	a^2	1	ac	a^2c	c
a^2	a^2	1	a	a^2c	c	ac
c	c	a^2c	ac	1	a^2	a
ac	ac	c	a^2c	a	1	a^2
a^2c	a^2c	ac	c	a^2	a	1

To see this, consider, for example, $(ac)a^2 = a(ca^2) = a(ac) = a^2c$.

More generally, we can say that S_3 has a *presentation* given by

$$S_3 = \langle a, c; a^3 = c^2 = 1, ac = ca^2 \rangle.$$

By this we mean that S_3 is *generated by* a, c, or that S_3 has *generators* a, c and the whole group and its multiplication table can be generated by using the *relations* $a^3 = c^2 = 1, ac = ca^2$.

A theorem of Cayley actually shows that every group is a permutation group. A group G is a permutation group on the group G itself considered as a set. This result however does not give much information about the group.

Theorem 9.3.6 (Cayley's theorem). *Let G be a group. Consider the set of elements of G. Then the group G is a permutation group on the set G, that is G is a subgroup of S_G.*

Proof. We show that to each $g \in G$ we can associate a permutation of the set G. If $g \in G$ let π_g be the map given by

$$\pi_g : g_1 \to gg_1 \quad \text{for each } g_1 \in G.$$

It is straightforward to show that each π_g is a permutation on G. □

9.4 Cosets and Lagrange's Theorem

In this section given a group G and a subgroup H we define an equivalence relation on G. The equivalence classes all have the same size and are called the (left) or (right) cosets of H in G.

Definition 9.4.1. Let G be a group and $H \subset G$ a subgroup. For $a, b \in G$ define $a \sim b$ if $a^{-1}b \in H$.

Lemma 9.4.2. *Let G be a group and $H \subset G$ a subgroup. Then the relation defined above is an equivalence relation on G. The equivalence classes all have the form aH for $a \in G$ and are called the left cosets of H in G. Clearly G is a disjoint union of its left cosets.*

Proof. Let us show, first of all, that this is an equivalence relation. Now $a \sim a$ since $a^{-1}a = e \in H$. Therefore the relation is reflexive. Further $a \sim b$ implies $a^{-1}b \in H$, but since H is a subgroup of G we have $b^{-1}a = (a^{-1}b)^{-1} \in H$ and so $b \sim a$. Therefore the relation is symmetric. Finally suppose that $a \sim b$ and $b \sim c$. Then $a^{-1}b \in H$ and $b^{-1}c \in H$. Since H is a subgroup $a^{-1}b \cdot b^{-1}c = a^{-1}c \in H$, and hence $a \sim c$. Therefore the relation is transitive and hence is an equivalence relation.

For $a \in G$ the equivalence class is

$$[a] = \{g \in G : a \sim g\} = \{a \in G : a^{-1}g \in H\}.$$

But then clearly $g \in aH$. It follows that the equivalence class for $a \in G$ is precisely the set

$$aH = \{g \in G : g = ah \text{ for some } h \in H\}.$$

These classes, aH, are called *left cosets* of H and since they are equivalence classes they partition G. This means that every element of g is in one and only one left coset. In particular, $bH = H = eH$ if and only if $b \in H$. □

If aH is a left coset then we call the element a a *coset representative* representative. A complete collection

$$\{a \in G : \{aH\} \text{ is the set of all distinct left cosets of } H\}$$

is called a (left) *transversal* of H in G.

One could define another equivalence relation by defining $a \sim b$ if and only if $ba^{-1} \in H$. Again this can be shown to be an equivalence relation on G, and the equivalence classes here are sets of the form

$$Ha = \{g \in G : g = ha \text{ for some } h \in H\},$$

called *right cosets* of H. Also, of course, G is the (disjoint) union of distinct right cosets.

It is easy to see that any two left (right) cosets have the same order (number of elements). To demonstrate this consider the mapping $aH \to bH$ via $ah \mapsto bh$ where $h \in H$. It is not hard to show that this mapping is 1-1 and onto (see exercises). Thus we have $|aH| = |bH|$. (This is also true for right cosets and can be established in a similar manner.) Letting $b \in H$ in the above discussion, we see $|aH| = |H|$, for any $a \in G$, that is the size of each left or right coset is exactly the same as the subgroup H.

One can also see that the collection $\{aH\}$ of all distinct left cosets has the same number of elements as the collection $\{Ha\}$ of all distinct right cosets. In other words, the number of left cosets equals the number of right cosets (this number may be infinite). For consider the map

$$f : aH \to Ha^{-1}.$$

This mapping is well-defined: for if $aH = bH$, then $b = ah$ where $h \in H$. Thus $f(bH) = Hb^{-1} = Hh^{-1}a^{-1} = f(aH)$. It is not hard to show that this mapping is 1-1 and onto (see exercises). Hence the number of left cosets equals the number of right cosets.

Definition 9.4.3. Let G be a group and $H \subset G$ a subgroup. The number of distinct left cosets, which is the same as the number of distinct right cosets, is called the *index* of H in G, denoted by $[G : H]$.

Now let us consider the case where the group G is finite. Each left coset has the same size as the subgroup H and here both are finite. Hence $|aH| = |H|$ for each coset. Further the group G is a disjoint union of the left cosets, that is

$$G = H \cup g_1 H \cup \cdots \cup g_n H.$$

Since this is a disjoint union we have

$$|G| = |H| + |g_1 H| + \cdots + |g_n H| = |H| + |H| + \cdots + |H| = |H|[G : H].$$

This establishes the following extremely important theorem.

Theorem 9.4.4 (Lagrange's theorem). *Let G be a group and $H \subset G$ a subgroup. Then*

$$|G| = |H|[G : H].$$

If G is a finite group this implies that both the order of a subgroup and the index of a subgroup are divisors of the order of the group.

This theorem plays a crucial role in the structure theory of finite groups since it greatly restricts the size of subgroups. For example in a group of order 10 there can be proper subgroups only of orders 1, 2 and 5.

As an immediate corollary, we have the following result.

Corollary 9.4.5. *The order of any element $g \in G$, where G is a finite group, divides the order of the group. In particular if $|G| = n$ and $g \in G$ then $o(g)|n$ and $g^n = 1$.*

Proof. Let $g \in G$ and $o(g) = m$. Then m is the size of the cyclic subgroup generated by g and hence divides m from Lagrange's theorem. Then $n = mk$ and so

$$g^n = g^{mk} = (g^m)^k = 1^k = 1. \qquad \square$$

Before leaving this section we consider some results concerning general subsets of a group.

Suppose that G is a group and S is an arbitrary nonempty subset of G, $S \subset G$ and $S \neq \emptyset$; such a set S is usually called a *complex* of G.

If U and V are two complexes of G, the product UV is defined as follows:

$$UV = \{g_1 g_2 \in G : u \in U, v \in V\}.$$

Now suppose that U, V are subgroups of G. When is the complex UV again a subgroup of G?

Theorem 9.4.6. *The product UV of two subgroups U, V of a group G is itself a subgroup if and only if U and V commute, that is, if and only if $UV = VU$.*

Proof. We note first that when we say U and V commute, we do *not* demand that this is so elementwise. In other words, it is not required that $uv = vu$ for all $u \in U$ and all $v \in V$; all that is required is that for any $u \in U$ and $v \in V$ $uv = v_1 u_1$, for some elements $u_1 \in U$ and $v_1 \in V$.

Assume that UV is a subgroup of G. Let $u \in U$ and $v \in V$. Then $u \in U \cdot 1 \subset UV$ and $v \in 1 \cdot V \subset UV$. But since UV is assumed itself to be a subgroup, it follows that $vu \in UV$. Hence each product $vu \in UV$ and so $VU \subset UV$. In an identical manner $UV \subset VU$ and so $UV = VU$.

Conversely, suppose that $UV = VU$. Let $g_1 = u_1 v_1 \in UV$, $g_2 = u_2 v_2 \in UV$. Then

$$g_1 g_1 = (u_1 v_1)(u_2 v_2) = u_1(v_1 u_2)v_2 = u_1 u_3 v_3 v_2 = (u_1 u_3)(v_3 v_2) \in UV$$

since $v_1 u_2 = u_3 v_3$ for some $u_3 \in U$ and $v_3 \in V$. Further

$$g_1^{-1} = (u_1 v_1)^{-1} = v_1^{-1} u_1^{-1} = u_4 v_4.$$

It follows that UV is a subgroup. $\square$

Theorem 9.4.7 (product formula). *Let U, V be subgroups of G and let R be a left transversal of the intersection $U \cap V$ in U. Then*

$$UV = \bigcup_{r \in R} rV$$

where this is a disjoint union.

In particular if U, V are finite then

$$|UV| = \frac{|U||V|}{|U \cap V|}.$$

Proof. Since $R \subset U$ we have that

$$\bigcup_{r \in R} rV \subset UV.$$

In the other direction let $uv \in UV$. Then

$$U = \bigcup_{r \in R} r(U \cap V).$$

It follows that $u = rv'$ with $r \in R$ and $v' \in U \cap V$. Hence

$$uv = rv'v \in rV.$$

The union of cosets of V is disjoint so

$$uv \in \bigcup_{r \in R} rV.$$

Therefore $UV \subset \bigcup_{r \in R} rV$ proving the equality.

Now suppose that $|U|$ and $|V|$ are finite. Then we have

$$|UV| = |R||V| = |U : U \cap V||V| = \frac{|U|}{|U \cap V|}|V| = \frac{|U||V|}{|U \cap V|}. \qquad \square$$

We now show that index is multiplicative. Later we will see how this fact is related to the multiplicativity of the degree of field extensions.

Theorem 9.4.8. *Suppose G is a group and U and V are subgroups with $U \subset V \subset G$. Then if G is the disjoint union*

$$G = \bigcup_{r \in R} rV$$

and V is the disjoint union

$$V = \bigcup_{s \in S} sU$$

then we get a disjoint union for G as

$$G = \bigcup_{r \in R, s \in S} rsU.$$

In particular if $[G : V]$ *and* $[V : U]$ *are finite then*

$$[G : U] = [G : V][V : U].$$

Proof. Now

$$G = \bigcup_{r \in R} rV = \bigcup_{r \in R} \Big(\bigcup_{s \in S} sU \Big) = \bigcup_{r \in R, s \in S} rsU.$$

Suppose that $r_1 s_1 U = r_2 s_2 U$. Then $r_1 s_1 UV = r_2 s_2 UV$. But $s_1 UV = V$ and $s_2 UV = V$ so $r_1 V = r_2 V$ which implies that $r_1 = r_2$. Then $s_1 U = s_2 U$ which implies that $s_1 = s_2$. Therefore the union is disjoint.

The index formula now follows directly. □

The next result says that the intersection of subgroups of finite index must again be of finite index.

Theorem 9.4.9 (Poincaré). *Suppose that* U, V *are subgroups of finite index in* G. *Then* $U \cap V$ *is also of finite index. Further*

$$[G : U \cap V] \leq [G : U][G : V].$$

If $[G : U], [G : V]$ *are relatively prime then equality holds.*

Proof. Let r be the number of left cosets of U in G that are contained in UV. r is finite since the index $[G : U]$ is finite. From Theorem 9.4.7 we then have

$$|V : U \cap V| = r \leq [G : U].$$

Then from Theorem 9.4.8

$$[G : U \cap V] = [G : V][V : U \cap V] \leq [G : V][G : U].$$

Since both $[G : U]$ and $[G : V]$ are finite so is $[G : U \cap V]$.

Now $[G : U] | [G : U \cap V]$ and $[G : V] | [G : U \cap V]$. If $[G : U]$ and $[G : V]$ are relatively prime then

$$[G : U][G : V] | [G : U \cap V] \implies [G : U][G : V] \leq [G : U \cap V]$$

Therefore we must have equality. □

Corollary 9.4.10. *Suppose that* $[G : U]$ *and* $[G : V]$ *are finite and relatively prime. Then* $G = UV$.

Proof. From Theorem 9.4.9 we have

$$[G : U \cap V] = [G : U][G : V].$$

From Theorem 9.4.8

$$[G : U \cap V] = [G : V][V : U \cap V].$$

Combing these we have

$$[V : U \cap V] = [G : U].$$

The number of left cosets of U in G that are contained in VU is equal to the number of all left cosets of U in G. It follows then that we must have $G = UV$. □

9.5 Generators and Cyclic Groups

We saw that if G is any group and $g \in G$ then the powers of g generate a subgroup of G called the cyclic subgroup generated by g. Here we explore more fully the idea of generating a group or subgroup. We first need the following.

Lemma 9.5.1. *If U and V are subgroups of a group G then their intersection $U \cap V$ is also a subgroup.*

Proof. Since the identity of G is in both U and V we have that $U \cap V$ is nonempty. Suppose that $g_1, g_2 \in U \cap V$. Then $g_1, g_2 \in U$ and hence $g_1^{-1}g_2 \in U$ since U is a subgroup. Analogously $g_1^{-1}g_2 \in V$. Hence $g^{-1}g_2 \in U \cap V$ and therefore $U \cap V$ is a subgroup. □

Now let S be a subset of a group G. The subset S is certainly contained in at least one subgroup of G, namely G itself. Let $\{U_\alpha\}$ be the collection of all subgroups of G containing S. Then $\bigcap_\alpha U_\alpha$ is again a subgroup of G from Lemma 9.5.1. Further it is the smallest subgroup of G containing S (see the exercises). We call $\bigcap_\alpha U_\alpha$ the subgroup of G generated by S and denote it by $\langle S \rangle$ or $\text{grp}(S)$. We call the set S a set of *generators* for $\langle S \rangle$.

Definition 9.5.2. A subset M of a group G is a *set of generators* for G if $G = \langle M \rangle$, that is the smallest subgroup of G containing M is all of G. We say that G is *generated* by M and that M is a set of *generators* for G.

Notice that any group G has at least one set of generators, namely G itself. If $G = \langle M \rangle$ and M is a finite set then we say that G is *finitely generated.* Clearly any finite group is finitely generated. Shortly we will give an example of a finitely generated infinite group.

Example 9.5.3. The set of all reflections forms a set of generators for the Euclidean group $\mathcal{E}$. Recall that any $T \in \mathcal{E}$ is either a translation, a rotation, a reflection or a glide reflection. It can be shown (see exercises) that any one of these can be expressed as a product of 3 or fewer reflections.

We now consider the case where a group G has a single generator.

Definition 9.5.4. A group G is *cyclic* if there exists a $g \in G$ such that $G = \langle g \rangle$.

In this case $G = \{g^n : n \in \mathbb{Z}\}$, that is G consists of all the powers of the element g. If there exists an integer m such that $g^m = 1$, then there exists a smallest such positive integer say n. It follows that $g^k = g^l$ if and only if $k \equiv l \mod n$. In this situation the distinct powers of g are precisely

$$\{1 = g^0, g, g^2, \dots, g^{n-1}\}.$$

It follows that $|G| = n$. We then call G a *finite cyclic group*. If no such power exists then all the powers of G are distinct and G is an infinite cyclic group.

We show next that any two cyclic groups of the same order are isomorphic.

Theorem 9.5.5. (a) *If $G = \langle g \rangle$ is an infinite cyclic group then $G \cong (\mathbb{Z}, +)$ that is the integers under addition.*

(b) *If $G = \langle g \rangle$ is a finite cyclic group of order n then $G \cong (\mathbb{Z}_n, +)$ that is the integers modulo n under addition.*

It follows that for a given order there is only one cyclic group up to isomorphism.

Proof. Let G be an infinite cyclic group with generator g. Map g onto $1 \in (\mathbb{Z}, +)$. Since g generates G and 1 generates $\mathbb{Z}$ under addition this can be extended to a homomorphism. It is straightforward to show that this defines an isomorphism.

Now let G be a finite cyclic group of order n with generator g. As above map g to $1 \in \mathbb{Z}_n$ and extend to a homomorphism. Again it is straightforward to show that this defines an isomorphism.

Now let G and H be two cyclic groups of the same order. If both are infinite then both are isomorphic to $(\mathbb{Z}, +)$ and hence isomorphic to each other. If both are finite of order n then both are isomorphic to $(\mathbb{Z}_n, +)$ and hence isomorphic to each other. □

Theorem 9.5.6. *Let $G = \langle g \rangle$ be a finite cyclic group of order n. Then every subgroup of G is also cyclic. Further if $d \mid n$ there exists a unique subgroup of G of order d.*

Proof. Let $G = \langle g \rangle$ be a finite cyclic group of order n and suppose that H is a subgroup of G. Notice that if $g^m \in H$ then g^{-m} is also in H since H is a subgroup. Hence H must contain positive powers of the generator g. Let t be the smallest positive power of g such that $g^t \in H$. We claim that $H = \langle g^t \rangle$ the cyclic subgroup of G generated by g^t. Let $h \in H$ then $h = g^m$ for some positive integer $m \geq t$. Divide m by t to get

$$m = qt + r \quad \text{where } r = 0 \text{ or } 0 < r < t.$$

If $r \neq 0$ then $r = m - qt > 0$. Now $g^m \in H$, $g^t \in H$ so $g^{-qt} \in H$ for any q since H is a subgroup. It follows that $g^m g^{-qt} = g^{m-qt} \in H$. This implies that $g^r \in H$. However this is a contradiction since $r < t$ and t is the least positive power in H. It follows that $r = 0$ so $m = qt$. This implies that $g^m = g^{qt} = (g^t)^q$, that is g^m is a multiple of g^t. Therefore every element of H is a multiple of g^t and therefore g^t generates H and hence H is cyclic.

Now suppose that $d|n$ so that $n = kd$. Let $H = \langle g^k \rangle$, that is the subgroup of G generated by g^k. We claim that H has order d and that any other subgroup H_1 of G with order d coincides with H. Now $(g^k)^d = g^{kd} = g^n = 1$ so the order of g^k divides d and hence is $\leq d$. Suppose that $(g^k)^{d_1} = g^{kd_1} = 1$ with $d_1 < d$. Then since the order of g is n we have $n = kd|kd_1$ with $d_1 < d$ which is impossible. Therefore the order of g^k is d and $h = \langle g^k \rangle$ is a subgroup of G of order d.

Now let H_1 be a subgroup of G of order d. We must show that $H_1 = H$. Let $h \in H_1$ so $h = g^t$ and hence $g^{td} = 1$. It follows that $n|td$ and so $kd|td$ and hence $k|t$, that is $t = qk$ for some positive integer q. Therefore $g^t = (g^k)^q \in H$. Therefore $H_1 \subset H$ and since they are of the same size $H = H_1$. □

Theorem 9.5.7. *Let $G = \langle g \rangle$ be an infinite cyclic group. Then a subgroup H is of the form $H = \langle g^t \rangle$ for a positive integer t. Further if t_1, t_2 are positive integers with $t_1 \neq t_2$ then $\langle g^{t_1} \rangle$ and $\langle g^{t_2} \rangle$ are distinct.*

Proof. Let $G = \langle g \rangle$ be an infinite cyclic group and H a subgroup of G. As in the proof of Theorem 9.5.6 H must contain positive powers of the generator g. Let t be the smallest positive power of g such that $g^t \in H$. We claim that $H = \langle g^t \rangle$ the cyclic subgroup of G generated by g^t. Let $h \in H$ then $h = g^m$ for some positive integer $m \geq t$. Divide m by t to get

$$m = qt + r \quad \text{where } r = 0 \text{ or } 0 < r < t.$$

If $r \neq 0$ then $r = m - qt > 0$. Now $g^m \in H$, $g^t \in H$ so $g^{-qt} \in H$ for any q since H is a subgroup. It follows that $g^m g^{-qt} = g^{m-qt} \in H$. This implies that $g^r \in H$. However this is a contradiction since $r < t$ and t is the least positive power in H. It follows that $r = 0$ so $m = qt$. This implies that $g^m = g^{qt} = (g^t)^q$, that is g^m is a multiple of g^t. Therefore every element of H is a multiple of g^t and therefore g^t generates H and hence $H = \langle g^t \rangle$.

From the proof above in the subgroup $\langle g^t \rangle$ the integer t is the smallest positive power of g in $\langle g^t \rangle$. Therefore if t_1, t_2 are positive integers with $t_1 \neq t_2$ then $\langle g^{t_1} \rangle$ and $\langle g^{t_2} \rangle$ are distinct. □

Theorem 9.5.8. *Let $G = \langle g \rangle$ be a cyclic group. Then*

(a) *If $G = \langle g \rangle$ is finite of order n then g^k is also a generator if and only if $(k,n) = 1$. That is the generators of G are precisely those powers g^k where k is relatively prime to n.*

(b) *If $G = \langle g \rangle$ is infinite then the only generators are g, g^{-1}.*

Proof. (a) Let $G = \langle g \rangle$ be a finite cyclic group of order n and suppose that $(k,n) = 1$. Then there exist integers x, y with $kx + ny = 1$. It follows that

$$g = g^{kx+ny} = (g^k)^x (g^n)^y = (g^k)^x$$

since $g^n = 1$. Hence g is a power of g^k that implies every element of G is also a power of g^k. Therefore g^k is also a generator.

Conversely suppose that g^k is also a generator. Then g is a power of g^k so there exists an x such that $g = g^{kx}$. It follows that $kx \equiv 1$ modulo n and so there exists a y such that

$$kx + ny = 1.$$

This then implies that $(k,n) = 1$.

(b) If $G = \langle g \rangle$ is infinite then any power of g other than g^{-1} generates a proper subgroup. If g is a power of g^n for some n so that $g = g^{nx}$ it follows that $g^{nx-1} = 1$ so that g has finite order contradicting that G is infinite cyclic. □

Recall that for positive integers n the Euler phi-function is defined as follows.

Definition 9.5.9. For any $n > 0$, let

$$\phi(n) = \text{number of integers less than or equal to } n \text{ and relatively prime to } n.$$

Example 9.5.10. $\phi(6) = 2$ since among $1, 2, 3, 4, 5, 6$ only $1, 5$ are relatively prime to 6.

Corollary 9.5.11. *If $G = \langle g \rangle$ is finite of order n then there are $\phi(n)$ generators for G where ϕ is the Euler phi-function.*

Proof. From Theorem 9.5.8 the generators of G are precisely the powers g^k where $(k,n) = 1$. The numbers relatively prime to n are counted by the Euler phi-function. □

Recall that in an arbitrary group G, if $g \in G$, then the order of g, denoted $o(g)$, is the order of the cyclic subgroup generated by g. Given two elements $g, h \in G$ in general there is no relationship between $o(g), o(h)$ and the order of the product gh. However if they commute there is a very direct relationship.

Lemma 9.5.12. *Let G be an arbitrary group and $g, h \in G$ both of finite order $o(g), o(h)$. If g and h commute, that is $gh = hg$, then $o(gh)$ divides* $\text{lcm}(o(g), o(h))$. *In particular if G is an abelian group then $o(gh) | \text{lcm}(o(g), o(h))$ for all $g, h \in G$ of finite order. Further if $\langle g \rangle \cap \langle h \rangle = \{1\}$ then $o(gh) = \text{lcm}(o(g), o(h))$.*

Proof. Suppose $o(g) = n$ and $o(h) = m$ are finite. If g, h commute then for any k we have $(gh)^k = g^k h^k$. Let $t = \text{lcm}(n, m)$ then $t = k_1 m, t = k_2 n$. Hence

$$(gh)^t = g^t h^t = (g^m)^{k_1} (h^n)^{k_2} = 1.$$

Therefore the order of gh is finite and divides t. Suppose that $\langle g \rangle \cap \langle h \rangle = \{1\}$ that is the cyclic subgroup generated by g intersects trivially with the cyclic subgroup generated by h. Let $k = o(gh)$ which we know is finite from the first part of the lemma. Let $t = \text{lcm}(n, m)$. We then have $(gh)^k = g^k h^k = 1$ which implies that $g^k = h^{-k}$. Since the cyclic subgroups have only trivial intersection this implies that $g^k = 1$ and $h^k = 1$. But then $n|k$ and $m|k$ and hence $t|k$. Since $k|t$ it follows that $k = t$. □

Recall that if m and n are relatively prime then $\text{lcm}(m, n) = mn$. Further if the orders of g and h are relatively prime it follows from Lagrange's theorem that $\langle g \rangle \cap \langle h \rangle = \{1\}$. We then get the following.

Corollary 9.5.13. *If g, h commute and $o(g)$ and $o(h)$ are finite and relatively prime then $o(gh) = o(g)o(h)$.*

Definition 9.5.14. If G is a finite abelian group then the *exponent* of G is the lcm of the orders of all elements of G. That is

$$\exp(G) = \text{lcm}\{o(g) : g \in G\}.$$

As a consequence of Lemma 9.5.12 we obtain

Lemma 9.5.15. *Let G be a finite abelian group. Then G contains an element of order* $\exp(G)$.

Proof. Suppose that $\exp(G) = p_1^{e_1} \cdots p_k^{e_k}$ with p_i distinct primes. By the definition of $\exp(G)$ there is a $g_i \in G$ with $o(g_i) = p_i^{e_i} r_i$ with p_i and r_i relatively prime. Let $h_i = g_i^{r_i}$. Then from Lemma 9.5.12 we get $o(h_i) = p_i^{e_i}$. Now let $g = h_1 h_2 \cdots h_k$. From the corollary to Lemma 9.5.12 we have $o(g) = p_1^{e1} \cdots p_k^{e_k} = \exp(G)$. □

If K is a field then the multiplicative subgroup of nonzero elements of K is an abelian group $K^\star$. The above results lead to the fact that a finite subgroup of $K^\star$ must actually be cyclic.

Theorem 9.5.16. *Let K be a field. Then any finite subgroup of $K^\star$ is cyclic.*

Proof. Let $A \subset K^\star$ with $|A| = n$. Suppose that $m = \exp(A)$. Consider the polynomial $f(x) = x^m - 1 \in K[x]$. Since the order of each element in A divides m it follows that $a^m = 1$ for all $a \in A$ and hence each $a \in A$ is a zero of the polynomial $f(x)$. Hence $f(x)$ has at least n zeros. Since a polynomial of degree m over a field can have at most m zeros it follows that $n < m$. From Lemma 9.5.15 there is an element $a \in A$ with $o(a) = m$. Since $|A| = n$ it follows that $m|n$ and hence $m < n$. Therefore $m = n$ and hence $A = \langle a \rangle$ showing that A is cyclic. □

We close this section with two other results concerning cyclic groups. The first proves, using group theory, a very interesting number theoretic result concerning the Euler phi-function.

Theorem 9.5.17. *For $n > 1$ and for $d \geq 1$*

$$\sum_{d|n} \phi(d) = n.$$

Proof. Consider a cyclic group G of order n. For each $d|n$, $d \geq 1$ there is a unique cyclic subgroup H of order d. H then has $\phi(d)$ generators. Each element in G generates its own cyclic subgroup H_1, say of order d and hence must be included in the $\phi(d)$ generators of H_1. Therefore

$$\sum_{d|n} \phi(d) = \text{sum of the numbers of generators of the cyclic subgroups of } G.$$

But this must be the whole group and hence this sum is n. □

We shall make use of the above theorem directly in the following theorem.

Theorem 9.5.18. *If $|G| = n$ and if for each positive d such that $d|n$, G has at most one cyclic subgroup of order d, then G is cyclic (and consequently, has exactly one cyclic subgroup of order d).*

Proof. For each $d|n$, $d > 0$, let $\psi(d) =$ the number of elements of G of order d. Then

$$\sum_{d|n} \psi(d) = n.$$

Now suppose that $\psi(d) \neq 0$ for a given $d|n$. Then there exists an $a \in G$ of order d which generates a cyclic subgroup, $\langle a \rangle$, of order d of G. We claim that all elements of G of order d are in $\langle a \rangle$. Indeed, if $b \in G$ with $o(b) = d$ and $b \notin \langle a \rangle$, then $\langle b \rangle$ is a second cyclic subgroup of order d, distinct from $\langle a \rangle$. This contradicts the hypothesis, so the claim is proved. Thus, if $\psi(d) \neq 0$, then $\psi(d) = \phi(d)$. In general, we have $\psi(d) \leq \phi(d)$, for all positive $d|n$. But $n = \sum_{d|n} \psi(d) \leq \sum_{d|n} \phi(d)$, by the previous theorem. It follows, clearly, from this that $\psi(d) = \phi(d)$ for all $d|n$. In particular, $\psi(n) = \phi(n) \geq 1$. Hence, there exists at least one element of G of order n; hence G is cyclic. This completes the proof. □

Corollary 9.5.19. *If in a group G of order n, for each $d|n$, the equation $x^d = 1$ has at most d solutions in G, then G is cyclic.*

Proof. The hypothesis clearly implies that G can have at most one cyclic subgroup of order d since all elements of such a subgroup satisfy the equation. So Theorem 9.5.17 applies to give our result. □

If H is a subgroup of a group G then G operates as a group of permutations on the set $\{aH : a \in R\}$ of left cosets of H in G where R is a left transversal of H in G. This we can use to show that a finitely generated group has only finitely many subgroups of a given finite index.

Theorem 9.5.20. *Let G be a finitely generated group. The number of subgroups of index $n < \infty$ is finite.*

Proof. Let H be a subgroup of index n. We choose a left transversal $\{c_1, \dots, c_n\}$ for H in G where $c_1 = 1$ represents H. G permutes the set of cosets $c_i H$ by multiplication from the left. This induces a homomorphism ψ_H from G to S_n as follows. For each $g \in G$ let $\psi_H(g)$ be the permutation which maps i to j if $gc_i H = c_j H$. $\psi_H(g)$ fixes the number 1 if and only if $g \in H$ because $c_1 H = H$. Now, let H and L be two different subgroups of index n in G. Then there exists $g \in H$ with $g \notin L$ and $\psi_H(g) \neq \psi_L(g)$, and hence ψ_H and ψ_L are different. Since G is finitely generated there are only finitely many homomorphisms from G to S_n. Therefore the number of subgroups of index $n < \infty$ is finite. □

9.6 Exercises

1. Prove Lemma 9.1.4.

2. Suppose that $g \in G$ and $g^m = 1$ for some positive integer m. Let n be the smallest positive integer such that $g^n = 1$. Show the set of elements $\{1, g, g^2, \dots, g^{n-1}\}$ are all distinct but for any other power g^k we have $g^k = g^t$ for some $k = 0, 1, \dots, n-1$.

3. Let G be a group and U_1, U_2 be finite subgroups of G. If $|U_1|$ and $|U_2|$ are relatively prime, then $U_1 \cap U_2 = \{e\}$.

4. Let A, B be subgroups of a finite group G. If $|A| \cdot |B| > |G|$ then $A \cap B \neq \{e\}$.

5. Let G be the set of all real matrices of the form $\left(\begin{smallmatrix} a & -b \\ b & a \end{smallmatrix}\right)$, where $a^2 + b^2 \neq 0$. Show:

   (a) G is a group.

   (b) For each $n \in \mathbb{N}$ there is at least one element of order n in G.

6. Let p be a prime, and let $G = \mathrm{SL}(2, p)$. Show:

   (a) G has at least $2p - 2$ elements of order p (their exact number is $p^2 - 1$).

   (b) If p is odd, then $x^p = 1$ for all $x \in G$.

7. Let p be a prime and $a \in \mathbb{Z}$. Show that $a^p \equiv a \bmod p$.

8. Let F be a field. Show that the set of $n \times n$ matrices of determinant 1 over F forms a group.

9. Here we outline a proof that every planar Euclidean congruence motion is either a rotation, translation, reflection or glide reflection. An isometry in this problem is a planar Euclidean congruence motion. Show:

   (a) If T is an isometry then it is completely determined by its action on a triangle – equivalent to showing that if T fixes three noncollinear points then it must be the identity.

   (b) If an isometry T has exactly one fixed point then it must be a rotation with that point as center.

   (c) If an isometry T has two fixed points then it fixes the line joining them. Then show that if T is not the identity it must be a reflection through this line.

   (d) If an isometry T has no fixed point but preserves orientation then it must be a translation.

   (e) If an isometry T has no fixed point but reverses orientation then it must be a glide reflection.

10. Let P_n be a regular n-gon and D_N its group of symmetries. Show that $|D_n| = 2n$. (Hint: First show that $|D_n| \leq 2n$ and then exhibit $2n$ distinct symmetries.)

11. If A, B have the same cardinality, then there exists a bijection $\sigma : A \to B$. Define a map $F : S_A \to S_B$ in the following manner: if $f \in S_A$, let $F(f)$ be the permutation on B given by $F(f)(b) = \sigma(f(\sigma^{-1}(b)))$. Show that F is an isomorphism.

12. Prove Lemma 9.3.3.

Chapter 10
Normal Subgroups, Factor Groups and Direct Products

10.1 Normal Subgroups and Factor Groups

In rings we saw that there were certain special types of subrings, called *ideals*, that allowed us to define factor rings. The analogous object for groups is called a *normal subgroup* which we will define and investigate in this section.

Definition 10.1.1. Let G be an arbitrary group and suppose that H_1 and H_2 are subgroups of G. We say that H_2 is *conjugate* to H_1 if there exists an element $a \in G$ such that $H_2 = aH_1a^{-1}$. H_1, H_2 are the called *conjugate subgroups* of G.

Lemma 10.1.2. *Let G be an arbitrary group. Then the relation of conjugacy is an equivalence relation on the set of subgroups of G.*

Proof. We must show that conjugacy is reflexive, symmetric and transitive. If H is a subgroup of G then $1^{-1}H1 = H$ and hence H is conjugate to itself and therefore the relation is reflexive.

Suppose that H_1 is conjugate to H_2. Then there exists a $g \in G$ with $g^{-1}H_1g = H_2$. This implies that $gH_2g^{-1} = H_1$. However $(g^{-1})^{-1} = g$ and hence letting $g^{-1} = g_1$ we have $g_1^{-1}H_2g_1 = H_1$. Therefore H_2 is conjugate to H_1 and conjugacy is symmetric.

Finally suppose that H_1 is conjugate to H_2 and H_2 is conjugate to H_3. Then there exist $g_1, g_2 \in G$ with $H_2 = g_1^{-1}H_1g_1$ and $H_3 = g_2^{-1}H_2g_2$. Then

$$H_3 = g_2^{-1}g_1^{-1}H_1g_1g_2 = (g_1g_2)^{-1}H_1(g_1g_2).$$

Therefore H_3 is conjugate to H_2 and conjugacy is transitive. □

Lemma 10.1.3. *Let G be an arbitrary group. Then for $g \in G$ the map $g : a \to g^{-1}ag$ is an automorphism on G.*

Proof. For a fixed $g \in G$ define the map $f : G \to G$ by $f(a) = g^{-1}ag$ for $a \in G$. We must show that this is a homomorphism and that it is one-to-one and onto.

Let $a_1, a_2 \in G$. Then

$$f(a_1a_2) = g^{-1}a_1a_2g = (g^{-1}a_1g)(g^{-1}a_2g) = f(a_1)f(a_2).$$

Hence f is a homomorphism.

If $f(a_1) = f(a_2)$ then $g^{-1}a_1g = g^{-1}a_2g$. Clearly by the cancellation law we then have $a_1 = a_2$ and hence f is one-to-one.

Finally let $a \in G$ and let $a_1 = gag^{-1}$. Then $a = g^{-1}a_1g$ and hence $f(a_1) = a$. It follows that f is onto and therefore f is an automorphism on G. □

In general a subgroup H of a group G may have many different conjugates. However in certain situations the only conjugate of a subgroup H is H itself. If this is the case we say that H is a *normal subgroup*. We will see shortly that this is precisely the analog for groups of the concept of an ideal in rings.

Definition 10.1.4. Let G be an arbitrary group. A subgroup H is a *normal subgroup* of G, which we denote by $H \lhd G$, if $g^{-1}Hg = H$ for all $g \in G$.

Since the conjugation map is an isomorphism it follows that if $g^{-1}Hg \subset H$ then $g^{-1}Hg = H$. Hence in order to show that a subgroup is normal we need only show inclusion.

Lemma 10.1.5. *Let N be a subgroup of a group G. Then if $aNa^{-1} \subset N$ for all $a \in G$, then $aNa^{-1} = N$. In particular, $aNa^{-1} \subset N$ for all $a \in G$ implies that N is a normal subgroup.*

Notice that if $g^{-1}Hg = H$ then $Hg = gH$. That is as sets the left coset gH is equal to the right coset Hg. Hence for each $h_1 \in H$ there is an $h_2 \in H$ with $gh_1 = h_2g$. If $H \lhd G$ this is true for all $g \in G$. Further if H is normal then for the product of two cosets g_1H and g_2H we have

$$(g_1H)(g_2H) = g_1(Hg_2)H = g_1g_2(HH) = g_1g_2H.$$

If $(g_1H)(g_2H) = (g_1g_2)H$ for all $g_1, g_2 \in G$ we necessarily have $gHg^{-1} = H$ for all $g \in G$.

Hence we have proved:

Lemma 10.1.6. *Let H be a subgroup of a group G. Then the following are equivalent:*

(1) *H is a normal subgroup of G.*

(2) *$g^{-1}Hg = H$ for all $g \in G$.*

(3) *$gH = Hg$ for all $g \in G$.*

(4) *$(g_1H)(g_2H) = (g_1g_2)H$ for all $g_1, g_2 \in G$.*

This is precisely the condition needed to construct factor groups. First we give some examples of normal subgroups.

Lemma 10.1.7. *Every subgroup of an abelian group is normal.*

Proof. Let G be abelian and H a subgroup of G. Suppose $g \in G$ then $gh = hg$ for all $h \in H$ since G is abelian. It follows that $gH = Hg$. Since this is true for every $g \in G$ it follows that H is normal. □

Lemma 10.1.8. *Let $H \subset G$ be a subgroup of index 2, that is $[G : H] = 2$. Then H is normal in G.*

Proof. Suppose that $[G : H] = 2$. We must show that $gH = Hg$ for all $g \in G$. If $g \in H$ then clearly $H = gH = Hg$. Therefore we may assume that g is not in H. Then there are only 2 left cosets and 2 right cosets. That is,

$$G = H \cup gH = H \cup Hg.$$

Since the union is a disjoint union we must have $gH = Hg$ and hence H is normal. □

Lemma 10.1.9. *Let K be any field. Then the group $\mathrm{SL}(n, K)$ is a normal subgroup of $\mathrm{GL}(n, K)$ for any positive integer n.*

Proof. Recall that $\mathrm{GL}(n, K)$ is the group of $n \times n$ matrices over the field K with nonzero determinant while $\mathrm{SL}(n, K)$ is the subgroup of $n \times n$ matrices over the field K with determinant equal to 1. Let $U \in \mathrm{SL}(n, K)$ and $T \in \mathrm{GL}(n, K)$. Consider $T^{-1}UT$. Then

$$\begin{aligned}\det(T^{-1}UT) &= \det(T^{-1})\det(U)\det(T) = \det(U)\det(T^{-1}T)\\ &= \det(U)\det(I) = \det(U) = 1.\end{aligned}$$

Hence $T^{-1}UT \in \mathrm{SL}(n, K)$ for any $U \in \mathrm{SL}(n, K)$ and any $T \in \mathrm{GL}(n, K)$. It follows that $T^{-1}\,\mathrm{SL}(n, K)T \subset \mathrm{SL}(n, K)$ and therefore $\mathrm{SL}(n, K)$ is normal in $\mathrm{GL}(n, K)$. □

The intersection of normal subgroups is again normal and the product of normal subgroups is normal.

Lemma 10.1.10. *Let N_1, N_2 be normal subgroups of the group G. Then*

(1) *$N_1 \cap N_2$ is a normal subgroup of G.*

(2) *N_1N_2 is a normal subgroup of G.*

(3) *If H is any subgroup of G then $N_1 \cap H$ is a normal subgroup of H and $N_1H = HN_1$.*

Proof. (a) Let $n \in N_1 \cap N_2$ and $g \in G$. Then $g^{-1}ng \in N_1$ since N_1 is normal. Similarly $g^{-1}ng \in N_2$ since N_2 is normal. Hence $g^{-1}ng \in N_1 \cap N_2$. It follows that $g^{-1}(N_1 \cap N_2)g \subset N_1 \cap N_2$ and therefore $N_1 \cap N_2$ is normal.

(b) Let $n_1 \in N_1, n_2 \in N_2$. Since N_1, N_2 are both normal $N_1N_2 = N_2N_1$ as sets and the complex N_1N_2 forms a subgroup of G. Let $g \in G$ and $n_1n_2 \in N_1N_2$. Then

$$g^{-1}(n_1n_2)g = (g^{-1}n_1g)(g^{-1}n_2g) \in N_1N_2$$

since $g^{-1}n_1g \in N_1$ and $g^{-1}n_2g \in N_2$. Therefore N_1N_2 is normal in G.

(c) Let $h \in H$ and $n \in N \cap H$. Then as in part (a) $h^{-1}nh \in N \cap H$ and therefore $N \cap H$ is a normal subgroup of H. The same argument as in part (b) shows that $N_1H = HN_1$. □

We now construct *factor groups* or *quotient groups* of a group modulo a normal subgroup.

Definition 10.1.11. Let G be an arbitrary group and H a normal subgroup of G. Let G/H denote the set of distinct left (and hence also right) cosets of H in G. On G/H define the multiplication

$$(g_1H)(g_2H) = g_1g_2H$$

for any elements g_1H, g_2H in G/H.

Theorem 10.1.12. *Let G be a group and H a normal subgroup of G. Then G/H under the operation defined above forms a group. This group is called the factor group or quotient group of G modulo H. The identity element is the coset $1H = H$ and the inverse of a coset gH is $g^{-1}H$.*

Proof. We first show that the operation on G/N is well-defined. Suppose that $a'N = aN$ and $b'N = bN$, then $b' \in bN$ and so $b' = bn_1$. Similarly $a' = an_2$ where $n_1, n_2 \in N$. Therefore

$$a'b'N = an_2bn_1N = an_2bN$$

since $n_1 \in N$. But $b^{-1}n_2b = n_3 \in N$, since N is normal, so the right-hand side of the equation can be written as

$$an_2bN = abN.$$

Thus we have shown that if $N \lhd G$ then $a'b'N = abN$, and the operation on G/N is indeed, well-defined.

The associative law is true because coset multiplication as defined above uses the ordinary group operation which is by definition associative.

The coset N serves as the identity element of G/N. Notice that

$$aN \cdot N = aN^2 = aN$$

and

$$N \cdot aN = aN^2 = aN.$$

The inverse of aN is $a^{-1}N$ since

$$aNa^{-1}N = aa^{-1}N^2 = N. \qquad \square$$

We emphasize that the elements of G/N are cosets and thus subsets of G. If $|G| < \infty$, then $|G/N| = [G : N]$, the member of cosets of N in G. It is also to be emphasized that in order for G/N to be a group N must be a normal subgroup of G.

In some cases properties of G are preserved in factor groups.

Lemma 10.1.13. *If G is abelian then any factor group of G is also abelian. If G is cyclic then any factor group of G is also cyclic.*

Proof. Suppose that G is abelian and H is a subgroup of G. H is necessarily normal from Lemma 10.1.7 so that we can form the factor group G/H. Let $g_1H, g_2H \in G/H$. Since G is abelian we have $g_1g_2 = g_2g_1$. Then in G/H,

$$(g_1H)(g_2H) = (g_1g_2)H = (g_2g_1)H = (g_2H)(g_1H).$$

Therefore G/H is abelian.

We leave the proof of the second part to the exercises. $\square$

An extremely important concept is when a group contains no proper normal subgroups other than the identity subgroup $\{1\}$.

Definition 10.1.14. A group $G \neq \{1\}$ is *simple* provided that $N \lhd G$ implies $N = G$ or $N = \{1\}$.

One of the most outstanding problems in group theory has been to give a complete classification of all finite simple groups. In other words, this is the program to discover all finite simple groups and to prove that there are no more to be found. This was accomplished through the efforts of many mathematicians. The proof of this magnificent result took thousands of pages. We refer the reader to [18] for a complete discussion of this. We give one elementary example.

Lemma 10.1.15. *Any finite group of prime order is simple and cyclic.*

Proof. Suppose that G is a finite group and $|G| = p$ where p is a prime. Let $g \in G$ with $g \neq 1$. Then $\langle g \rangle$ is a nontrivial subgroup of G so its order divides the order of G by Lagrange's theorem. Since $g \neq 1$ and p is a prime we must have $|\langle g \rangle| = p$. Therefore $\langle g \rangle$ is all of G, that is $G = \langle g \rangle$ and hence G is cyclic.

The argument above shows that G has no nontrivial proper subgroups and therefore no nontrivial normal subgroups. Therefore G is simple. $\square$

In the next chapter we will examine certain other finite simple groups.

10.2 The Group Isomorphism Theorems

In Chapter 1 we saw that there was a close relationship between ring homomorphisms and factor rings. In particular to each ideal, and consequently to each factor ring, there is a ring homomorphism that has that ideal as its kernel. Conversely to each ring homomorphism its kernel is an ideal and the corresponding factor ring is isomorphic to the image of the homomorphism. This was formalized in Theorem 1.5.7 which we called the *ring isomorphism theorem*. We now look at the group theoretical analog of this result, called the *group isomorphism theorem*. We will then examine some consequences of this result that will be crucial in the Galois theory of fields.

Definition 10.2.1. If G_1 and G_2 are groups and $f : G_1 \to G_2$ is a group homomorphism then the *kernel* of f, denoted $\ker(f)$, is defined as

$$\ker(f) = \{g \in G_1 : f(g) = 1\}.$$

That is the kernel is the set of the elements of G_1 that map onto the identity of G_2. The *image* of f, denoted $\operatorname{im}(f)$, is the set of elements of G_2 mapped onto by f from elements of G_1. That is

$$\operatorname{im}(f) = \{g \in G_2 : f(g_1) = g_2 \text{ for some } g_1 \in G_1\}.$$

Note that if f is a surjection then $\operatorname{im}(f) = G_2$.

As with ring homomorphisms the kernel measures how far a homomorphism is from being an injection, that is, a one-to-one mapping.

Lemma 10.2.2. *Let G_1 and G_2 are groups and $f : G_1 \to G_2$ a group homomorphism. Then f is injective if and only if* $\ker(f) = \{1\}$.

Proof. Suppose that f is injective. Since $f(1) = 1$ we always have $1 \in \ker(f)$. Suppose that $g \in \ker(f)$. Then $f(g) = f(1)$. Since f is injective this implies that $g = 1$ and hence $\ker(f) = \{1\}$.

Conversely suppose that $\ker(f) = \{1\}$ and $f(g_1) = f(g_2)$. Then

$$f(g_1)(f(g_2))^{-1} = 1 \implies f(g_1 g_2^{-1}) = 1 \implies g_1 g_2^{-1} \in \ker(f).$$

Then since $\ker(f) = \{1\}$ we have $g_1 g_2^{-1} = 1$ and hence $g_1 = g_2$. Therefore f is injective. □

We now state the *group isomorphism theorem*. This is entirely analogous to the ring isomorphism theorem replacing ideals by normal subgroups. We note that this theorem is sometimes called the *first group isomorphism theorem*.

Theorem 10.2.3 (group isomorphism theorem). (a) *Let G_1 and G_2 be groups and $f : G_1 \to G_2$ a group homomorphism. Then* $\ker(f)$ *is a normal subgroup of G_1,* $\operatorname{im}(f)$ *is a subgroup of G_2 and*

$$G/\ker(f) \cong \operatorname{im}(f).$$

(b) *Conversely suppose that N is a normal subgroup of a group G. Then there exists a group H and a homomorphism $f : G \to H$ such that* $\ker(f) = N$ *and* $\operatorname{im}(f) = H$.

Proof. (a) Since $1 \in \ker(f)$ the kernel is nonempty. Suppose that $g_1, g_2 \in \ker(f)$. Then $f(g_1) = f(g_2) = 1$. It follows that $f(g_1g_2^{-1}) = f(g_1)(f(g_2))^{-1} = 1$. Hence $g_1g_2^{-1} \in \ker(f)$ and therefore $\ker(f)$ is a subgroup of G_1. Further for any $g \in G_1$ we have

$$\begin{aligned} f(g^{-1}g_1g) &= (f(g))^{-1}f(g_1)f(g) \\ &= (f(g))^{-1} \cdot 1 \cdot f(g) = f(g^{-1}g) = f(1) = 1. \end{aligned}$$

Hence $g^{-1}g_1g \in \ker(f)$ and $\ker(f)$ is a normal subgroup.

It is straightforward to show that $\operatorname{im}(f)$ is a subgroup of G_2.

Consider the map $\hat{f} : G/\ker(f) \to \operatorname{im}(f)$ defined by

$$\hat{f}(g\ker(f)) = f(g).$$

We show that this is an isomorphism.

Suppose that $g_1\ker(f) = g_2\ker(f)$ then $g_1g_2^{-1} \in \ker(f)$ so that $f(g_1g_2^{-1}) = 1$. This implies that $f(g_1) = f(g_2)$ and hence the map $\hat{f}$ is well-defined. Now

$$\begin{aligned} \hat{f}(g_1\ker(f)g_2\ker(f)) &= \hat{f}(g_1g_2\ker(f)) = f(g_1g_2) \\ &= f(g_1)f(g_2) = \hat{f}(g_1\ker(f))\hat{f}(g_2\ker(f)) \end{aligned}$$

and therefore $\hat{f}$ is a homomorphism.

Suppose that $\hat{f}(g_1\ker(f)) = \hat{f}(g_2\ker(f))$ then $f(g_1) = f(g_2)$ and hence $g_1\ker(f) = g_2\ker(f)$. It follows that $\hat{f}$ is injective.

Finally suppose that $h \in \operatorname{im}(f)$. Then there exists a $g \in G_1$ with $f(g) = h$. Then $\hat{f}(g\ker(f)) = h$ and $\hat{f}$ is a surjection onto $\operatorname{im}(f)$. Therefore $\hat{f}$ is an isomorphism completing the proof of part (a).

(b) Conversely suppose that N is a normal subgroup of G. Define the map $f : G \to G/N$ by $f(g) = gN$ for $g \in G$. By the definition of the product in the quotient group G/N it is clear that f is a homomorphism with $\operatorname{im}(f) = G/N$. If $g \in \ker(f)$ then $f(g) = gN = N$ since N is the identity in G/N. However this implies that $g \in N$ and hence it follows that $\ker(f) = N$ completing the proof. □

There are two related theorems that are called the second isomorphism theorem and the third isomorphism theorem.

Theorem 10.2.4 (second isomorphism theorem). *Let N be a normal subgroup of a group G and U a subgroup of G. Then $U \cap N$ is normal in U and*

$$(UN)/N \cong U/(U \cap N).$$

Proof. From Lemma 10.1.10 we know that $U \cap N$ is normal in U. Define the map

$$\alpha : UN \to U/U \cap N$$

by $\alpha(un) = u(U \cap N)$. If $un = u'n'$ then $u'^{-1}u = n'n^{-1} \in U \cap N$. Therefore $u'(U \cap N) = u(U \cap N)$ and hence the map α is well-defined.

Suppose that $un, u'n' \in UN$ Since N is normal in G we have that $unu'n' \in uu'N$. Hence $unu'n' = uu'n''$ with $n'' \in N$. Then

$$\alpha(unu'n') = \alpha(uu'n) = uu'(U \cap N).$$

However $U \cap N$ is normal in U so

$$uu'(U \cap N) = u(U \cap N)u'(U \cap N) = \alpha(un)\alpha(u'n').$$

Therefore α is a homomorphism.

We have $\operatorname{im}(\alpha) = U/(U \cap N)$ by definition. Suppose that $un \in \ker(\alpha)$. Then $\alpha(un) = U \cap N \subset N$ which implies $u \in N$. Therefore $\ker(f) = N$. From the group isomorphism theorem we then have

$$UN/N \cong U/(U \cap N)$$

proving the theorem. □

Theorem 10.2.5 (third isomorphism theorem). *Let N and M be normal subgroups of a group G with N a subgroup of M. Then M/N is a normal subgroup in G/N and*

$$(G/N)/(M/N) \cong G/M.$$

Proof. Define the map $\beta : G/N \to G/M$ by

$$\beta(gN) = gM.$$

It is straightforward that β is well-defined and a homomorphism. If $gN \in \ker(\beta)$ then $\beta(gN) = gM = M$ and hence $g \in M$. It follows that $\ker(\beta) = M/N$. In particular this shows that M/N is normal in G/N. From the group isomorphism theorem then

$$(G/N)/(M/N) \cong G/M.$$ □

For a normal subgroup N in G the homomorphism $f : G \to G/N$ provides a one-to-one correspondence between subgroups of G containing N and the subgroups of G/N. This correspondence will play a fundamental role in the study of subfields of a field.

Theorem 10.2.6 (correspondence theorem). *Let N be a normal subgroup of a group G and let f be the corresponding homomorphism $f : G \to G/N$. Then the mapping*

$$\phi : H \to f(H)$$

where H is a subgroup of G containing N provides a one-to-one correspondence between all the subgroups of G/N and the subgroups of G containing N.

Proof. We first show that the mapping ϕ is surjective. Let H_1 be a subgroup of G/N and let

$$H = \{g \in G : f(g) \in H_1\}.$$

We show that H is a subgroup of G and that $N \subset H$.

If $g_1, g_2 \in H$ then $f(g_1) \in H_1$ and $f(g_2) \in H_1$. Therefore $f(g_1)f(g_2) \in H_1$ and hence $f(g_1g_2) \in H_1$. Therefore $g_1g_2 \in H$. In an identical fashion $g_1^{-1} \in H$. Therefore H is a subgroup of G. If $n \in N$ then $f(n) = 1 \in H_1$ and hence $n \in H$. Therefore $N \subset H$ showing that the map ϕ is surjective.

Suppose that $\phi(H_1) = \phi(H_2)$ where H_1 and H_2 are subgroups of G containing N. This implies that $f(H_1) = f(H_2)$. Let $g_1 \in H_1$. Then $f(g_1) = f(g_2)$ for some $g_2 \in H_2$. Then $g_1g_2^{-1} \in \ker(f) = N \subset H_2$. It follows that $g_1g_2^{-1} \in H_2$ so that $g_1 \in H_2$. Hence $H_1 \subset H_2$. In a similar fashion $H_2 \subset H_1$ and therefore $H_1 = H_2$. It follows that ϕ is injective. □

10.3 Direct Products of Groups

In this section we look at a very important construction, the direct product, which allows us to build new groups out of existing groups. This construction is the analog for groups of the direct sum of rings. As an application of this construction, in the next section we present a theorem which completely describes the structure of finite abelian groups.

Let G_1, G_2 be groups and let G be the Cartesian product of G_1 and G_2. That is

$$G = G_1 \times G_2 = \{(a, b) : a \in G_1, b \in G_2\}.$$

On G define

$$(a_1, b_1) \cdot (a_2, b_2) = (a_1a_2, b_1b_2).$$

With this operation it is direct to verify the groups axioms for G and hence G becomes a group.

Theorem 10.3.1. *Let G_1, G_2 be groups and G the Cartesian product $G_1 \times G_2$ with the operation defined above. Then G forms a group called the direct product of G_1 and G_2. The identity element is $(1, 1)$ and $(g, h)^{-1} = (g^{-1}, h^{-1})$.*

This construction can be iterated to any finite number of groups (also to an infinite number but we won't consider that here) $G_1, \dots, G_n$ to form the direct product $G_1 \times G_2 \times \cdots \times G_n$.

Theorem 10.3.2. *For groups G_1 and G_2 we have $G_1 \times G_2 \cong G_2 \times G_1$ and $G_1 \times G_2$ is abelian if and only if each G_i, $i = 1, 2$, is abelian.*

Proof. The map $(a, b) \to (b, a)$ where $a \in G_1, b \in G_2$ provides an isomorphism from $G_1 \times G_2 \to G_2 \times G_1$.

Suppose that both G_1, G_2 are abelian. Then if $a_1, a_2 \in G_1, b_1, b_2 \in G_2$ we have

$$(a_1, b_1)(a_2, b_2) = (a_1 a_2, b_1 b_2) = (a_2 a_1, b_2 b_1) = (a_2, b_2)(a_1, b_1)$$

and hence $G_1 \times G_2$ is abelian.

Conversely suppose $G_1 \times G_2$ is abelian and suppose that $a_1, a_2 \in G_1$. Then for the identity $1 \in G_2$ we have

$$(a_1 a_2, 1) = (a_1, 1)(a_2, 1) = (a_2, 1)(a_1, 1) = (a_2 a_1, 1).$$

Therefore $a_1 a_2 = a_2 a_1$ and G_1 is abelian. Identically G_2 is abelian. □

We show next that in $G_1 \times G_2$ there are normal subgroups H_1, H_2 with $H_1 \cong G_1$ and $H_2 \cong G_2$.

Theorem 10.3.3. *Let $G = G_1 \times G_2$. Let $H_1 = \{(a, 1) : a \in G_1\}$ and $H_2 = \{(1, b) : b \in G_2\}$. Then both H_1 and H_2 are normal subgroups of G with $G = H_1 H_2$ and $H_1 \cap H_2 = \{1\}$. Further $H_1 \cong G_1$, $H_2 \cong G_2$ and $G/H_1 \cong G_2$ and $G/H_2 \cong G_1$.*

Proof. Map $G_1 \times G_2$ onto G_2 by $(a, b) \to b$. It is clear that this map is a homomorphism and that the kernel is $H_1 = \{(a, 1) : a \in G_1\}$. This establishes that H_1 is a normal subgroup of G and that $G/H_1 \cong G_2$. In an identical fashion we get that $G/H_2 \cong G_1$. The map $(a, 1) \to a$ provides the isomorphism from H_1 onto G_1. □

If the factors are finite it is easy to find the order of $G_1 \times G_2$. The size of the Cartesian product is just the product of the sizes of the factors.

Lemma 10.3.4. *If $|G_1|$ and $|G_2|$ are finite then $|G_1 \times G_2| = |G_1||G_2|$.*

Now suppose that G is a group with normal subgroups G_1, G_2 such that $G = G_1 G_2$ and $G_1 \cap G_2 = \{1\}$. Then we will show that G is isomorphic to the direct product $G_1 \times G_2$. In this case we say that G is the *internal direct product* of its subgroups and that G_1, G_2 are *direct factors* of G.

Theorem 10.3.5. *Suppose that G is a group with normal subgroups G_1, G_2 such that $G = G_1G_2$ and $G_1 \cap G_2 = \{1\}$. Then G is isomorphic to the direct product $G_1 \times G_2$.*

Proof. Since $G = G_1G_2$ each element of G has the form ab with $a \in G_1, b \in G_2$. We first show that each $a \in G_1$ commutes with each $b \in G_2$. Consider the element $aba^{-1}b^{-1}$. Since G_1 is normal $ba^{-1}b^{-1} \in G_1$ which implies that $abab^{-1} \in G_1$. Since G_2 is normal $aba^{-1} \in G_2$ which implies that $aba^{-1}b^{-1} \in G_2$. Therefore $aba^{-1}b^{-1} \in G_1 \cap G_2 = \{1\}$ and hence $aba^{-1}b^1 = 1$ so that $ab = ba$.

Now map G onto $G_1 \times G_2$ by $f(ab) \to (a,b)$. We claim that this is an isomorphism. It is clearly onto. Now

$$\begin{aligned} f((a_1b_1)(a_2b_2)) &= f(a_1a_2b_1b_2) = (a_1a_2, b_1b_2) \\ &= (a_1, b_1)(a_2, b_2) = f((a_1, b_1))f(a_2, b_2)) \end{aligned}$$

so that f is a homomorphism. The kernel is $G_1 \cap G_2 = \{1\}$ and so f is an isomorphism. □

Although the end resulting groups are isomorphic we call $G_1 \times G_2$ an external direct product if we started with the groups G_1, G_2 and constructed $G_1 \times G_2$ and call $G_1 \times G_2$ an internal direct product if we started with a group G having normal subgroups as in the theorem.

10.4 Finite Abelian Groups

We now use the results of the last section to present a theorem that completely provides the structure of finite abelian groups. This theorem is a special case of a general result on modules that we will examine in detail later in the book.

Theorem 10.4.1 (basis theorem for finite abelian groups). *Let G be a finite abelian group. Then G is a direct product of cyclic groups of prime power order.*

Before giving the proof we give two examples showing how this theorem leads to the classification of finite abelian groups.

Since all cyclic groups of order n are isomorphic to $(\mathbb{Z}_n, +)$ we will denote a cyclic group of order n by $\mathbb{Z}_n$.

Example 10.4.2. Classify all abelian groups of order 60. Let G be an abelian group of order 60. From Theorem 10.4.1 G must be a direct product of cyclic groups of prime power order. Now $60 = 2^2 \cdot 3 \cdot 5$ so the only primes involved are 2, 3 and 5. Hence the cyclic group involved in the direct product decomposition of G have order either 2, 4, 3 or 5 (by Lagrange's theorem they must be divisors of 60). Therefore G

must be of the form

$$G \cong \mathbb{Z}_4 \times \mathbb{Z}_3 \times \mathbb{Z}_5$$
$$G \cong \mathbb{Z}_2 \times \mathbb{Z}_2 \times \mathbb{Z}_3 \times \mathbb{Z}_5.$$

Hence up to isomorphism there are only two abelian groups of order 60.

Example 10.4.3. Classify all abelian groups of order 180. Now $180 = 2^2 \cdot 3^2 \cdot 5$ so the only primes involved are 2, 3 and 5. Hence the cyclic group involved in the direct product decomposition of G have order either 2, 4, 3, 9 or 5 (by Lagrange's theorem they must be divisors of 180). Therefore G must be of the form

$$G \cong \mathbb{Z}_4 \times \mathbb{Z}_9 \times \mathbb{Z}_5$$
$$G \cong \mathbb{Z}_2 \times \mathbb{Z}_2 \times \mathbb{Z}_9 \times \mathbb{Z}_5$$
$$G \cong \mathbb{Z}_4 \times \mathbb{Z}_3 \times \mathbb{Z}_3 \times \mathbb{Z}_5$$
$$G \cong \mathbb{Z}_2 \times \mathbb{Z}_2 \times \mathbb{Z}_3 \times \mathbb{Z}_3 \times \mathbb{Z}_5.$$

Hence up to isomorphism there are four abelian groups of order 180.

The proof of Theorem 10.4.1 involves the following lemmas.

Lemma 10.4.4. *Let G be a finite abelian group and let $p \mid |G|$ where p is a prime. Then all the elements of G whose orders are a power of p form a normal subgroup of G. This subgroup is called the p-primary component of G, which we will denote by G_p.*

Proof. Let p be a prime with $p \mid |G|$ and let a and b be two elements of G of order a power of p. Since G is abelian the order of ab is the lcm of the orders which is again a power of p. Therefore $ab \in G_p$. The order of a^{-1} is the same as the order of a so $a^{-1} \in G_p$ and therefore G_p is a subgroup. □

Lemma 10.4.5. *Let G be a finite abelian group of order n. Suppose that $n = p_1^{e_1} \cdots p_k^{e_k}$ with $p_1, \dots, p_k$ distinct primes. Then*

$$G \cong G_{p_1} \times \cdots \times G_{p_k}$$

where G_{p_i} is the p_i-primary component of G.

Proof. Each G_{p_i} is normal since G is abelian and since distinct primes are relatively prime the intersection of the G_{p_i} is the identity. Therefore Lemma 10.4.5 will follow by showing that each element of G is a product of elements in the G_{p_1}.

Let $g \in G$. Then the order of g is $p_1^{f_1} \cdots p_k^{f_k}$. We write this as $p_i^{f_i} m$ with $(m, p_i) = 1$. Then g^m has order $p_i^{f_i}$ and hence is in G_{p_i}. Now since $p_1, \dots, p_k$ are relatively prime there exists $m_1, \dots, m_k$ with

$$m_1 p_1^{f_1} + \cdots + m_k p_k^{f_k} = 1$$

and hence

$$g = (g^{p_1^{f_1}})^{m_1} \cdots (g^{p_k^{f_k}})^{m_k}.$$

Therefore g is a product of elements in the G_{p_i}. □

We next need the concept of a *basis*. Let G be any finitely generated abelian group (finite or infinite) and let $g_1, \dots, g_n$ be a set of generators for G. The generators $g_1, \dots, g_n$ form a *basis* if

$$G = \langle g_1 \rangle \times \cdots \times \langle g_n \rangle,$$

that is G is the direct product of the cyclic subgroups generated by the g_i. The basis theorem for finite abelian groups says that any finite abelian group has a basis.

Suppose that G is a finite abelian group with a basis $g_1, \dots, g_k$ so that $G = \langle g_1 \rangle \times \cdots \times \langle g_k \rangle$. Since G is finite each g_i has finite order say m_i. It follows than from the fact that G is a direct product that each $g \in G$ can be expressed as

$$g = g_1^{n_1} \cdots g_k^{n_k}$$

and further the integers $n_1, \dots, n_k$ are unique modulo the order of g_i. Hence each integer n_i can be chosen in the range $0, 1, \dots, m_i - 1$ and within this range for the element g the integer n_i is unique.

From the previous lemma each finite abelian group splits into a direct product of its p-primary components for different primes p. Hence to complete the proof of the basis theorem we must show that any finite abelian group of order p^m for some prime p has a basis. We call an abelian group of order p^m an abelian p-group.

Consider an abelian group G of order p^m for a prime p. It is somewhat easier to complete the proof if we consider the group using additive notation. That is the operation is considered $+$, the identity as 0 and powers are given by multiples. Hence if an element $g \in G$ has order p^k then in additive notation $p^k g = 0$. A set of elements $g_1, \dots, g_k$ is then a basis for G if each $g \in G$ can be expressed uniquely as $g = m_1 g_1 + \cdots + m_k g_k$ where the m_i are unique modulo the order of g_i. We say that the $g_1, \dots, g_k$ are *independent* and this is equivalent to the fact that whenever $m_1 g_1 + \cdots + m_k g_k = 0$ then $m_i \equiv 0$ modulo the order of g_i. We now prove that any abelian p-group has a basis.

Lemma 10.4.6. *Let G be a finite abelian group of prime power order p^n for some prime p. Then G is a direct product of cyclic groups.*

Notice that in the group G we have $p^n g = 0$ for all $g \in G$ as a consequence of Lagrange's theorem. Further every element has order a power of p. The smallest power of p say p^r such that $p^r g = 0$ for all $g \in G$ is called the *exponent* of G. Any finite abelian p-group must have some exponent p^r.

Proof. The proof of this lemma is by induction on the exponent.

The lowest possible exponent is p so suppose first that $pg = 0$ for all $g \in G$. Since G is finite it has a finite system of generators. Let $S = \{g_1, \dots, g_k\}$ be a minimal set of generators for G. We claim that this is a basis. Since this is a set of generators to show its a basis we must show that they are independent. Hence suppose that we have

$$m_1 g_1 + \cdots + m_k g_k = 0 \tag{1}$$

for some set of integers m_i. Since the order of each g_i is p, as explained above we may assume that $0 \leq m_i < p$ for $i = 1, \dots, k$. Suppose that one $m_i \neq 0$. Then $(m_i, p) = 1$ and hence there exists an x_i with $m_i x_i \equiv 1 \bmod p$ (see Chapter 4). Multiplying the equation (1) by x_i we get modulo p,

$$m_1 x_i g_1 + \cdots + g_i + \cdots + m_k x_i g_k,$$

and rearranging

$$g_i = -m_1 x_i g_1 - \cdots - m_k x_k g_k,$$

But then g_i can be expressed in terms of the other g_j and therefore the set $\{g_1, \dots, g_k\}$ is not minimal. It follows that $g_1, \dots, g_k$ constitute a basis and the lemma is true for the exponent p.

Now suppose that any finite abelian group of exponent p^{n-1} has a basis and assume that G has exponent p^n. Consider the set $\overline{G} = pG = \{pg : g \in G\}$. It is straightforward that this forms a subgroup (see exercises). Since $p^n g = 0$ for all $g \in G$ it follows that $p^{n-1} g = 0$ for all $g \in \overline{G}$ and so the exponent of $\overline{G} \leq p^{n-1}$. By the inductive hypothesis $\overline{G}$ has a basis

$$S = \{pg_1, \dots, pg_k\}.$$

Consider the set $\{g_1, \dots, g_k\}$ and adjoin to this set the set of all elements $h \in G$ satisfying $ph = 0$. Call this set S_1 so that we have

$$S_1 = \{g_1, \dots, g_k, h_1, \dots, h_t\}.$$

We claim that S_1 is a set of generators for G. Let $g \in G$. Then $pg \in \overline{G}$ which has the basis $pg_1, \dots, pg_k$ so that

$$pg = m_1 pg_1 + \cdots + m_k pg_k.$$

This implies that

$$p(g - m_1 g_1 - \cdots - m_k g_k) = 0$$

so that $g_1 - m_1g_1 - \cdots - m_kg_k$ must be one of the h_i. Hence

$$g - m_1g_1 - \cdots - m_kg_k = h_i \quad \text{so that } g = m_1g_1 + \cdots + m_kg_k + h_i$$

proving the claim.

Now S_1 is finite so there is a minimal subset of S_1 that is still a generating system for G. Call this S_0 and suppose that S_0, renumbering if necessary, is

$$S_0 = \{g_1, \ldots, g_r, h_1, \ldots, h_s\} \quad \text{with } ph_i = 0 \text{ for } i = 1, \ldots, s.$$

The subgroup generated by $h_1, \ldots, h_s$ has exponent p so by inductive hypothesis has a basis. We may assume than that $h_1, \ldots, h_s$ is a basis for this subgroup and hence is independent. We claim now that $g_1, \ldots, g_r, h_1, \ldots, h_s$ are independent and hence form a basis for G.

Suppose that

$$m_1g_r + \cdots + m_rg_r + n_1h_1 + \cdots + n_sh_s = 0 \tag{2}$$

for some integers $m_1, \ldots, m_r, h_1, \ldots, h_s$. Each m_i, n_i must be divisible by p. Suppose for example that some m_i is not. Then $(m_i, p) = 1$ and then $(m_i, p^n) = 1$. This implies that there exists an x_i with $m_ix_i \equiv 1 \bmod p^n$. Multiplying through by x_i and rearranging we then obtain

$$g_i = -m_1x_ig_1 - \cdots - n_sx_ih_s.$$

Therefore g_i can be expressed in terms of the remaining elements of S_0 contradicting the minimality of S_0. An identical argument works if some n_i is not divisible by p. Therefore the relation (2) takes the form

$$a_1pg_1 + \cdots + a_rpg_r + b_1ph_1 + \cdots + b_sph_s = 0. \tag{3}$$

Each of the terms $ph_i = 0$ so that (3) becomes

$$a_pg_1 + \cdots + a_rpg_r = 0.$$

The $g_1, \ldots, g_r$ are independent and hence $a_ip = 0$ for each i and hence $a_i = 0$. Now (2) becomes

$$n_1h_1 + \cdots + n_sh_s = 0.$$

However $h_1, \ldots, h_s$ are independent so each $n_i = 0$ completing the claim.

Therefore the whole group G has a basis proving the lemma by induction. □

For more details see the proof of the general result on modules over principal ideal domains later in the book. There is also an additional elementary proof for the basis theorem for finitely generated abelian groups.

10.5 Some Properties of Finite Groups

Classification is an extremely important concept in algebra. A large part of the theory is devoted to classifying all structures of a given type, for example all UFD's. In most cases this is not possible. Since for a given finite n there are only finitely many group tables it is theoretically possible to classify all groups of order n. However even for small n this becomes impractical. We close the chapter by looking at some further results on finite groups and then using these to classify all the finite groups up to order 10.

Before stating the classification we give some further examples of groups that are needed.

Example 10.5.1. In Example 9.2.6 we saw that the symmetry group of an equilateral triangle had 6 elements and is generated by elements r and f which satisfy the relations $r^3 = f^2 = 1$, $f^{-1}rf = r^{-1}$, where r is a rotation of $120°$ about the center of the triangle and f is a reflection through an altitude. This was called the dihedral group D_3 of order 6.

This can be generalized to any regular n-gon. If D is a regular n-gon, then the symmetry group D_n has $2n$ elements and is called the *dihedral group* of order $2n$. It is generated by elements r and f which satisfy the relations $r^n = f^2 = 1$, $f^{-1}rf = r^{n-1}$, where r is a rotation of $\frac{2\pi}{n}$ about the center of the n-gon and f is a reflection.

Hence, D_4, the symmetries of a square, has order 8 and D_5, the symmetries of a regular pentagon, has order 10.

Example 10.5.2. Let i, j, k be the generators of the quaternions. Then we have

$$i^2 = j^2 = k^2 = -1, \quad (-1)^2 = 1 \quad \text{and} \quad ijk = 1.$$

These elements then form a group of order 8 called the *quaternion group* denoted by Q. Since $ijk = 1$ we have $ij = -ji$, and the generators i and j satisfy the relations $i^4 = j^4 = 1, i^2 = j^2, ij = i^2ji$.

We now state the main classification and then prove it in a series of lemmas.

Theorem 10.5.3. *Let G be a finite group.*

(a) *If $|G| = 2$ then $G \cong \mathbb{Z}_2$.*

(b) *If $|G| = 3$ then $G \cong \mathbb{Z}_3$.*

(c) *If $|G| = 4$ then $G \cong \mathbb{Z}_4$ or $G \cong \mathbb{Z}_2 \times \mathbb{Z}_2$.*

(d) *If $|G| = 5$ then $G \cong \mathbb{Z}_5$.*

(e) *If $|G| = 6$ then $G \cong \mathbb{Z}_6 \cong \mathbb{Z}_2 \times \mathbb{Z}_3$ or $G \cong D_3$, the dihedral group with 6 elements. (Note $D_3 \cong S_3$ the symmetric group on 3 symbols.)*

(f) *If $|G| = 7$ then $G \cong \mathbb{Z}_7$.*

(g) *If* $|G| = 8$ *then* $G \cong \mathbb{Z}_8$ *or* $G \cong \mathbb{Z}_4 \times \mathbb{Z}_2$ *or* $G \cong \mathbb{Z}_2 \times \mathbb{Z}_2 \times \mathbb{Z}_2$ *or* $G \cong D_4$, *the dihedral group of order* 8, *or* $G \cong Q$ *the quaternion group.*

(h) *If* $|G| = 9$ *then* $G \cong \mathbb{Z}_9$ *or* $G \cong \mathbb{Z}_3 \times \mathbb{Z}_3$.

(i) *If* $|G| = 10$ *then* $G \cong \mathbb{Z}_{10} \cong \mathbb{Z}_2 \times \mathbb{Z}_5$ *or* $G \cong D_5$, *the dihedral group with* 10 *elements.*

Recall from Section 10.1 that a finite group of prime order must be cyclic. Hence in the theorem the cases $|G| = 2, 3, 5, 7$ are handled. We next consider the case where G has order p^2 where p is a prime.

Definition 10.5.4. If G is a group then its *center* denoted $Z(G)$ is the set of elements in G which commute with everything in G. That is

$$Z(G) = \{g \in G : gh = hg \text{ for any } h \in G\}.$$

Lemma 10.5.5. *For any group G the following hold:*

(a) *The center* $Z(G)$ *is a normal subgroup.*

(b) $G = Z(G)$ *if and only if G is abelian.*

(c) *If* $G/Z(G)$ *is cyclic then G is abelian.*

Proof. (a) and (b) are direct and we leave them to the exercises. Consider the case where $G/Z(G)$ is cyclic. Then each coset of $Z(G)$ has the form $g^m Z(G)$ where $g \in G$. Let $a, b \in G$. Then since a, b are in cosets of the center we have $a = g^m u$ and $b = g^n v$ with $u, v \in Z(G)$. Then

$$ab = (g^m u)(g^n v) = (g^m g^n)(uv) = (g^n g^m)(vu) = (g^n v)(g^m u) = ba$$

since u, v commute with everything. Therefore G is abelian. □

A *p-group* is any finite group of prime power order. We need the following. The proof of this is based on what is called the *class equation* which we will prove in Chapter 13.

Lemma 10.5.6. *A finite* p*-group has a nontrivial center of order at least* p.

Lemma 10.5.7. *If* $|G| = p^2$ *with* p *a prime then* G *is abelian and hence* $G \cong \mathbb{Z}_{p^2}$ *or* $G \cong \mathbb{Z}_p \times \mathbb{Z}_p$.

Proof. Suppose that $|G| = p^2$. Then from the previous lemma G has a nontrivial center and hence $|Z(G)| = p$ or $|Z(G)| = p^2$. If $|Z(G)| = p^2$ then $G = Z(G)$ and G is abelian. If $|Z(G)| = p$ then $|G/Z(G)| = p$. Since p is a prime this implies that $G/Z(G)$ is cyclic and hence from Lemma 10.5.5 G is abelian. □

Lemma 10.5.7 handles the cases $n = 4$ and $n = 9$. Therefore if $|G| = 4$ we must have $G \cong \mathbb{Z}_4$ or $G \cong \mathbb{Z}_2 \times \mathbb{Z}_2$ and if $|G| = 9$ we must have $G \cong \mathbb{Z}_9$ or $G \cong \mathbb{Z}_3 \times \mathbb{Z}_3$.

This leaves $n = 6, 8, 10$. We next handle 6 and 10.

Lemma 10.5.8. *If G is any group where every nontrivial element has order 2 then G is abelian.*

Proof. Suppose that $g^2 = 1$ for all $g \in G$. This implies that $g = g^{-1}$ for all $g \in G$. Let a, b be arbitrary elements of G. Then

$$(ab)^2 = 1 \implies abab = 1 \implies ab = b^{-1}a^{-1} = ba.$$

Therefore a, b commute and G is abelian. □

Lemma 10.5.9. *If* $|G| = 6$ *then* $G \cong \mathbb{Z}_6$ *or* $G \cong D_3$.

Proof. Since $6 = 2 \cdot 3$ if G was abelian then $G \cong \mathbb{Z}_2 \times \mathbb{Z}_3$. Notice that if an abelian group has an element of order m and an element of order n with $(n, m) = 1$ then it has an element of order mn. Therefore for 6 if G is abelian there is an element of order 6 and hence $G \cong \mathbb{Z}_2 \times \mathbb{Z}_3 \cong \mathbb{Z}_6$.

Now suppose that G is nonabelian. The nontrivial elements of G have orders 2, 3 or 6. If there is an element of order 6 then G is cyclic and hence abelian. If every element has order 2 then G is abelian. Therefore there is an element of order 3 say $g \in G$. The cyclic subgroup $\langle g \rangle = \{1, g, g^2\}$ then has index 2 in G and is therefore normal. Let $h \in G$ with $h \notin \langle g \rangle$. Since g, g^2 both generate $\langle g \rangle$ we must have $\langle g \rangle \cap \langle h \rangle = \{1\}$. If h also had order 3 then $|\langle g, h \rangle| = \frac{|\langle g \rangle||\langle h \rangle|}{|\langle g \rangle \cap \langle h \rangle|} = 9$ which is impossible. Therefore h must have order 2. Since $\langle g \rangle$ is normal we have $h^{-1}gh = g^t$ for $t = 1, 2$. If $h^{-1}gh = g$ then g, h commute and the group G is abelian. Therefore $h^{-1}gh = g^2 = g^{-1}$. It follows that g, h generate a subgroup of G satisfying

$$g^3 = h^2 = 1, \quad h^1 gh = g^{-1}.$$

This defines a subgroup of order 6 isomorphic to D_3 and hence must be all of G. □

Lemma 10.5.10. *If* $|G| = 10$ *then* $G \cong \mathbb{Z}_{10}$ *or* $G \cong D_5$.

Proof. The proof is almost identical to that for $n = 6$. Since $10 = 2 \cdot 5$ if G were abelian $G \cong \mathbb{Z}_2 \times \mathbb{Z}_5 = \mathbb{Z}_{10}$.

Now suppose that G is nonabelian. As for $n = 6$, G must contain a normal cyclic subgroup of order 5 say $\langle g \rangle = \{1, g, g^2, g^3, g^4\}$. If $h \notin \langle g \rangle$ then exactly as for $n = 6$ it follows that h must have order 2 and $h^{-1}gh = g^t$ for $t = 1, 2, 3, 4$. If

$h^{-1}gh = g$ then g, h commute and G is abelian. Notice that $h^{-1} = h$. Suppose that $h^{-1}gh = hgh = g^2$. Then

$$(hgh)^3 = (g^2)^3 = g^6 = g \implies g = h^2gh^2 = hg^2h = g^4 \implies g = 1$$

which is a contradiction. Similarly $hgh = g^3$ leads to a contradiction. Therefore $h^{-1}gh = g^4 = g^{-1}$ and g, h generate a subgroup of order 10 satisfying

$$g^5 = h^2 = 1; \quad h^{-1}gh = g^{-1}.$$

Therefore this is all of G and is isomorphic to D_5. □

This leaves the case $n = 8$ that is the most difficult. If $|G| = 8$ and G is abelian then clearly $G \cong \mathbb{Z}_8$ or $G \cong \mathbb{Z}_4 \times \mathbb{Z}_2$ or $G \cong \mathbb{Z}_2 \times \mathbb{Z}_2 \times \mathbb{Z}_2$. The proof of Theorem 10.5.3 is then completed with the following.

Lemma 10.5.11. *If G is a nonabelian group of order 8 then $G \cong D_4$ or $G \cong Q$.*

Proof. The nontrivial elements of G have orders 2, 4 or 8. If there is an element of order 8 then G is cyclic and hence abelian while if every element has order 2 then G is abelian. Hence we may assume that G has an element of order 4 say g. Then $\langle g \rangle$ has index 2 and is a normal subgroup. Suppose first that G has an element $h \notin \langle g \rangle$ of order 2. Then

$$h^{-1}gh = g^t \quad \text{for some } t = 1, 2, 3.$$

If $h^{-1}gh = g$ then as in the cases 6 and 10, $\langle g, h \rangle$ defines an abelian subgroup of order 8 and hence G is abelian. If $h^{-1}gh = g^2$ then

$$(h^{-1}gh)^2 = (g^2)^2 = g^4 = 1 \implies g = h^{-2}gh^2 = h^{-1}g^2h = g^4 \implies g^3 = 1$$

contradicting the fact that g has order 4. Therefore $h^{-1}gh = g^3 = g^{-1}$. It follows that g, h define a subgroup of order 8 isomorphic to D_4. Since $|G| = 8$ this must be all of G and $G \cong D_4$.

Therefore we may now assume that every element $h \in G$ with $h \notin \langle g \rangle$ has order 4. Let h be such an element. Then h^2 has order 2 so $h^2 \in \langle g \rangle$ which implies that $h^2 = g^2$. This further implies that g^2 is central, that is commutes with everything. Identifying g with i, h with j and g^2 with -1 we get that G is isomorphic to Q completing Lemma 10.5.11 and the proof of Theorem 10.5.3. □

In principle this type of analysis can be used to determine the structure of any finite group although it quickly becomes impractical. A major tool in this classification is the following important result known as the Sylow theorem which now we just state. We will prove this theorem in Chapter 13. If $|G| = p^m n$ with p a prime and $(n, p) = 1$ then a subgroup of G of order p^m is called a *p-Sylow subgroup*. It is not clear at first that a group will contain p-Sylow subgroups.

Theorem 10.5.12 (Sylow theorem). *Let $|G| = p^m n$ with p a prime and $(n, p) = 1$.*

(a) *G contains a p-Sylow subgroup.*

(b) *All p-Sylow subgroups of G are conjugate G.*

(c) *Any p-subgroup of G is contained in a p-Sylow subgroup.*

(d) *The number of p-Sylow subgroups of G is of the form $1 + pk$ and divides n.*

10.6 Exercises

1. Prove that if G is cyclic then any factor group of G is also cyclic.
2. Prove that for any group G the center $Z(G)$ is a normal subgroup and $G = Z(G)$ if and only if G is abelian.
3. Let U_1 and U_2 be subgroups of a group G. Let $x, y \in G$. Show:
   (i) If $xU_1 = yU_2$ then $U_1 = U_2$.
   (ii) Give an example to show that $xU_1 = U_2x$ does not imply $U_1 = U_2$.
4. Let U, V be subgroups of a group G. Let $x, y \in G$. If $UxV \cap UyV \neq \emptyset$ then $UxV = UyV$.
5. Let N be a cyclic normal subgroup of the group G. Then all subgroups of N are normal subgroups of G. Give an example to show that the statement is not correct if N is not cyclic.
6. Let N_1 and N_2 be normal subgroups of G. Show:
   (i) If all elements in N_1 and N_2 have finite order, then also the elements of N_1N_2.
   (ii) Let $e_1, e_2 \in \mathbb{N}$. If $n_i^{e_i} = 1$ for all $n_i \in N_i$ $(i = 1, 2)$, then $x^{e_1e_2} = 1$ for all $x \in N_1N_2$.
7. Find groups N_1, N_2 and G with $N_1 \lhd N_2 \lhd G$, but N_1 is not a normal subgroup of G.
8. Let G be a group generated by a and b and let $bab^{-1} = a^r$ and $a^n = 1$ for suitable $r \in \mathbb{Z}, n \in \mathbb{N}$. Show:
   (i) The subgroup $A := \langle a \rangle$ is a normal subgroup of G.
   (ii) $G/A = \langle bA \rangle$.
   (iii) $G = \{b^j a^i : i, j \in \mathbb{Z}\}$.
9. Prove that any group of order 24 cannot be simple.

Chapter 11
Symmetric and Alternating Groups

11.1 Symmetric Groups and Cycle Decomposition

Groups most often appear as groups of transformations or permutations on a set. In Galois Theory groups will appear as permutation groups on the zeros of a polynomial. In Section 9.3 we introduced permutation groups and the symmetric group S_n. In this chapter we look more carefully at the structure of S_n and for each n introduce a very important normal subgroup A_n of S_n called the *alternating group* on n symbols.

Recall that if A is a set, a *permutation* on A is a one-to-one mapping of A onto itself. The set S_A of all permutations on A forms a group under composition called the *symmetric group* on A. If $|A| > 2$ then S_A is nonabelian. Further if A, B have the same cardinality, then $S_A \cong S_B$.

If $|A| = n$ then $|S_A| = n!$ and in this case we denote S_A by S_n, called the *symmetric group on n symbols*. For example $|S_3| = 6$. In Example 9.3.5 we showed that the six elements of S_3 can be given by:

$$1 = \begin{pmatrix} 1 & 2 & 3 \\ 1 & 2 & 3 \end{pmatrix}, \quad a = \begin{pmatrix} 1 & 2 & 3 \\ 2 & 3 & 1 \end{pmatrix}, \quad b = \begin{pmatrix} 1 & 2 & 3 \\ 3 & 1 & 2 \end{pmatrix}$$

$$c = \begin{pmatrix} 1 & 2 & 3 \\ 2 & 1 & 3 \end{pmatrix}, \quad d = \begin{pmatrix} 1 & 2 & 3 \\ 3 & 2 & 1 \end{pmatrix}, \quad e = \begin{pmatrix} 1 & 2 & 3 \\ 1 & 3 & 2 \end{pmatrix}.$$

Further we saw that S_3 has a *presentation* given by

$$S_3 = \langle a, c; a^3 = c^2 = 1, ac = ca^2 \rangle.$$

By this we mean that S_3 is *generated by* a, c, or that S_3 has *generators* a, c and the whole group and its multiplication table can be generated by using the *relations* $a^3 = c^2 = 1, ac = ca^2$.

In general a *permutation* group is any subgroup of S_A for a set A.

For the remainder of this chapter we will only consider finite symmetric groups S_n and always consider the set A as $A = \{1, 2, 3, \ldots, n\}$.

Definition 11.1.1. Suppose that f is a permutation of $A = \{1, 2, \ldots, n\}$, which has the following effect on the elements of A: There exists an element $a_1 \in A$ such that $f(a_1) = a_2$, $f(a_2) = a_3, \ldots, f(a_{k-1}) = a_k$, $f(a_k) = a_1$, and f leaves all other elements (if there are any) of A fixed, i.e., $f(a_j) = a_j$ for $a_j \neq a_i, i = 1, 2, \ldots, k$. Such a permutation f is called a *cycle* or a *k-cycle*.

We use the following notation for a k-cycle, f, as given above:

$$f = (a_1, a_2, \ldots, a_k).$$

The cycle notation is read from left to right, it says f takes a_1 into a_2, a_2 into a_3, etc., and finally a_k, the last symbol, into a_1, the first symbol. Moreover, f leaves all the other elements not appearing in the representation above fixed.

Note that one can write the same cycle in many ways using this type of notation; e.g., $f = (a_2, a_3, \ldots, a_k, a_1)$. In fact any cyclic rearrangement of the symbols gives the same cycle. The integer k is the *length* of the cycle. Note we allow a cycle to have length 1, i.e., $f = (a_1)$, for instance, this is just the identity map. For this reason, we will usually designate the identity of S_n by (1) or just 1. (Of course, it also could be written as (a_i) where $a_i \in A$.)

If f and g are two cycles, they are called *disjoint cycles* if the elements moved by one are left fixed by the other, that is, their representations contain different elements of the set A (their representations are disjoint as sets).

Lemma 11.1.2. *If f and g are disjoint cycles, then they must commute, that is, $fg = gf$.*

Proof. Since the cycles f and g are disjoint, each element moved by f is fixed by g and vice versa. First suppose $f(a_i) \neq a_i$. This implies that $g(a_i) = a_i$ and $f^2(a_i) \neq f(a_i)$. But since $f^2(a_i) \neq f(a_i)$, $g(f(a_i)) = f(a_i)$. Thus $(fg)(a_i) = f(g(a_i)) = f(a_i)$ while $(gf)(a_i) = g(f(a_i)) = f(a_i)$. Similarly if $g(a_j) \neq a_j$, then $(fg)(a_j) = (gf)(a_j)$. Finally, if $f(a_k) = a_k$ and $g(a_k) = a_k$ then clearly $(fg)(a_k) = a_k = (gf)(a_k)$. Thus $gf = fg$. □

Before proceeding further with the theory, let us consider a specific example. Let $A = \{1, 2, \ldots, 8\}$ and let

$$f = \begin{pmatrix} 1 & 2 & 3 & 4 & 5 & 6 & 7 & 8 \\ 2 & 4 & 6 & 5 & 1 & 7 & 3 & 8 \end{pmatrix}.$$

We pick an arbitrary number from the set A, say 1. Then $f(1) = 2$, $f(2) = 4$, $f(4) = 5$, $f(5) = 1$. Now select an element from A not in the set $\{1, 2, 4, 5\}$, say 3. Then $f(3) = 6$, $f(6) = 7$, $f(7) = 3$. Next select any element of A not occurring in the set $\{1, 2, 4, 5\} \cup \{3, 6, 7\}$. The only element left is 8, and $f(8) = 8$. It is clear that we can now write the permutation f as a product of cycles:

$$f = (1, 2, 4, 5)(3, 6, 7)(8)$$

where the order of the cycles is immaterial since they are disjoint and therefore commute. It is customary to omit such cycles as (8) and write f simply as

$$f = (1245)(367)$$

with the understanding that the elements of A not appearing are left fixed by f.

It is not difficult to generalize what was done here for a specific example, and show that any permutation f can be written uniquely, except for order, as a product of disjoint cycles. Thus let f be a permutation on the set $A = \{1, 2, \dots, n\}$, and let $a_1 \in A$. Let $f(a_1) = a_2$, $f^2(a_1) = f(a_2) = a_3$, etc., and continue until a repetition is obtained. We claim that this first occurs for a_1, that is, the first repetition is say $f^k(a_1) = f(a_k) = a_{k+1} = a_1$. For suppose the first repetition occurs at the k-th iterate of f and

$$f^k(a_1) = f(a_k) = a_{k+1},$$

and $a_{k+1} = a_j$, where $j < k$. Then

$$f^k(a_1) = f^{j-1}(a_1),$$

and so $f^{k-j+1}(a_1) = a_1$. However, $k - j + 1 < k$ if $j \neq 1$, and we assumed that the first repetition occurred for k. Thus, $j = 1$ and so f does cyclically permute the set $\{a_1, a_2, \dots, a_k\}$. If $k < n$, then there exists $b_1 \in A$ such that $b_1 \notin \{a_1, a_2, \dots, a_k\}$ and we may proceed similarly with b_1. We continue in this manner until all the elements of A are accounted for. It is then seen that f can be written in the form

$$f = (a_1, \dots, a_k)(b_1, \dots, b_\ell)(c_1, \dots, c_m) \cdots (h_1, \dots, h_t).$$

Note that all powers $f^i(a_1)$ belong to the set $\{a_1 = f^0(a_1) = f^k(a_1),\ a_2 = f^1(a_1), \dots, a_k = f^{k-1}(a_1)\}$, all powers $f^i(b_1)$ belong to the set $\{b_1 = f^0(b_1) = f^\ell(b_1),\ b_2 = f^1(b_1), \dots, b_\ell = f^{\ell-1}(b_1)\}, \dots$. Here, by definition, b_1 is the smallest element in $\{1, 2, \dots, n\}$ which does not belong to $\{a_1 = f^0(a_1) = f^k(a_1),\ a_2 = f^1(a_1), \dots, a_k = f^{k-1}(a_1)\}$, c_1 is the smallest element in $\{1, 2, \dots, n\}$ which does not belong to

$$\begin{aligned}&\{a_1 = f^0(a_1) = f^k(a_1),\ a_2 = f^1(a_1),\ \dots,\ a_k = f^{k-1}(a_1)\}\\ &\cup \{b_1 = f^0(b_1) = f^\ell(b_1),\ b_2 = f^1(b_1),\ \dots,\ b_\ell = f^{\ell-1}(b_1)\}.\end{aligned}$$

Therefore by construction, all the cycles are disjoint. From this it follows that $k + \ell + m + \cdots + t = n$. It is clear that this factorization is unique except for the order of the factors since it tells explicitly what effect f has on each element of A.

In summary we have proven the following result.

Theorem 11.1.3. *Every permutation of S_n can be written uniquely as a product of disjoint cycles (up to order).*

Example 11.1.4. The elements of S_3 can be written in cycle notation as $1 = (1)$, $(1,2), (1,3), (2,3), (1,2,3), (1,3,2)$. This is the largest symmetric group which consists entirely of cycles.

In S_4, for example, the element $(1,2)(3,4)$ is not a cycle.

Suppose we multiply two elements of S_3 say $(1,2)$ and $(1,3)$. In forming the product or composition here, we read from right to left. Thus to compute $(1,2)(1,3)$: We note the permutation $(1,3)$ takes 1 into 3 and then the permutation $(1,2)$ takes 3 into 3 so the composite $(1,2)(1,3)$ takes 1 into 3. Continuing the permutation $(1,3)$ takes 3 into 1 and then the permutation $(1,2)$ takes 1 into 2, so the composite $(1,2)(1,3)$ takes 3 into 2. Finally $(1,3)$ takes 2 into 2 and then $(1,2)$ takes 2 into 1 so $(1,2)(1,3)$ takes 2 into 1. Thus we see

$$(1,2)(1,3) = (1,3,2).$$

As another example of this *cycle multiplication* consider the product in S_5,

$$(1,2)(2,4,5)(1,3)(1,2,5).$$

Reading from right to left $1 \mapsto 2 \mapsto 2 \mapsto 4 \mapsto 4$ so $1 \mapsto 4$. Now $4 \mapsto 4 \mapsto 4 \mapsto 5 \mapsto 5$ so $4 \mapsto 5$. Next $5 \mapsto 1 \mapsto 3 \mapsto 3 \mapsto 3$ so $5 \mapsto 3$. Then $3 \mapsto 3 \mapsto 1 \mapsto 1 \mapsto 2$ so $3 \mapsto 2$. Finally $2 \mapsto 5 \mapsto 5 \mapsto 2 \mapsto 1$, so $2 \mapsto 1$. Since all the elements of $A = \{1,2,3,4,5\}$ have been accounted for, we have

$$(1,2)(2,4,5)(1,3)(1,2,5) = (1,4,5,3,2).$$

Let $f \in S_n$. If f is a cycle of length 2, i.e., $f = (a_1, a_2)$ where $a_1, a_2 \in A$, then f is called a *transposition*. Any cycle can be written as a product of transpositions, namely

$$(a_1, \dots, a_k) = (a_1, a_k)(a_1, a_{k-1}) \cdots (a_1, a_2).$$

From Theorem 11.1.3 any permutation can be written in terms of cycles, but from the above any cycle can be written as a product of transpositions. Thus we have the following result.

Theorem 11.1.5. *Let $f \in S_n$ be any permutation of degree n. Then f can be written as a product of transpositions.*

11.2 Parity and the Alternating Groups

If f is a permutation with a cycle decomposition

$$(a_1, \dots, a_k)(b_1, \dots, b_j) \cdots (m_1, \dots, m_t)$$

then f can be written as a product of

$$W(f) = (k-1) + (j-1) + \cdots + (t-1)$$

transpositions. The number $W(f)$ is uniquely associated with the permutation f since f is uniquely represented (up to order) as a product of disjoint cycles. However, there is nothing unique about the number of transpositions occurring in an arbitrary representation of f as a product of transpositions. For example in S_3

$$(1,3,2) = (1,2)(1,3) = (1,2)(1,3)(1,2)(1,2),$$

since $(1,2)(1,2) = (1)$, the identity permutation of S_3.

Although the number of transpositions is not unique in the representation of a permutation, f, as a product of transpositions, we shall show, however, that the parity (even or oddness) of that number *is* unique. Moreover, this depends solely on the number $W(f)$ uniquely associated with the representation of f. More explicitly, we have the following result.

Theorem 11.2.1. *If f is a permutation written as a product of disjoint cycles and if $W(f)$ is the associated integer given above, then if $W(f)$ is even (odd) any representation of f as a product of transpositions must contain an even (odd) number of transpositions.*

Proof. We first observe the following:

$$(a,b)(b,c_1,\dots,c_t)(a,b_1,\dots,b_k) = (a,b_1,\dots,b_k,b,c_1,\dots,c_t),$$
$$(a,b)(a,b_1,\dots,b_k,b,c_1,\dots,c_t) = (a,b_1,\dots,b_k)(b,c_1,\dots,c_t).$$

Suppose now that f is represented as a product of disjoint cycles, where we include all the 1-cycles of elements of A which f fixes, if any. If a and b occur in the same cycle in this representation for f,

$$f = \cdots(a,b_1,\dots,b_k,b,c_1,\dots,c_t)\cdots,$$

then in the computation of $W(f)$ this cycle contributes $k+t+1$. Now consider $(a,b)f$. Since the cycles are disjoint and disjoint cycles commute,

$$(a,b)f = \cdots(a,b)(a,b_1,\dots,b_k,b,c_1,\dots,c_t)\cdots$$

since neither a nor b can occur in any factor of f other than $(a,b_1,\dots,b_k,b,c_1,\dots,c_t)$. So that (a,b) cancels out and we find that $(a,b)f = \cdots(b,c_1,\dots,c_t)(a,b_1,\dots,b_k)\cdots$. Since $W((b,c_1,\dots,c_t)(a,b_1,\dots,b_k)) = k+t$ but $W(a,b_1,\dots,b_k,b,c_1,\dots,c_t) = k+t+1$, we have $W((a,b)f) = W(f)-1$.

A similar analysis shows that in the case where a and b occur in different cycles in the representation of f, then $W((a,b)f) = W(f)+1$. Combining both cases, we have

$$W((a,b)f) = W(f) \pm 1.$$

Now let f be written as a product of m transpositions, say

$$f = (a_1, b_1)(a_2, b_2) \cdots (a_m, b_m).$$

Then

$$(a_m, b_m) \cdots (a_2, b_2)(a_1, b_1) f = 1.$$

Iterating this, together with the fact that $W(1) = 0$, shows that

$$W(f)(\pm 1)(\pm 1)(\pm 1) \cdots (\pm 1) = 0,$$

where there are m terms of the form ± 1. Thus

$$W(f) = (\pm 1)(\pm 1) \cdots (\pm 1),$$

m times. Note if exactly p are $+$ and $q = m - p$ are $-$ then $m = p + q$ and $W(f) = p - q$. Hence $m \equiv W(f) \pmod 2$. Thus, $W(f)$ is even if and only if m is even and this completes the proof. □

It now makes sense to state the following definition since we know that the parity is indeed unique.

Definition 11.2.2. A permutation $f \in S_n$ is said to be *even* if it can be written as a product of an even number of transpositions. Similarly, f is called *odd* if it can be written as a product of an odd number of transpositions.

Definition 11.2.3. On the group S_n we define the *sign function* $\text{sgn} : S_n \to (\mathbb{Z}_2, +)$ by $\text{sgn}(\pi) = 0$ if π is an even permutation and $\text{sgn}(\pi) = 1$ if π is an odd permutation.

We note that if f and g are even permutations then so are fg and f^{-1} and also the identity permutation is even. Further if f is even and g is odd it is clear that fg is odd. From this it is straightforward to establish the following.

Lemma 11.2.4. sgn *is a homomorphism from* S_n *onto* $(\mathbb{Z}_2, +)$.

We now let

$$A_n = \{\pi \in S_n : \text{sgn}(\pi) = 0\}.$$

That is, A_n is precisely the set of even permutations in S_n.

Theorem 11.2.5. *For each* $n \in \mathbb{N}$ *the set* A_n *forms a normal subgroup of index* 2 *in* S_n *called the alternating group on* n *symbols. Further* $|A_n| = \frac{n!}{2}$.

Proof. By Lemma 11.2.4 $\text{sgn} : S_n \to (\mathbb{Z}_2, +)$ is a homomorphism. Then $\ker(\text{sgn}) = A_n$ and therefore A_n is a normal subgroup of S_n. Since $\text{im}(\text{sgn}) = \mathbb{Z}_2$ we have $|\text{im}(\text{sgn})| = 2$ and hence $|S_n/A_n| = 2$. Therefore $[S_n : A_n] = 2$. Since $|S_n| = n!$ then $|A_n| = \frac{n!}{2}$ follows from Lagrange's theorem. □

11.3 Conjugation in S_n

Recall that in a group G two elements $x, y \in G$ are *conjugates* if there exists a $g \in G$ with $g^{-1}xg = y$. Conjugacy is an equivalence relation on G. In the symmetric groups S_n it is easy to determine if two elements are conjugates. We say that two permutations in S_n have the *same cycle structure* if they have the same number of cycles and the lengths are the same. Hence for example in S_8 the permutations

$$\pi_1 = (1, 3, 6, 7)(2, 5) \quad \text{and} \quad \pi_2 = (2, 3, 5, 6)(1, 8)$$

have the same cycle structure. In particular if π_1, π_2 are two permutations in S_n then π_1, π_2 are conjugates if and only if they have the same cycle structure. Therefore in S_8 the permutations

$$\pi_1 = (1, 3, 6, 7)(2, 5) \quad \text{and} \quad \pi_2 = (2, 3, 5, 6)(1, 8)$$

are conjugates.

Lemma 11.3.1. *Let*

$$\pi = (a_{11}, a_{12}, \dots, a_{1k_1}) \cdots (a_{s1}, a_{s2}, \dots, a_{sk_s})$$

be the cycle decomposition of $\pi \in S_n$. Let $\tau \in S_n$ and denote by a_{ij}^{τ} the image of a_{ij} under τ. Then

$$\tau\pi\tau^{-1} = (a_{11}^{\tau}, a_{12}^{\tau}, \dots, a_{1k_1}^{\tau}) \cdots (a_{s1}^{\tau}, a_{s2}^{\tau}, \dots, a_{sk_s}^{\tau}).$$

Proof. (a) Consider a_{11} then operating on the left like functions we have

$$\tau\pi\tau^{-1}(a_{11}^{\tau}) = \tau\pi(a_{11}) = \tau(a_{12}) = a_{12}^{\tau}.$$

The same computation then follows for all the symbols a_{ij} proving the lemma. □

Theorem 11.3.2. *Two permutations $\pi_1, \pi_2 \in S_n$ are conjugates if and only if they are of the same cycle structure.*

Proof. Suppose that $\pi_2 = \tau\pi_1\tau^{-1}$. Then from Lemma 11.3.1 we have that π_1 and π_2 are of the same cycle structure.

Conversely suppose that π_1 and π_2 are of the same cycle structure. Let

$$\pi_1 = (a_{11}, a_{12}, \dots, a_{1k_1}) \cdots (a_{s1}, a_{s2}, \dots, a_{sk_s})$$
$$\pi_2 = (b_{11}, b_{12}, \dots, b_{1k_1}) \cdots (b_{s1}, b_{s2}, \dots, b_{sk_s})$$

where we place the cycles of the same length under each other. Let τ be the permutation in S_n that maps each symbol in π_1 to the digit below it in π_2. Then from Lemma 11.3.1 we have $\tau\pi_1\tau^{-1} = \pi_2$ and hence π_1 and π_2 are conjugate. □

11.4 The Simplicity of A_n

A *simple group* is a group G with no nontrivial proper normal subgroups. Up to this point the only examples we have of simple groups are cyclic groups of prime order. In this section we prove that if $n \geq 5$ each alternating group A_n is a simple group.

Theorem 11.4.1. *For each $n \geq 3$ each $\pi \in A_n$ is a product of cycles of length* 3.

Proof. Let $\pi \in A_n$. Since π is a product of an even number of transpositions to prove the theorem it suffices to show that if τ_1, τ_2 are transpositions then $\pi_1\pi_2$ is a product of 3-cycles.

Suppose that a, b, c, d are different digits in $\{1, \dots, n\}$. There are three cases to consider. First:

$$\text{Case (1):} \quad (a, b)(a, b) = 1 = (1, 2, 3)^0$$

and hence it is true here.

Next:

$$\text{Case (2):} \quad (a, b)(b, c) = (c, a, b)$$

and hence it is true here also.

Finally:

$$\text{Case (3):} \quad (a, b)(c, d) = (a, b)(b, c)(b, c)(c, d) = (c, a, b)(c, d, b)$$

since $(b, c)(b, c) = 1$. Therefore it is true here also proving the theorem. □

Now our main result:

Theorem 11.4.2. *For $n \geq 5$ the alternating group A_n is a simple nonabelian group.*

Proof. Suppose that N is a nontrivial normal subgroup of A_n with $n \geq 5$. We show that $N = A_n$ and hence that A_n is simple.

We claim first that N must contain a 3-cycle. Let $1 \neq \pi \in N$ then π is not a transposition since $\pi \in A_n$. Therefore π moves at least 3 digits. If π moves exactly 3 digits then it is a 3-cycle and we are done. Suppose then that π moves at least 4 digits. Let $\pi = \tau_1 \cdots \tau_r$ with τ_i disjoint cycles.

Case (1): There is a $\tau_i = (\dots, a, b, c, d)$. Set $\sigma = (a, b, c) \in A_n$. Then

$$\pi\sigma\pi^{-1} = \tau_i\sigma\tau_i^{-1} = (b, c, d).$$

However from Lemma 11.3.1 $(b, c, d) = (a^{\tau_i}, b^{\tau_i}, c^{\tau_i})$. Further since $\pi \in N$ and N is normal we have

$$\pi(\sigma\pi^{-1}\sigma^{-1}) = (b, c, d)(a, c, b) = (a, d, b).$$

Therefore in this case N contains a 3-cycle.

Case (2): There is a τ_i which is a 3-cycle. Then

$$\pi = (a, b, c)(d, e, \ldots).$$

Now set $\sigma = (a, b, d) \in A_n$ and then

$$\pi\sigma\pi^{-1} = (b, c, e) = (a^\pi, b^\pi, d^\pi)$$

and

$$\sigma^{-1}\pi\sigma\pi^{-1} = (a, b, d)(b, c, e) = (b, c, e, d, a) \in N.$$

Now use Case (1). Therefore in this case N has a 3-cycle.

In the final case π is a disjoint product of transpositions.

Case (3): $\pi = (a, b)(c, d) \cdots$. Since $n \geq 5$ there is an $e \neq a, b, c, d$. Let $\sigma = (a, c, e) \in A_n$. Then

$$\pi\sigma\pi^{-1} = (b, d, e_1) \quad \text{with } e_1 = e^\pi \neq b, d.$$

However $(a^\pi, c^\pi, e^\pi) = (b, d, e_1)$. Let $\gamma = (\sigma^{-1}\pi\sigma)\pi^{-1}$. This is in N since N is normal. If $e = e_1$ then $\gamma = (e, c, a)(b, d, e) = (a, e, b, d, c)$ and we can use Case (1) to get that N contains a 3-cycle. If $e \neq e_1$ then $\Gamma = (e, c, a)(b, d, e_1) \in N$ and then we can use Case (2) to obtain that N contains a 3-cycle.

These three cases show that N must contain a 3-cycle.

If N is normal in A_n then from the argument above N contains a 3-cycle τ. However from Theorem 11.3.2 any two 3-cycles in S_n are conjugate. Hence τ is conjugate to any other 3-cycle in S_n. Since N is normal and $\tau \in N$ each of these conjugates must also be in N. Therefore N contains all 3-cycles in S_n. From Theorem 11.4.1 each element of A_n is a product of 3-cycles. It follows then that each element of A_n is in N. However since $N \subset A_n$ this is only possible if $N = A_n$ completing the proof. □

Theorem 11.4.3. *Let p be a prime and $U \subset S_p$ a subgroup. Let τ be a transposition and α a p-cycle with $\alpha, \tau \in U$. Then $U = S_p$.*

Proof. Suppose without loss of generality that $\tau = (1, 2)$. Since $\alpha, \alpha^2, \ldots, \alpha^{p-1}$ are p-cycles with no fixed points then there is an i with $\alpha^i(1) = 2$. Without loss of generality we may then assume that $\alpha = (1, 2, a_3, \ldots, a_p)$. Let

$$\pi = \begin{pmatrix} 1 & 2 & a_3 & \cdots & a_p \\ 1 & 2 & 3 & \cdots & p \end{pmatrix}.$$

Then from Lemma 10.3.4 we have

$$\pi\alpha\pi^{-1} = (1, 2, \ldots, p).$$

Further $\pi(1, 2)\pi^{-1} = (1, 2)$. Hence $U_1 = \pi U \pi^{-1}$ contains $(1, 2)$ and $(1, 2, \ldots, p)$.

Now we have

$$(1,2,\dots,p)(1,2)(1,2,\dots,p)^{-1} = (2,3) \in U_1.$$

Analogously

$$(1,2,\dots,p)(2,3)(1,2,\dots,p)^{-1} = (3,4) \in U_1,$$

and so on until

$$(1,2,\dots,p)(p-2,p-1)(1,2,\dots,p)^{-1} = (p-1,p) \in U_1.$$

Hence the transpositions $(1,2),(2,3),\dots,(p-1,p) \in U_1$. Moreover

$$(1,2)(2,3)(1,2) = (1,3) \in U_1.$$

In an identical fashion each $(1,k) \in U_1$. Then for any digits s,t we have

$$(1,s)(1,t)(1,s) = (s,t) \in U_1.$$

Therefore U_1 contains all the transpositions of S_p and hence $U_1 = S_p$. Since $U = \pi U_1 \pi^{-1}$ we must have $U = S_p$ also. □

11.5 Exercises

1. Show that for $n \geq 3$ the group A_n is generated by $\{(12k) : k \geq 3\}$.
2. Let $\sigma = (k_1,\dots,k_s) \in S_n$ be a permutation. Show that the order of σ is the least common multiple of $k_1,\dots,k_s$. Compute the order of $\tau = \begin{pmatrix} 1&2&3&4&5&6&7 \\ 2&6&5&1&3&4&7 \end{pmatrix} \in S_7$.
3. Let $G = S_4$.
   (i) Determine a noncyclic subgroup H of order 4 of G.
   (ii) Show that H is normal.
   (iii) Show that $f(g)(h) := ghg^{-1}$ defines an epimorphism $f : G \to \operatorname{Aut}(H)$ for $g \in G$ and $h \in H$. Determine its kernel.
4. Show that all subgroups of order 6 of S_4 are conjugate.

Chapter 12
Solvable Groups

12.1 Solvability and Solvable Groups

The original motivation for Galois theory grew out of a famous problem in the theory of equations. This problem was to determine the solvability or insolvability of a polynomial equation of degree 5 or higher in terms of a formula involving the coefficients of the polynomial and only using algebraic operations and radicals. This question arose out of the well-known quadratic formula.

The ability to solve quadratic equations and in essence the quadratic formula was known to the Babylonians some 3600 years ago. With the discovery of imaginary numbers, the quadratic formula then says that any degree two polynomial over $\mathbb{C}$ can be solved by radicals in terms of the coefficients. In the sixteenth century the Italian mathematician Niccolo Tartaglia discovered a similar formula in terms of radicals to solve cubic equations. This *cubic formula* is now known erroneously as *Cardano's formula* in honor of Cardano, who first published it in 1545. An earlier special version of this formula was discovered by Scipione del Ferro. Cardano's student Ferrari extended the formula to solutions by radicals for fourth degree polynomials. The combination of these formulas says that polynomial equations of degree four or less over the complex numbers can be solved by radicals.

From Cardano's work until the very early nineteenth century, attempts were made to find similar formulas for degree five polynomials. In 1805 Ruffini proved that fifth degree polynomial equations are insolvable by radicals in general. Therefore there exists no comparable formula for degree 5. Abel in 1825–1826 and Galois in 1831 extended Ruffini's result and proved the insolubility by radicals for all degrees five or greater. In doing this, Galois developed a general theory of field extensions and its relationship to group theory. This has come to be known as *Galois theory* and is really the main focus of this book.

The solution of the insolvability of the quintic and higher involved a translation of the problem into a group theory setting. For a polynomial equation to be solvable by radicals its corresponding Galois group (a concept we will introduce in Chapter 16) must be a *solvable group*. This is a group with a certain defined structure. In this chapter we introduce and discuss this class of groups.

12.2 Solvable Groups

A *normal series* for a group G is a finite chain of subgroups beginning with G and ending with the identity subgroup $\{1\}$

$$G = G_0 \supset G_1 \supset G_2 \supset \cdots \supset \{1\}$$

in which each G_{i+1} is a proper normal subgroup of G_i. The factor groups G_i/G_{i+1} are called the *factors* of the series and n is the length of the series.

Definition 12.2.1. A group G is *solvable* if it has a normal series with abelian factors, that is G_i/G_{i+1} is abelian for all $i = 0, 1, \dots, n$. Such a normal series is called a *solvable series.*

If G is an abelian group then $G = G_0 \supset \{1\}$ provides a solvable series. Hence any abelian group is solvable. Further the symmetric group S_3 on 3-symbols is also solvable however nonabelian. Consider the series

$$S_3 \supset A_3 \subset \{1\}.$$

Since $|S_3| = 6$ we have $|A_3| = 3$ and hence A_3 is cyclic and therefore abelian. Further $|S_3/A_3| = 2$ and hence the factor group S_3/A_3 is also cyclic and hence abelian. Therefore the series above gives a solvable series for S_3.

Lemma 12.2.2. *If G is a finite solvable group then G has a normal series with cyclic factors.*

Proof. If G is a finite solvable group then by definition it has a normal series with abelian factors. Hence to prove the lemma it suffices to show that a finite abelian group has a normal series with cyclic factors.

Let A be a nontrivial finite abelian group. We do an induction on the order of A. If $|A| = 2$ then A itself is cyclic and the result follows. Suppose that $|A| > 2$. Choose an $1 \neq a \in A$. Let $N = \langle a \rangle$ so that N is cyclic. Then we have the normal series $A \supset N \supset \{1\}$ with A/N abelian. Further A/N has order less than A so A/N has a normal series with cyclic factors and the result follows. □

Solvability is preserved under subgroups and factor groups.

Theorem 12.2.3. *Let G be a solvable group. Then:*

(1) *Any subgroup H of G is also solvable.*

(2) *Any factor group G/N of G is also solvable.*

Proof. (1) Let G be a solvable group and suppose that

$$G = G_0 \supset G_1 \supset \cdots \supset G_r \supset \{1\}$$

is a solvable series for G. Hence G_{i+1} is a normal subgroup of G_i for each i and the factor group G_i/G_{i+1} is abelian.

Now let H be a subgroup of G and consider the chain of subgroups

$$H = H \cap G_0 \supset H \cap G_1 \supset \cdots \supset H \cap G_r \supset \{1\}.$$

Since G_{i+1} is normal in G_i we know that $H \cap G_{i+1}$ is normal in $H \cap G_i$ and hence this gives a finite normal series for H. Further from the second isomorphism theorem we have for each i,

$$\begin{aligned}(H \cap G_i)/(H \cap G_{i+1}) &= (H \cap G_i)/((H \cap G_i) \cap G_{i+1}) \\ &\cong (H \cap G_i)G_{i+1}/G_{i+1} \subset G_i/G_{i+1}.\end{aligned}$$

However G_i/G_{i+1} is abelian so each factor in the normal series for H is abelian. Therefore the above series is a solvable series for H and hence H is also solvable.

(2) Let N be a normal subgroup of G. Then from (1) N is also solvable. As above let

$$G = G_0 \supset G_1 \supset \cdots \supset G_r \supset \{1\}$$

be a solvable series for G. Consider the chain of subgroups

$$G/N = G_0N/N \supset G_1N/N \supset \cdots \supset G_rN/N \supset N/N = \{1\}.$$

Let $m \in G_{i-1}, n \in N$. Then since N is normal in G,

$$\begin{aligned}(mn)^{-1}G_iN(mn) &= n^{-1}m^{-1}G_imnN = n^{-1}G_inN \\ &= n^{-1}NG_i = NG_i = G_iN.\end{aligned}$$

It follows that $G_{i+1}N$ is normal in G_iN for each i and therefore the series for G/N is a normal series.

Further, again from the isomorphism theorems

$$\begin{aligned}(G_iN/N)/(G_{i+1}N/N) &\cong G_i/(G_i \cap G_{i+1}N) \\ &\cong (G_i/G_{i+1}/((G_i \cap G_{i+1}N)/G_{i+1}).\end{aligned}$$

However the last group $(G_i/G_{i+1}/((G_i \cap G_{i+1}N)/G_{i+1})$ is a factor group of the group G_i/G_{i+1} which is abelian. Hence this last group is also abelian and therefore each factor in the normal series for G/N is abelian. Hence this series is a solvable series and G/N is solvable. □

The following is a type of converse of the above theorem.

Theorem 12.2.4. *Let G be group and H a normal subgroup of G. If both H and G/H are solvable then G is solvable.*

Proof. Suppose that

$$H = H_0 \supset H_1 \supset \cdots \supset H_r \supset \{1\}$$
$$G/H = G_0/H \supset G_1/H \supset \cdots \supset G_s/H \supset H/H = \{1\}$$

are solvable series for H and G/H respectively. Then

$$G = G_0 \supset G_1 \supset \cdots \supset G_s = H \supset H_1 \supset \cdots \supset \{1\}$$

gives a normal series for G. Further from the isomorphism theorems again

$$G_i/G_{i+1} \cong (G_i/H)/(G_{i+1}/H)$$

and hence each factor is abelian. Therefore this is a solvable series for G and hence G is solvable. □

This theorem allows us to prove that solvability is preserved under direct products.

Corollary 12.2.5. *Let G and H be solvable groups. Then their direct product $G \times H$ is also solvable.*

Proof. Suppose that G and H are solvable groups and $K = G \times H$. Recall from Chapter 10 that G can be considered as a normal subgroup of K with $K/G \cong H$. Therefore G is a solvable subgroup of K and G/K is a solvable quotient. It follows then from Theorem 12.2.4 that K is solvable. □

We saw that the symmetric group S_3 is solvable. However the following theorem shows that the symmetric group S_n is not solvable for $n \geq 5$. This result will be crucial to the proof of the insolvability of the quintic and higher.

Theorem 12.2.6. *For $n \geq 5$ the symmetric group S_n is not solvable.*

Proof. For $n \geq 5$ we saw that the alternating group A_n is simple. Further A_n is nonabelian. Hence A_n cannot have a nontrivial normal series and so no solvable series. Therefore A_n is not solvable. If S_n were solvable for $n \geq 5$ then from Theorem 12.2.3 A_n would also be solvable. Therefore S_n must also be nonsolvable for $n \geq 5$. □

In general for a simple, solvable group we have the following.

Lemma 12.2.7. *If a group G is both simple and solvable then G is cyclic of prime order.*

Proof. Suppose that G is a nontrivial simple, solvable group. Since G is simple the only normal series for G is $G = G_0 \supset \{1\}$. Since G is solvable the factors are abelian and hence G is abelian. Again since G is simple G must be cyclic. If G were infinite then $G \cong (\mathbb{Z}, +)$. However then $2\mathbb{Z}$ is a proper normal subgroup, a contradiction. Therefore G must be finite cyclic. If the order were not prime then for each proper divisor of the order there would be a nontrivial proper normal subgroup. Therefore G must be of prime order. □

In general a finite p-group is solvable.

Theorem 12.2.8. *A finite p-group G is solvable.*

Proof. Suppose that $|G| = p^n$. We do this by induction on n. If $n = 1$ then $|G| = p$ and G is cyclic, hence abelian and therefore solvable. Suppose that $n > 1$. Then as used previously G has a nontrivial center $Z(G)$. If $Z(G) = G$ then G is abelian and hence solvable. If $Z(G) \neq G$ then $Z(G)$ is a finite p-group of order less than p^n. From our inductive hypothesis $Z(G)$ must be solvable. Further $G/Z(G)$ is then also a finite p-group of order less than p^n so it is also solvable. Hence $Z(G)$ and $G/Z(G)$ are both solvable so from Theorem 12.2.4 G is solvable. □

12.3 The Derived Series

Let G be a group and let $a, b \in G$. The product $aba^{-1}b^{-1}$ is called the *commutator* of a and b. We write $[a, b] = aba^{-1}b^{-1}$.

Clearly $[a, b] = 1$ if and only if a and b commute.

Definition 12.3.1. Let G' be the subgroup of G which is generated by the set of all commutators

$$G' = gp(\{[x, y] : x, y \in G\}).$$

G' is called the *commutator* or (*derived*) *subgroup* of G. We sometimes write $G' = [G, G]$.

Theorem 12.3.2. *For any group G the commutator subgroup G' is a normal subgroup of G and G/G' is abelian. Further if H is a normal subgroup of G then G/H is abelian if and only if $G' \subset H$.*

Proof. The commutator subgroup G' consists of all finite products of commutators and inverses of commutators. However

$$[a, b]^{-1} = (aba^{-1}b^{-1})^{-1} = bab^{-1}a^{-1} = [b, a]$$

and so the inverse of a commutator is once again a commutator. It then follows that G' is precisely the set of all finite products of commutators, i.e., G' is the set of all elements of the form

$$h_1 h_2 \cdots h_n$$

where each h_i is a commutator of elements of G.

If $h = [a, b]$ for $a, b \in G$, and $x \in G$, $xhx^{-1} = [xax^{-1}, xbx^{-1}]$ is again a commutator of elements of G. Now from our previous comments, an arbitrary element of G' has the form $h_1 h_2 \cdots h_n$, where each h_i is a commutator. Thus $x(h_1 h_2 \cdots h_n)x^{-1} = (xh_1x^{-1})(xh_2x^{-1}) \cdots (xh_nx^{-1})$ and, since by the above each xh_ix^{-1} is a commutator, $x(h_1 h_2 \cdots h_n)x^{-1} \in G'$. It follows that G' is a normal subgroup of G.

Consider the factor group G/G'. Let aG' and bG' be any two elements of G/G'. Then

$$\begin{aligned}[aG', bG'] &= aG' \cdot bG' \cdot (aG')^{-1} \cdot (bG')^{-1} \\ &= aG' \cdot bG' \cdot a^{-1}G' \cdot b^{-1}G' = aba^{-1}b^{-1}G' = G'\end{aligned}$$

since $[a, b] \in G'$. In other words, any two elements of G/G' commute and therefore G/G' is abelian.

Now let N be a normal subgroup of G with G/N abelian. Let $a, b \in G$ then aN and bN commute since G/N is abelian. Therefore

$$[aN, bN] = aNbNa^{-1}Nb^{-1}N = aba^{-1}b^{-1}N = N.$$

It follows that $[a, b] \in N$. Therefore all commutators of elements in G lie in N and therefore the commutator subgroup $G' \subset N$. □

From the second part of Theorem 12.3.2 we see that G' is the minimal normal subgroup of G such that G/N is abelian. We call $G/G' = G_{\text{ab}}$ the *abelianization* of G.

We consider next the following inductively defined sequence of subgroups of an arbitrary group G called the *derived series*.

Definition 12.3.3. For an arbitrary group G define $G^{(0)} = G$ and $G^{(1)} = G'$ and then inductively $G^{(n+1)} = (G^{(n)})'$. That is $G^{(n+1)}$ is the commutator subgroup or derived group of $G^{(n)}$. The chain of subgroups

$$G = G^{(0)} \supset G^{(1)} \supset \cdots \supset G^{(n)} \supset \cdots$$

is called the *derived series* for G.

Notice that since $G^{(i+1)}$ is the commutator subgroup of $G^{(i)}$ we have $G^{(i)}/G^{(i+1)}$ is abelian. If the derived series was finite then G would have a normal series with abelian factors and hence be solvable. The converse is also true and characterizes solvable groups in terms of the derived series.

Theorem 12.3.4. *A group G is solvable if and only if its derived series is finite. That is there exists an n such that $G^{(n)} = \{1\}$.*

Proof. If $G^{(n)} = \{1\}$ for some n then as explained above the derived series provides a solvable series for G and hence G is solvable.

Conversely suppose that G is solvable and let

$$G = G_0 \supset G_1 \supset \cdots \supset G_r = \{1\}$$

be a solvable series for G. We claim first that $G_i \supset G^{(i)}$ for all i. We do this by induction on r. If $r = 0$ then $G = G_0 = G^{(0)}$. Suppose that $G_i \supset G^{(i)}$. Then $G_i' \supset (G^{(i)})' = G^{(i+1)}$. Since G_i/G_{i+1} is abelian it follows from Theorem 12.3.2 that $G_{i+1} \supset G_i'$. Therefore $G_{i+1} \supset G^{(i+1)}$ establishing the claim.

Now if G is solvable from the claim we have that $G_r \supset G^{(r)}$. However $G_r = \{1\}$ and therefore $G^{(r)} = \{1\}$ proving the theorem. □

The length of the derived series is called the *solvability length* of a solvable group G. The class of solvable groups of *class* c consists of those solvable groups of solvability length c or less.

12.4 Composition Series and the Jordan–Hölder Theorem

The concept of a normal series is extremely important in the structure theory of groups. This is especially true for finite groups. If

$$G = G_0 \supset G_1 \supset \cdots \supset \{1\}$$
$$G = H_0 \supset H_1 \supset \cdots \supset \{1\}$$

are two normal series for the group G then the second is a *refinement* of the first if all the terms of the second occur in the first series. Further, two normal series are called *equivalent* or (*isomorphic*) if there exists a 1-1 correspondence between the factors (hence the length must be the same) of the two series such that the corresponding factors are isomorphic.

Theorem 12.4.1 (Schreier's theorem). *Any two normal series for a group G have equivalent refinements.*

Proof. Consider two normal series for G

$$G = G_0 \supset G_1 \supset \cdots \supset G_s \supset \{1\} = G_{s+1}$$
$$G = H_0 \supset H_1 \supset \cdots \supset H_t \supset \{1\} = G_{t+1}.$$

Now define

$$G_{ij} = (G_i \cap H_j)G_{i+1}, \quad j = 0, 1, 2, \dots, t+1,$$
$$H_{ji} = (G_i \cap H_j)H_{j+1}, \quad i = 0, 1, 2, \dots, s+1.$$

Then we have

$$G = G_{00} \supset G_{01} \supset \cdots \supset G_{0,s+1} = G_1$$
$$= G_{10} \supset \cdots \supset G_{1,s+1} = G_2 \supset \cdots \supset G_{t,s+1} = \{e\},$$

and

$$G = H_{00} \supset H_{01} \supset \cdots \supset H_{0,t+1} = H_1$$
$$= H_{10} \supset \cdots \supset H_{1,t+1} = H_2 \supset \cdots \supset H_{s,t+1} = \{e\}.$$

Now applying the third isomorphism theorem to the groups G_i, H_j, G_{i+1}, H_{j+1}, we have that $G_{i,j+1} = (G_i \cap H_{j+1})G_{i+1}$ is a normal subgroup of $G_{i,j} = (G_i \cap H_j)G_{i+1}$ and $H_{j,i+1} = (G_{i+1} \cap H_j)H_{j+1}$ is a normal subgroup of $H_{j,i} = (G_i \cap H_j)H_{j+1}$. Furthermore, also

$$G_{ij}/G_{i,j+1} \cong H_{ji}/H_{j,i+1}.$$

Thus the above two are normal series which are refinements of the two given series and they are equivalent. □

A proper normal subgroup N of a group G is called *maximal* in G if there does not exist any normal subgroup $N \subset M \subset G$ with all inclusions proper. This is the group theoretic analog of a maximal ideal. An alternative characterization is the following. N is a maximal normal subgroup of G if and only if G/N is simple.

A normal series where each factor is simple can have no refinements.

Definition 12.4.2. A *composition series* for a group G is a normal series where all the inclusions are proper and such that G_{i+1} is maximal in G_i. Equivalently a normal series where each factor is simple.

It is possible that an arbitrary group does not have a composition series or even if it does have one, a subgroup of it may not have one. Of course, a finite group does have a composition series.

In the case in which a group, G, does have a composition series the following important theorem, called the Jordan–Hölder theorem, provides a type of unique factorization.

Theorem 12.4.3 (Jordan–Hölder theorem). *If a group G has a composition series, then any two composition series are equivalent, that is the composition factors are unique.*

Proof. Suppose we are given two composition series. Applying Theorem 12.4.1 we get that the two composition series have equivalent refinements. But the only refinement of a composition series is one obtained by introducing repetitions. If in the 1-1 correspondence between the factors of these refinements, the paired factors equal to $\{e\}$ are disregarded, that is if we drop the repetitions, we get clearly that the original composition series are equivalent. □

We remarked in Chapter 10 that the simple groups are important because they play a role in finite group theory somewhat analogous to that of the primes in number theory. In particular, an arbitrary finite group, G, can be broken down into simple components. These uniquely determined simple components are, according to the Jordan–Hölder theorem, the factors of a composition series for G.

12.5 Exercises

1. Let K be a field and

$$G = \left\{ \begin{pmatrix} a & x & y \\ 0 & b & z \\ 0 & 0 & c \end{pmatrix} : a, b, c, x, y, z \in K, abc \neq 0 \right\}.$$

   Show that G is solvable.

2. A group G is called *polycyclic* if it has a normal series with cyclic factors. Show:
   (i) Each subgroup and each factor group of a polycyclic group is polycyclic.
   (ii) In a polycyclic group each normal series has the same number of infinite cyclic factors.

3. Let G be a group. Show:
   (i) If G is finite and solvable, then G is polycyclic.
   (ii) If G is polycyclic, then G is finitely generated.
   (iii) The group $(\mathbb{Q}, +)$ is solvable, but not polycyclic.

4. Let N_1 and N_2 be normal subgroups of G. Show:
   (i) If N_1 and N_2 are solvable, then also $N_1 N_2$ is a solvable normal subgroup of G.
   (ii) Is (i) still true, if we replace "solvable" by "abelian"?

5. Let $N_1, \dots, N_t$ be normal subgroups of a group G. If all factor groups G/N_i are solvable, then also $G/(N_1 \cap \cdots \cap N_t)$ is solvable.

Chapter 13
Groups Actions and the Sylow Theorems

13.1 Group Actions

A *group action* of a group G on a set A is a homomorphism from G into S_A the symmetric group on A. We say that G *acts* on A. Hence G acts on A if to each $g \in G$ corresponds a permutation

$$\pi_g : A \to A$$

such that

(1) $\pi_{g_1}(\pi_{g_2}(a)) = \pi_{g_1 g_2}(a)$ for all $g_1, g_2 \in G$ and for all $a \in A$

(2) $1(a) = a$ for all $a \in A$.

For the remainder of this chapter if $g \in G$ and $a \in A$ we will write ga for $\pi_g(a)$.

Group actions are an extremely important idea and we use this idea in the present chapter to prove several fundamental results in group theory.

If G acts on the set A then we say that two elements $a_1, a_2 \in A$ are *congruent* under G if there exists a $g \in G$ with $ga_1 = a_2$. The set

$$G_a = \{a_1 \in A : a_1 = ga \text{ for some } g \in G\}$$

is called the *orbit* of a. It consists of elements congruent to a under G.

Lemma 13.1.1. *If G acts on A then congruence under G is an equivalence relation on A.*

Proof. Any element $a \in A$ is congruent to itself via the identity map and hence the relation is reflexive. If $a_1 \sim a_2$ so that $ga_1 = a_2$ for some $g \in G$ then $g^{-1}a_2 = a_1$ and so $a_2 \sim a_1$ and the relation is symmetric. Finally is $g_1 a_1 = a_2$ and $g_2 a_2 = a_3$ then $g_2 g_1 a_1 = a_3$ and relation is transitive. □

Recall that the equivalence classes under an equivalence relation partition a set. For a given $a \in A$ its equivalence class under this relation is precisely its orbit as defined above.

Corollary 13.1.2. *If G acts on the set A then the orbits under G partition the set A.*

We say that G *acts transitively* on A if any two elements of A are congruent under G. That is the action is transitive if for any $a_1, a_2 \in A$ there is some $g \in G$ such that $ga_1 = a_2$.

If $a \in A$ the *stabilizer* of a consists of those $g \in G$ that *fix* a. Hence

$$\mathrm{Stab}_G(a) = \{g \in G : ga = a\}.$$

The following lemma is easily proved and left to the exercises.

Lemma 13.1.3. *If G acts on A then for any $a \in A$ the stabilizer* $\mathrm{Stab}_G(a)$ *is a subgroup of G.*

We now prove the crucial theorem concerning group actions.

Theorem 13.1.4. *Suppose that G acts on A and $a \in A$. Let G_a be the orbit of a under G and* $\mathrm{Stab}_G(a)$ *its stabilizer. Then*

$$|G : \mathrm{Stab}_G(a)| = |G_a|.$$

That is the size of the orbit of a is the index of its stabilizer in G.

Proof. Suppose that $g_1, g_2 \in G$ with $g_1 \mathrm{Stab}_G(a) = g_2 \mathrm{Stab}_G(a)$, that is they define the same left coset of the stabilizer. Then $g_2^{-1}g_1 \in \mathrm{Stab}_G(a)$. This implies that $g_2^{-1}g_1a = a$ so that $g_2a = g_1a$. Hence any two elements in the same left coset of the stabilizer produce the same image of a in G_a. Conversely if $g_1a = g_2$ then g_1, g_2 define the same left coset of $\mathrm{Stab}_G(a)$. This shows that there is a one-to-one correspondence between left cosets of $\mathrm{Stab}_G(a)$ and elements of G_a. It follows that the size of G_a is precisely the index of the stabilizer. □

We will use this theorem repeatedly with different group actions to obtain important group theoretic results.

13.2 Conjugacy Classes and the Class Equation

In Section 10.5 we introduced the *center* of a group

$$Z(G) = \{g \in G : gg_1 = g_1g \text{ for all } g_1 \in G\},$$

and showed that it is a normal subgroup of G. We then used this normal subgroup in conjunction with what we called the *class equation* to show that any finite p-group has a nontrivial center. In this section we use group actions to derive the class equation and prove the result for finite p-groups.

Recall that if G is a group then two elements $g_1, g_2 \in G$ are *conjugate* if there exists a $g \in G$ with $g^{-1}g_1g = g_2$. We saw that conjugacy is an equivalence relation on G. For $g \in G$ its equivalence class is called its *conjugacy class* that we will denote by $\mathrm{Cl}(g)$. Thus

$$\mathrm{Cl}(g) = \{g_1 \in G : g_1 \text{ is conjugate to } g\}.$$

If $g \in G$ then its *centralizer* $C_G(g)$ is the set of elements in G that commute with g:

$$C_G(g) = \{g_1 \in G : gg_1 = g_1 g\}.$$

Theorem 13.2.1. *Let G be a finite group and $g \in G$. Then the centralizer of g is a subgroup of G and*

$$|G : C_G(g)| = |\mathrm{Cl}(g)|.$$

That is the index of the centralizer of g is the size of its conjugacy class.

In particular for a finite group the size of each conjugacy class divides the order of the group.

Proof. Let the group G act on itself by conjugation. That is $g(g_1) = g^{-1}g_1 g$. It is easy to show that this is an action on the set G (see exercises). For $g \in G$ its orbit under this action is precisely its conjugacy class $\mathrm{Cl}(g)$ and the stabilizer is its centralizer $C_G(g)$. The statements in the theorem then follow directly from Theorem 13.1.4. □

For any group G, since conjugacy is an equivalence relation, the conjugacy classes partition G. Hence

$$G = \dot{\bigcup_{g \in G}} \mathrm{Cl}(g)$$

where this union is a disjoint union. It follows that

$$|G| = \sum_{g \in G} |\mathrm{Cl}(g)|$$

where this sum is taken over distinct conjugacy classes.

If $\mathrm{Cl}(g) = \{g\}$ that is the conjugacy class of g is g alone then $C_G(g) = G$ so that g commutes with all of G. Therefore in this case $g \in Z(G)$. This is true for every element of the center and therefore

$$G = Z(G) \cup \dot{\bigcup_{g \notin Z(G)}} \mathrm{Cl}(g)$$

where again the second union is a disjoint union. The size of G is then the sum of these disjoint pieces so

$$|G| = |Z(G)| + \sum_{g \notin Z(G)} |\mathrm{Cl}(g)|.$$

However from Theorem 13.2.1 $|\mathrm{Cl}(g)| = |G : C_G(g)|$ so the equation above becomes

$$|G| = |Z(G)| + \sum_{g \notin Z(G)} |G : C_G(g)|.$$

This is known as the *class equation*.

Theorem 13.2.2 (class equation). *Let G be a finite group. Then*

$$|G| = |Z(G)| + \sum_{g \notin Z(G)} |G : C_G(g)|$$

where the sum is taken over the distinct centralizers.

As a first application we prove the result that finite p-groups have nontrivial centers.

Theorem 13.2.3. *Let G be a finite p-group. Then G has a nontrivial center.*

Proof. Let G be a finite p-group so that $|G| = p^n$ for some n and consider the class equation

$$|G| = |Z(G)| + \sum_{g \notin Z(G)} |G : C_G(g)|.$$

Since $|G : C_G(g)|$ divides $|G|$ for each $g \in G$ we must have that $p||G : C_G(g)|$ for each $g \in G$. Further $p||G|$. Therefore p must divide $|Z(G)|$ and hence $|Z(G)| = p^m$ for some $m \geq 1$ and therefore $Z(G)$ is nontrivial. □

The idea of conjugacy and the centralizer of an element can be extended to subgroups. If H_1, H_2 are subgroups of a group G then H_1, H_2 are conjugate if there exists a $g \in G$ such that $g^{-1}H_1g = H_2$. As for elements conjugacy is an equivalence relation on the set of subgroups of G.

If $H \subset G$ is a subgroup then its *conjugacy class* consists of all the subgroups of G conjugate to it. The *normalizer* of H is

$$N_G(H) = \{g \in G : g^{-1}Hg = H\}.$$

As for elements let G act on the set of subgroups of G by conjugation. That is for $g \in G$ the map is given by $H \mapsto g^{-1}Hg$. For $H \subset G$ the stabilizer under this action is precisely the normalizer. Hence exactly as for elements we obtain the following theorem.

Theorem 13.2.4. *Let G be a group and $H \subset G$ a subgroup. Then the normalizer $N_G(H)$ of H is a subgroup of G, H is normal in $N_G(H)$ and*

$$|G : N_G(H)| = \textit{number of conjugates of } H \textit{ in } G.$$

13.3 The Sylow Theorems

If G is a finite group and $H \subset G$ is a subgroup then Lagrange's theorem guarantees that the order of H divides the order of G. However the converse of Lagrange's theorem is false. That is, if G is a finite group of order n and if $d|n$, then G need not contain a subgroup of order d. If d is a prime p or a power of a prime p^e, however, then we shall see that G must contain subgroups of that order. In particular, we shall see that if p^d is the highest power of p that divides n, than all subgroups of that order are actually conjugate, and we shall finally get a formula concerning the number of such subgroups. These theorems constitute the Sylow theorems which we will examine in this section. First we give an example where the converse of Lagrange's theorem is false.

Lemma 13.3.1. *The alternating group on* 4 *symbols* A_4 *has order* 12 *but has no subgroup of order* 6.

Proof. Suppose that there exists a subgroup $U \subset A_4$ with $|U|=6$. Then $|A_4 : U|=2$ since $|A_4| = 12$ and hence U is normal in A_4.

Now id, $(1,2)(3,4)$, $(1,3)(2,4)$, $(1,4)(2,3)$ are in A_4. These each have order 2 and commute so they form a subgroup $V \subset A_4$ of order 4. This subgroup $V \cong \mathbb{Z}_2 \times \mathbb{Z}_2$. Then

$$12 = |A_4| \geq |VU| = \frac{|V||U|}{|V \cap U|} = \frac{4 \cdot 6}{|V \cap U|}.$$

It follows that $V \cap U \neq \{1\}$ and since U is normal we have that $V \cap U$ is also normal in A_4.

Now $(1,2)(3,4) \in V$ and by renaming the entries in V if necessary we may assume that it is also in U so that $(1,2)(3,4) \in V \cap U$. Since $(1,2,3) \in A_4$ we have

$$(3,2,1)(1,2)(3,4)(1,2,3) = (1,3)(2,4) \in V \cap U$$

and then

$$(3,2,1)(1,4)(2,3)(1,2,3) = (1,2)(3,4) \in V \cap U.$$

But then $V \subset V \cap U$ and so $V \subset U$. But this is impossible since $|V| = 4$ which doesn't divide $|U| = 6$. □

Definition 13.3.2. Let G be a finite group with $|G| = n$ and let p be a prime such that $p^a | n$ but no higher power of p divides n. A subgroup of G of order p^a is called a *p-Sylow subgroup*.

It is not a clear that a p-Sylow subgroup must exist. We will prove that for each $p | n$ a p-Sylow subgroup exists.

We first consider and prove a very special case.

Theorem 13.3.3. *Let* G *be a finite abelian group and let* p *be a prime such that* $p \,|\, |G|$. *Then* G *contains at least one element of order* p.

Proof. Suppose that G is a finite abelian group of order pn. We use induction on n. If $n = 1$ then G has order p and hence is cyclic and so has an element of order p. Suppose that the theorem is true for all abelian groups of order pm with $m < n$ and suppose that G has order pn. Suppose that $g \in G$. If the order of g is pt for some integer t then $g^t \neq 1$ and g^t has order p proving the theorem in this case. Hence we may suppose that $g \in G$ has order prime to p and we show that there must be an element whose order is a multiple of p and then use the above argument to get an element of exact order p.

Hence we have $g \in G$ with order m where $(m, p) = 1$. Since $m \| G| = pn$ we must have $m|n$. Since G is abelian $\langle g \rangle$ is normal and the factor group $G/\langle g \rangle$ is abelian of order $p(\frac{n}{m}) < pn$. By the inductive hypothesis $G/\langle g \rangle$ has an element $h\langle g \rangle$ of order p, $h \in G$, and hence $h^p = g^k$ for some k. g^k has order $m_1|m$ and therefore h has order pm_1. Now as above h^{m_1} has order p proving the theorem. □

Therefore if G is an abelian group and if $p|n$, then G contains a subgroup of order p, the cyclic subgroup of order p generated by an element $a \in G$ of order p whose existence is guaranteed by the above theorem. We now present the first Sylow theorem.

Theorem 13.3.4 (first Sylow theorem). *Let G be a finite group and let $p \| G|$, then G contains a p-Sylow subgroup, that is a p-Sylow subgroup exists.*

Proof. Let G be a finite group of order pn and as above we do induction on n. If $n = 1$ then G is cyclic and G is its own maximal p-subgroup and hence all of G is a p-Sylow subgroup. We assume then that if $|G| = pm$ with $m < n$ then G has a p-Sylow subgroup.

Assume that $|G| = p^t m$ with $(m, p) = 1$. We must show that G contains a subgroup of order p^t. If H is a proper subgroup whose index is prime to p then $|H| = p^t m_1$ with $m_1 < m$. Therefore by the inductive hypothesis H has a p-Sylow subgroup of order p^t. This will also be a subgroup of G and hence a p-Sylow subgroup of G.

Therefore we may assume that the index of any proper subgroup H of G must have index divisible by p. Now consider the class equation for G,

$$|G| = |Z(G)| + \sum_{g \notin Z(G)} |G : C_G(g)|.$$

By assumption each of the indices are divisible by p and also $p \| G|$. Therefore $p \| Z(G)|$. It follows that $Z(G)$ is a finite abelian group whose order is divisible by p. From Theorem 13.3.3 there exists an element $g \in Z(G) \subset G$ of order p. Since $g \in Z(G)$ we must have $\langle g \rangle$ normal in G. The factor group $G/\langle g \rangle$ then has order $p^{t-1}m$ and by the inductive hypothesis must have a p-Sylow subgroup $\overline{K}$ of order p^{t-1} and hence of index m. By the correspondence theorem there is a subgroup K of G with $\langle g \rangle \subset K$ such that $K/H \cong \overline{K}$. Therefore $|K| = p^t$ and K is a p-Sylow subgroup of G. □

On the basis of this theorem, we can now strengthen the result obtained in Theorem 13.3.3.

Theorem 13.3.5 (Cauchy). *If G is a finite group and if p is a prime such that $p||G|$, then G contains at least one element of order p.*

Proof. Let P be a p-Sylow subgroup of G, and let $|P| = p^t$. If $g \in P$, $g \neq 1$, then the order of g is p^{t_1}. Then $g^{p^{t_1-1}}$ has order p. □

We have seen that p-Sylow subgroups exist. We now wish to show that any two p-Sylow subgroups are conjugate. This is the content of the second Sylow theorem.

Theorem 13.3.6 (second Sylow theorem). *Let G be a finite group and p a prime such that $p||G|$. Then any p-subgroup H of G is contained in a p-Sylow subgroup. Further all p-Sylow subgroups of G are conjugate. That is, if P_1 and P_2 are any two p-Sylow subgroups of G then there exists an $a \in G$ such that $P_1 = aP_2a^{-1}$.*

Proof. Let Ω be the set of p-Sylow subgroups of G and let G act on Ω by conjugation. This action will of course partition Ω into disjoint orbits. Let P be a fixed p-Sylow subgroup and Ω_P be its orbit under the conjugation action. The size of the orbit is the index of its stabilizer that is $|\Omega_P| = |G : \mathrm{Stab}_G(P)|$. Now $P \subset \mathrm{Stab}_G(P)$ and P is a maximal p-subgroup of G. It follows that the index of $\mathrm{Stab}_G(P)$ must be prime to p and so the number of p-Sylow subgroups conjugate to P is prime to p.

Now let H be a p-subgroup of G and let H act on Ω_P by conjugation. Ω_P will itself decompose into disjoint orbits under this actions. Further the size of each orbit is an index of a subgroup of H and hence must be a power of p. On the other hand the size of the whole orbit is prime to p. Therefore there must be one orbit that has size exactly 1. This orbit contains a p-Sylow subgroup P' and P' is fixed by H under conjugation, that is H normalizes P'. It follows that HP' is a subgroup of G and P' is normal in HP'. From the second isomorphism theorem we then obtain

$$HP'/P' \cong H/(H \cap P').$$

Since H is a p-group the size of $H/(H \cap P')$ is a power of p and therefore so is the size of HP'/P'. But P' is also a p-group so it follows that HP' also has order a power of p. Now $P' \subset HP'$ but P' is a maximal p-subgroup of G. Hence $HP' = P'$. This is possible only if $H \subset P'$ proving the first assertion in the theorem. Therefore any p-subgroup of G is obtained in a p-Sylow subgroup.

Now let H be a p-Sylow subgroup P_1 and let P_1 act on Ω_P. Exactly as in the argument above $P_1 \subset P'$ where P' is a conjugate of P. Since P_1 and P' are both p-Sylow subgroups they have the same size and hence $P_1 = P'$. This implies that P_1 is a conjugate of P. Since P_1 and P are arbitrary p-Sylow subgroups it follows that all p-Sylow subgroups are conjugate. □

We come now to the last of the three Sylow theorems. This one gives us information concerning the number of p-Sylow subgroups.

Theorem 13.3.7 (third Sylow theorem). *Let G be a finite group and p a prime such that $p||G|$. Then the number of p-Sylow subgroups of G is of the form $1 + pk$ and divides the order of $|G|$. It follows that if $|G| = p^a m$ with $(p,m) = 1$ then the number of p-Sylow subgroups divides m.*

Proof. Let P be a p-Sylow subgroup and let P act on Ω, the set of all p-Sylow subgroups, by conjugation. Now P normalizes itself so there is one orbit, namely P that has size exactly 1. Every other orbit has size a power of p since the size is the index of a nontrivial subgroup of P and therefore must be divisible by p. Hence the size of the Ω is $1 + pk$. □

13.4 Some Applications of the Sylow Theorems

We now give some applications of the Sylow theorems. First we show that the converse of Lagrange's theorem is true for both general p-groups and for finite abelian groups.

Theorem 13.4.1. *Let G be a group of order p^n. Then G contains at least one normal subgroup of order p^m, for each m such that $0 \le m \le n$.*

Proof. We use induction on n. For $n = 1$ the theorem is trivial. By Lemma 10.5.7 any group of order p^2 is abelian. This together with Theorem 13.3.3 establishes the claim for $n = 2$.

We now assume the theorem is true for all groups G of order p^k where $1 \le k < n$, where $n > 2$. Let G be a group of order p^n. From Lemma 10.3.4 G has a nontrivial center of order at least p and hence an element $g \in Z(G)$ of order p. Let $N = \langle g \rangle$. Since $g \in Z(G)$ it follows that N is normal subgroup of order p. Then G/N is of order p^{n-1}, and therefore, contains (by the induction hypothesis) normal subgroups of orders p^{m-1}, for $0 \le m-1 \le n-1$. These groups are of the form H/N, where the normal subgroup $H \subset G$ contains N and is of order p^m, $1 \le m \le n$, because $|H| = |N|[H : N] = |N| \cdot |H/N|$. □

On the basis of the first Sylow theorem we see that if G is a finite group and if $p^k||G|$, then G must contain a subgroup of order p^k. One can actually show that, as in the case of Sylow p-groups, the number of such subgroups is of the form $1 + pt$, but we shall not prove this here.

Theorem 13.4.2. *Let G be a finite abelian group of order n. Suppose that $d|n$. Then G contains a subgroup of order d.*

Proof. Suppose that $n = p_1^{e_1} \cdots p_k^{e_k}$ is the prime factorization of n. Then $d = p_1^{f_1} \cdots p_k^{f_k}$ for some nonnegative $f_1, \dots, f_k$. Now G has p_1-Sylow subgroup H_1 of order $p_1^{e_1}$. Hence from Theorem 13.4.1 H_1 has a subgroup K_1 of order $p_1^{f_1}$. Similarly there are subgroups $K_2, \dots, K_k$ of G of respective orders $p_2^{f_2}, \dots, p_k^{f_k}$. Further since the orders are disjoint $K_i \cap K_j = \{1\}$ if $i \neq j$. It follows that $\langle K_1, K_2, \dots, K_k \rangle$ has order $|K_1||K_2| \cdots |K_k| = p_1^{f_1} \cdots p_k^{f_k} = d$. □

In Section 10.5 we examined the classification of finite groups of small orders. Here we use the Sylow theorems to extend some of this material further.

Theorem 13.4.3. *Let p, q be distinct primes with $p < q$ and q not congruent to 1 mod p. Then any group of order pq is cyclic. For example any group of order 15 must be cyclic.*

Proof. Suppose that $|G| = pq$ with $p < q$ and q not congruent to 1 mod p. The number of q-Sylow subgroups is of the form $1 + qk$ and divides p. Since q is greater than p this implies that there can be only one and hence there is a normal q-Sylow subgroup H. Since q is a prime, H is cyclic of order q and therefore there is an element g of order q.

The number of p-Sylow subgroups is of the form $1 + pk$ and divides q. Since q is not congruent to 1 mod p this implies that there also can be only one p-Sylow subgroup and hence there is a normal p-Sylow subgroup K. Since p is a prime K is cyclic of order p and therefore there is an element h of order p. Since p, q are distinct primes $H \cap K = \{1\}$. Consider the element $g^{-1}h^{-1}gh$. Since K is normal $g^{-1}hg \in K$. Then $g^{-1}h^{-1}gh = (g^{-1}h^{-1}g)h \in K$. But H is also normal so $h^{-1}gh \in H$. This then implies that $g^{-1}h^{-1}gh = g^{-1}(h^{-1}gh) \in H$ and therefore $g^{-1}h^{-1}gh \in K \cap H$. It follows then that $g^{-1}h^{-1}gh = 1$ or $gh = hg$. Since g, h commute the order of gh is the lcm of the orders of g and h which is pq. Therefore G has an element of order pq. Since $|G| = pq$ this implies that G is cyclic. □

In the above theorem since we assumed that q is not congruent to 1 mod p and hence $p \neq 2$. In the case when $p = 2$ we get another possibility.

Theorem 13.4.4. *Let p be an odd prime and G a finite group of order $2p$. Then either G is cyclic or G is isomorphic to the dihedral group of order $2p$, that is the group of symmetries of a regular p-gon. In this latter case G is generated by two elements g and h which satisfy the relations $g^p = h^2 = (gh)^2 = 1$.*

Proof. As in the proof of Theorem 13.4.3 G must have a normal cyclic subgroup of order p say $\langle g \rangle$. Since $2 \mid |G|$ the group G must have an element of order 2 say h. Consider the order of (gh). By Lagrange's theorem this element can have order $1, 2, p, 2p$. If the order is 1 then $gh = 1$ or $g = h^{-1} = h$. This is impossible since g has order p and h has order 2. If the order of gh is p then from the second Sylow theorem $gh \in \langle g \rangle$. But this implies that $h \in \langle g \rangle$ which is impossible since every nontrivial element of $\langle g \rangle$ has order p. Therefore the order of gh is either 2 or $2p$.

If the order of gh is $2p$ then since G has order $2p$ it must be cyclic.

If the order of gh is 2 then within G we have the relations $g^p = h^2 = (gh)^2 = 1$. Let $H = \langle g, h \rangle$ be the subgroup of G generated by g and h. The relations $g^p = h^2 = (gh)^2 = 1$ imply that H has order $2p$. Since $|G| = 2p$ we get that $H = G$. G is isomorphic to the dihedral group D_p of order $2p$ (see exercises).

In the above description g represents a rotation of $\frac{2\pi}{p}$ of a regular p-gon about its center while h represents any reflection across a line of symmetry of the regular p-gon. □

We have looked at the finite fields $\mathbb{Z}_p$. We give an example of a p-Sylow subgroup of a matrix group over $\mathbb{Z}_p$.

Example 13.4.5. Consider $\mathrm{GL}(n,p)$, the group of $n \times n$ invertible matrices over $\mathbb{Z}_p$. If $\{v_1, \dots, v_n\}$ is a basis for $(\mathbb{Z}_p)^n$ over $\mathbb{Z}_p$ then the size of $\mathrm{GL}(n,p)$ is the number of independent images $\{w_1, \dots, w_n\}$ of $\{v_1, \dots, v_n\}$. For w_1 there are $p^n - 1$ choices, for w_2 there are $p^n - p$ choices and so on. It follows that

$$|\mathrm{GL}(n,p)| = (p^n - 1)(p^n - p) \cdots (p^n - p^{n-1}) = p^{1+2+\cdots+(n-1)} m = p^{\frac{n(n-1)}{2}} m$$

with $(p,m) = 1$. Therefore a p-Sylow subgroup must have size $p^{\frac{n(n-1)}{2}}$.

Let P be the subgroup of upper triangular matrices with 1's on the diagonal. Then P has size $p^{1+2+\cdots+(n-1)} = p^{\frac{n(n-1)}{2}}$ and is therefore a p-Sylow subgroup of $\mathrm{GL}(n,p)$.

The final example is a bit more difficult. We mentioned that a major result on finite groups is the classification of the finite simple groups. This classification showed that any finite simple group is either cyclic of prime order, in one of several classes of groups such as the A_n or one of a number of special examples called sporadic groups. One of the major tools in this classification is the following famous result called the Feit–Thompson theorem that showed that any finite group G of odd order is solvable and, in addition, if G is not cyclic then G is nonsimple.

Theorem 13.4.6 (Feit–Thompson theorem). *Any finite group of odd order is solvable.*

The proof of this theorem, one of the major results in algebra in the twentieth century is way beyond the scope of this book – the proof is actually hundreds of pages in length when one counts the results used. However we look at the smallest nonabelian simple group.

Theorem 13.4.7. *Suppose that G is a simple group of order 60. Then G is isomorphic to A_5. Further A_5 is the smallest nonabelian finite simple group.*

Proof. Suppose that G is a simple group of order $60 = 2^2 \cdot 3 \cdot 5$. The number of 5-Sylow subgroups is of the form $1 + 5k$ and divides 12. Hence there is 1 or 6. Since G is assumed simple and all 5-Sylow subgroups are conjugate there cannot be only one and hence there are 6. Since each of these is cyclic of order 5 they intersect only in the identity. Hence these 6 subgroups cover 24 distinct elements.

The number of 3-Sylow subgroups is of the form $1 + 3k$ and divides 20. Hence there are $1, 4, 10$. We claim that there are 10. There can't be only 1 since G is simple. Suppose there were 4. Let G act on the set of 3-Sylow subgroups by conjugation. Since an action is a permutation this gives a homomorphism f from G into S_4. By the first isomorphism theorem $G/\ker(f) \cong \operatorname{im}(f)$. However since G is simple the kernel must be trivial and this implies that G would imbed into S_4. This is impossible since $|G| = 60 > 24 = |S_4|$. Therefore there are 10 3-Sylow subgroups. Since each of these is cyclic of order 3 they intersect only in the identity and therefore these 10 subgroups cover 20 distinct elements.

Hence together with the elements in the 5-Sylow subgroups we have 44 nontrivial elements.

The number of 2-Sylow subgroups is of the form $1 + 2k$ and divides 15. Hence there are $1, 3, 5, 15$. We claim that there are 5. As before there can't be only 1 since G is simple. There can't be 3 since as for the case of 3-Sylow subgroups this would imply an imbedding of G into S_3 which is impossible since $|S_3| = 6$. Suppose that there were 15 2-Sylow subgroups each of order 4. The intersections would have a maximum of 2 elements and therefore each of these would contribute at least 2 distinct elements. This gives a minimum of 30 distinct elements. However we already have 44 nontrivial elements from the 3-Sylow and 5-Sylow subgroups. Since $|G| = 60$ this is too many. Therefore G must have 5 2-Sylow subgroups.

Now let G act on the set of 2-Sylow subgroups. This then as above implies an imbedding of G into S_5 so we may consider G as a subgroup of S_5. However the only subgroup of S_5 of order 60 is A_5 and therefore $G \cong A_5$.

The proof that A_5 is the smallest nonabelian simple group is actually brute force. We show that any group G of order less than 60 either has prime order or is nonsimple. There are strong tools that we can use. By the Feit–Thompson theorem we must only consider groups of even order. From Theorem 13.4.4 we don't have to consider orders $2p$. The rest can be done by an analysis using Sylow theory. For example we show that any group of order 20 is nonsimple. Since $20 = 2^2 \cdot 5$ the number of 5-Sylow subgroups is $1 + 5k$ and divides 4. Hence there is only one and therefore it must be normal and so G is nonsimple. There is a strong theorem, whose proof is usually done with representation theory, which says that any group whose order is divisible by only two primes is solvable. Therefore for $|G| = 60$ we only have to show that groups of order $30 = 2 \cdot 3 \cdot 5$ and $42 = 2 \cdot 3 \cdot 7$ are nonsimple. This is done in the same manner as the first part of this proof. Suppose $|G| = 30$. The number of 5-Sylow subgroups is of the form $1 + 5k$ and divides 6 hence there are 1 or 6. If G were simple there would have to be 6 covering 24 distinct elements. The number of 3-Sylow subgroups is of the form $1 + 3k$ and divides 10 hence there are $1, 4, 10$. If there were 10 these would cover an additional 20 distinct elements which is impossible since we already have 24 and G has order 30. If there were 4 and G were simple then G would imbed into S_4, again impossible since $|G| = 30 > 24$. Therefore there is only one and hence a normal 3-Sylow subgroup. It follows that G cannot be simple. The case $|G| = 42$ is even simpler. There must be a normal 7-Sylow subgroup. □

13.5 Exercises

1. Prove Lemma 13.1.3.
2. Let the group G act on itself by conjugation. That is $g(g_1) = g^{-1}g_1g$. Prove that this is an action on the set G.
3. Show that the dihedral group D_n of order $2n$ has the presentation $\langle r, f; r^n = f^2 = (rf)^2 = 1\rangle$.
4. Show that each group of order ≤ 59 is solvable.
5. Let P_1 and P_2 be two different p-Sylow subgroups of a finite group G. Show that P_1P_2 is not a subgroup of G.
6. Let G be a finite group. For a prime p the following are equivalent:
   (i) G has exactly one p-Sylow subgroup.
   (ii) The product of two elements of order p has again order p.
7. Let P and Q be two p-Sylow subgroups of the finite group G. If $Z(P)$ is a normal subgroup of Q, then $Z(P) = Z(Q)$.
8. Let p be a prime and $G = \mathrm{SL}(2, p)$. Let $P = \langle a\rangle$, where $a = \left(\begin{smallmatrix} 1 & 1 \\ 0 & 1 \end{smallmatrix}\right)$.
   (i) Determine the normalizer $N_G(P)$ and the number of p-Sylow subgroups of G.
   (ii) Determine the centralizer $C_G(a)$. How many elements of order p does G have? In how many conjugacy classes can they be decomposed?
   (iii) Show that all subgroups of G of order $p(p-1)$ are conjugate.
   (iv) Show that G has no elements of order $p(p-1)$ for $p \ge 5$.

Chapter 14
Free Groups and Group Presentations

14.1 Group Presentations and Combinatorial Group Theory

In discussing the symmetric group on 3 symbols and then the various dihedral groups in Chapters 9, 10 and 11 we came across the concept of a *group presentation*. Roughly for a group G a presentation consists of a set of *generators* X for G, so that $G = \langle X \rangle$, and a set of *relations* between the elements of X from which in principle the whole group table can be constructed. In this chapter we make this concept precise. As we will see, every group G has a presentation but it is mainly in the case where the group is finite or countably infinite that presentations are most useful. Historically the idea of group presentations arose out of the attempt to describe the countably infinite fundamental groups that came out of low dimensional topology. The study of groups using group presentations is called *combinatorial group theory*.

Before looking at group presentations in general we revisit two examples of finite groups and then a class of infinite groups.

Consider the symmetric group on 3 symbols, S_3. We saw that it has the following 6 elements:

$$1 = \begin{pmatrix} 1 & 2 & 3 \\ 1 & 2 & 3 \end{pmatrix}, \quad a = \begin{pmatrix} 1 & 2 & 3 \\ 2 & 3 & 1 \end{pmatrix}, \quad b = \begin{pmatrix} 1 & 2 & 3 \\ 3 & 1 & 2 \end{pmatrix}$$

$$c = \begin{pmatrix} 1 & 2 & 3 \\ 2 & 1 & 3 \end{pmatrix}, \quad d = \begin{pmatrix} 1 & 2 & 3 \\ 3 & 2 & 1 \end{pmatrix}, \quad e = \begin{pmatrix} 1 & 2 & 3 \\ 1 & 3 & 2 \end{pmatrix}.$$

Notice that $a^3 = 1$, $c^2 = 1$ and that $ac = ca^2$. We claim that

$$\langle a, c; a^3 = c^2 = (ac)^2 = 1 \rangle$$

is a presentation for S_3. First it is easy to show that $S_3 = \langle a, c \rangle$. Indeed

$$1 = 1, \quad a = a, \quad b = a^2, \quad c = c, \quad d = ac, \quad e = a^2 c$$

and so clearly a, c generate S_3.

Now from $(ac)^2 = acac = 1$ we get that $ca = a^2 c$. This implies that if we write any sequence (or word in our later language) in a and c we can also rearrange it so that the only powers of a are a and a^2, the only powers of c are c and all a terms precede c terms. For example

$$aca^2cac = aca(acac) = a(ca) = a(a^2c) = (a^3)c = c.$$

Therefore using the three relations form the presentation above each element of S_3 can be written as $a^\alpha c^\beta$ with $\alpha = 0, 1, 2$ and $\beta = 0, 1$. From this the multiplication of any two elements can be determined.

Exactly this type of argument applies to all the dihedral groups D_n. We saw that in general $|D_n| = 2n$. Since these are the symmetry groups of a regular n-gon we always have a rotation r of angle $\frac{2\pi}{n}$ about the center of the n-gon. This element r would have order n. Let f be a reflection about any line of symmetry. Then $f^2 = 1$ and rf is a reflection about the rotated line which is also a line of symmetry. Therefore $(rf)^2 = 1$. Exactly as for S_3 the relation $(rf)^2 = 1$ implies that $fr = r^{-1}f = r^{n-1}f$. This allows us to always place r terms in front of f terms in any word on r and f. Therefore the elements of D_n are always of the form

$$r^\alpha f^\beta, \quad \alpha = 0, 1, 2, \dots, n-1, \quad \beta = 0, 1$$

and further the relations $r^n = f^2 = (rf)^2 = 1$ allow us to rearrange any word in r and f into this form. It follows that $|\langle r, f\rangle| = 2n$ and hence $D_n = \langle r, f\rangle$ together with the relations above. Hence we obtain:

Theorem 14.1.1. *If D_n is the symmetry group of a regular n-gon then a presentation for D_n is given by*

$$D_n = \langle r, f; r^n = f^2 = (rf)^2 = 1\rangle.$$

We now give one class of infinite examples. If G is an infinite cyclic group, so that $G \cong \mathbb{Z}$, then $G = \langle g; \rangle$ is a presentation for G. That is G has a single generator with no relations.

A direct product of n copies of $\mathbb{Z}$ is called a *free abelian group* of rank n. We will denote this by $\mathbb{Z}^n$. A presentation for $\mathbb{Z}^n$ is then given by

$$\mathbb{Z}^n = \langle x_1, x_2, \dots, x_n; x_i x_j = x_j x_i \text{ for all } i, j = 1, \dots, n\rangle.$$

14.2 Free Groups

Crucial to the concept of a group presentation is the idea of a *free group*.

Definition 14.2.1. A group F is *free on a subset* X if every map $f : X \to G$ with G a group can be extended to a unique homomorphism $f : F \to G$. X is called a *free basis* for F. In general a group F is a *free group* if it is free on some subset X. If X is a free basis for a free group F we write $F = F(X)$.

We first show that given any set X there does exist a free group with free basis X. Let $X = \{x_i\}_{i\in I}$ be a set (possibly empty). We will construct a group $F(X)$ which is free with free basis X. First let X^{-1} be a set disjoint from X but bijective to X. If $x_i \in X$ then the corresponding element of X^{-1} under the bijection we denote x_i^{-1} and say that x_i and x_i^{-1} are *associated.* The set X^{-1} is called the *set of formal inverses* from X and we call $X \cup X^{-1}$ the *alphabet*. Elements of the alphabet are called *letters*, hence a letter has the form $x_i^{\epsilon_1}$ where $\epsilon_i = \pm 1$. A *word* in X is a finite sequence of letters from the alphabet. That is a word has the form

$$w = x_{i_1}^{\epsilon_{i_1}} x_{i_2}^{\epsilon_{i_2}} \dots x_{i_n}^{\epsilon_{i_n}}$$

where $x_{i_j} \in X$ and $\epsilon_{i_j} = \pm 1$. If $n = 0$ we call it the *empty word* which we will denote e. The integer n is called the *length* of the word. Words of the form $x_i x_i^{-1}$ or $x_i^{-1} x_i$ are called *trivial words*. We let $W(X)$ be the set of all words on X.

If $w_1, w_2 \in W(X)$ we say that w_1 is *equivalent* to w_2, denoted $w_1 \sim w_2$, if w_1 can be converted to w_2 by a finite string of insertions and deletions of trivial words. For example if $w_1 = x_3 x_4 x_4^{-1} x_2 x_2$ and $w_2 = x_3 x_2 x_2$ then $w_1 \sim w_2$. It is straightforward to verify that this is an equivalence relation on $W(X)$ (see exercises). Let $F(X)$ denote the set of equivalence classes in $W(X)$ under this relation, hence $F(X)$ is a set of equivalence classes of words from X.

A word $w \in W(X)$ is said to be *freely reduced* or *reduced* if it has no trivial subwords (a subword is a connected sequence within a word). Hence in the example above $w_2 = x_3 x_2 x_2$ is reduced but $w_1 = x_3 x_4 x_4^{-1} x_2 x_2$ is not reduced. In each equivalence class in $F(X)$ there is a unique element of minimal length. Further this element must be reduced or else it would be equivalent to something of smaller length. Two reduced words in $W(X)$ are either equal or not in the same equivalence class in $F(X)$. Hence $F(X)$ can also be considered as the set of all reduced words from $W(X)$.

Given a word $w = x_{i_1}^{\epsilon_{i_1}} x_{i_2}^{\epsilon_{i_2}} \dots x_{i_n}^{\epsilon_{i_n}}$ we can find the unique reduced word $\overline{w}$ equivalent to w via the following *free reduction process*. Beginning from the left side of w we cancel each occurrence of a trivial subword. After all these possible cancellations we have a word w'. Now we repeat the process again starting from the left side. Since w has finite length eventually the resulting word will either be empty or reduced. The final reduced $\overline{w}$ is the *free reduction* of w.

Now we build a multiplication on $F(X)$. If

$$w_1 = x_{i_1}^{\epsilon_{i_1}} x_{i_2}^{\epsilon_{i_2}} \dots x_{i_n}^{\epsilon_{i_n}}, \quad w_2 = x_{j_1}^{\epsilon_{j_1}} x_{j_2}^{\epsilon_{j_2}} \dots x_{j_m}^{\epsilon_{j_m}}$$

are two words in $W(X)$ then their *concatenation* $w_1 \star w_2$ is simply placing w_2 after w_1,

$$w_1 \star w_2 = x_{i_1}^{\epsilon_{i_1}} x_{i_2}^{\epsilon_{i_2}} \dots x_{i_n}^{\epsilon_{i_n}} x_{j_1}^{\epsilon_{j_1}} x_{j_2}^{\epsilon_{j_2}} \dots x_{j_m}^{\epsilon_{j_m}}.$$

If $w_1, w_2 \in F(x)$ then we define their product as

$$w_1 w_2 = \text{equivalence class of } w_1 \star w_2.$$

That is we concatenate w_1 and w_2, and the product is the equivalence class of the resulting word. It is easy to show that if $w_1 \sim w_1'$ and $w_2 \sim w_2'$ then $w_1 \star w_2 \sim$

$w_1' \star w_2'$ so that the above multiplication is well-defined. Equivalently we can think of this product in the following way. If w_1, w_2 are reduced words then to find $w_1 w_2$ first concatenate and then freely reduce. Notice that if $x_{i_n}^{\epsilon_{i_n}} x_{j_1}^{\epsilon_{j_1}}$ is a trivial word then it is cancelled when the concatenation is formed. We say then that there is *cancellation* in forming the product $w_1 w_2$. Otherwise the product is formed without cancellation.

Theorem 14.2.2. *Let X be a nonempty set and let $F(X)$ be as above. Then $F(X)$ is a free group with free basis X. Further if $X = \emptyset$ then $F(X) = \{1\}$, if $|X| = 1$ then $F(X) \cong \mathbb{Z}$ and if $|X| \geq 2$ then $F(X)$ is nonabelian.*

Proof. We first show that $F(X)$ is a group and then show that it satisfies the universal mapping property on X. We consider $F(X)$ as the set of reduced words in $W(X)$ with the multiplication defined above. Clearly the empty word acts as the identity element 1. If $w = x_{i_1}^{\epsilon_{i_1}} x_{i_2}^{\epsilon_{i_2}} \dots x_{i_n}^{\epsilon_{i_n}}$ and $w_1 = x_{i_n}^{-\epsilon_{i_n}} x_{i_{n-1}}^{-\epsilon_{i_{n-1}}} \dots x_{i_1}^{-\epsilon_{i_1}}$ then both $w \star w_1$ and $w_1 \star w$ freely reduce to the empty word and so w_1 is the inverse of w. Therefore each element of $F(X)$ has an inverse. Therefore to show that $F(X)$ forms a group we must show that the multiplication is associative. Let

$$w_1 = x_{i_1}^{\epsilon_{i_1}} x_{i_2}^{\epsilon_{i_2}} \dots x_{i_n}^{\epsilon_{i_n}}, \quad w_2 = x_{j_1}^{\epsilon_{j_1}} x_{j_2}^{\epsilon_{j_2}} \dots x_{j_m}^{\epsilon_{j_m}}, \quad w_3 = x_{k_1}^{\epsilon_{k_1}} x_{k_2}^{\epsilon_{k_2}} \dots x_{k_p}^{\epsilon_{k_p}}$$

be three freely reduced words in $F(X)$. We must show that

$$(w_1 w_2) w_3 = w_1 (w_2 w_3).$$

To prove this we use induction on m, the length of w_2. If $m = 0$ then w_2 is the empty word and hence the identity and it is certainly true. Now suppose that $m = 1$ so that $w_2 = x_{j_1}^{\epsilon_{j_1}}$. We must consider exactly four cases.

Case (1): There is no cancellation in forming either $w_1 w_2$ or $w_2 w_3$. That is $x_{j_1}^{\epsilon_{j_1}} \neq x_{i_n}^{-\epsilon_{i_n}}$ and $x_{j_1}^{\epsilon_{j_1}} \neq x_{k_1}^{-\epsilon_{k_1}}$. Then the product $w_1 w_2$ is just the concatenation of the words and so is $(w_1 w_2) w_3$. The same is true for $w_1(w_2 w_3)$. Therefore in this case $w_1(w_2 w_3) = (w_1 w_2) w_3$.

Case (2): There is cancellation in forming $w_1 w_2$ but not in forming $w_2 w_3$. Then if we concatenate all three words the only cancellation occurs between w_1 and w_2 in either $w_1(w_2 w_3)$ or in $(w_1 w_2) w_3$ and hence they are equal. Therefore in this case $w_1(w_2 w_3) = (w_1 w_2) w_3$.

Case (3): There is cancellation in forming $w_2 w_3$ but not in forming $w_1 w_2$. This is entirely analogous to Case (2) so therefore in this case $w_1(w_2 w_3) = (w_1 w_2) w_3$.

Case (4): There is cancellation in forming $w_1 w_2$ and also in forming $w_2 w_3$. Then $x_{j_1}^{\epsilon_{j_1}} = x_{i_n}^{-\epsilon_{i_n}}$ and $x_{j_1}^{\epsilon_{j_1}} = x_{k_1}^{-\epsilon_{k_1}}$. Here

$$(w_1 w_2) w_3 = x_{i_1}^{\epsilon_{i_1}} \dots x_{i_{n-1}}^{\epsilon_{i_{n-1}}} x_{k_1}^{\epsilon_{k_1}} x_{k_2}^{\epsilon_{k_2}} \dots x_{k_p}^{\epsilon_{k_p}}.$$

On the other hand

$$w_1(w_2w_3) = x_{i_1}^{\epsilon_{i_1}} \dots x_{i_n}^{\epsilon_{in}} x_{k_2}^{\epsilon_{k_2}} \dots x_{k_p}^{\epsilon_{k_p}}.$$

However these are equal since $x_{i_n}^{\epsilon_{in}} = x_{k_1}^{\epsilon_{k_1}}$. Therefore in this final case $w_1(w_2w_3) = (w_1w_2)w_3$. It follows inductively from these four cases that the associative law holds in $F(X)$ and therefore $F(X)$ forms a group.

Now suppose that $f : X \to G$ is a map from X into a group G. By the construction of $F(X)$ as a set of reduced words this can be extended to a unique homomorphism. If $w \in F$ with $w = x_{i_1}^{\epsilon_{i_1}} \dots x_{i_n}^{\epsilon_{in}}$ then define $f(w) = f(x_{i_1})^{\epsilon_{i_1}} \cdots f(x_{i_n})^{\epsilon_{in}}$. Since multiplication in $F(X)$ is concatenation this defines a homomorphism and again form the construction of $F(X)$ its the only one extending f. This is analogous to constructing a linear transformation from one vector space to another by specifying the images of a basis. Therefore $F(X)$ satisfies the universal mapping property of Definition 14.2.1 and hence $F(X)$ is a free group with free basis X.

The final parts of Theorem 14.2.2 are straightforward. If X is empty the only reduced word is the empty word and hence the group is just the identity. If X has a single letter then $F(X)$ has a single generator and is therefore cyclic. It is easy to see that it must be torsion-free and therefore $F(X)$ is infinite cyclic, that is $F(x) \cong \mathbb{Z}$. Finally if $|X| \geq 2$ let $x_1, x_2 \in X$. Then $x_1x_2 \neq x_2x_1$ and both are reduced. Therefore $F(X)$ is nonabelian. □

The proof of Theorem 14.2.2 provides another way to look at free groups.

Theorem 14.2.3. *F is a free group if and only if there is a generating set X such that every element of F has a unique representation as a freely reduced word on X.*

The structure of a free group is entirely dependent on the cardinality of a free basis. In particular the cardinality of a free basis $|X|$ for a free group F is unique and is called the *rank of F*. If $|X| < \infty$, F is of *finite rank*. If F has rank n and $X = \{x_1, \dots, x_n\}$ we say that F is free on $\{x_1, \dots, x_n\}$. We denote this by $F(x_1, x_2, \dots, x_n)$.

Theorem 14.2.4. *If X and Y are sets with the same cardinality, that is $|X| = |Y|$, then $F(X) \cong F(Y)$, the resulting free groups are isomorphic. Further if $F(X) \cong F(Y)$ then $|X| = |Y|$.*

Proof. Suppose that $f : X \to Y$ is a bijection from X onto Y. Now $Y \subset F(Y)$ so there is a unique homomorphism $\phi : F(X) \to F(Y)$ extending f. Since f is a bijection it has an inverse $f^{-1} : Y \to X$ and since $F(Y)$ is free there is a unique homomorphism ϕ_1 from $F(Y)$ to $F(X)$ extending f^{-1}. Then $\phi\phi_1$ is the identity map on $F(Y)$ and $\phi_1\phi$ is the identity map on $F(X)$. Therefore ϕ, ϕ_1 are isomorphisms with $\phi = \phi_1^{-1}$.

Conversely suppose that $F(X) \cong F(Y)$. In $F(X)$ let $N(X)$ be the subgroup generated by all squares in $F(X)$ that is

$$N(X) = \langle\{g^2 : g \in F(X)\}\rangle.$$

Then $N(X)$ is a normal subgroup and the factor group $F(X)/N(X)$ is abelian where every nontrivial element has order 2 (see exercises). Therefore $F(X)/N(X)$ can be considered as a vector space over $\mathbb{Z}_2$, the finite field of order 2, with X as a vector space basis. Hence $|X|$ is the dimension of this vector space. Let $N(Y)$ be the corresponding subgroup of $F(Y)$. Since $F(X) \cong F(Y)$ we would have $F(X)/N(X) \cong F(Y)/N(Y)$ and therefore $|Y|$ is the dimension of the vector space $F(Y)/N(Y)$. Therefore $|X| = |Y|$ from the uniqueness of dimension of vector spaces. □

Expressing elements of $F(X)$ as a reduced word gives a *normal form* for elements in a free group F. As we will see in Section 14.5 this solves what is termed the *word problem* for free groups. Another important concept is the following: a freely reduced word $W = x_{v_1}^{e_1} x_{v_2}^{e_2} \ldots x_{v_n}^{e_n}$ is *cyclically reduced* if $v_1 \neq v_n$ or if $v_1 = v_n$ then $e_1 \neq -e_n$. Clearly then every element of a free group is conjugate to an element given by a cyclically reduced word. This provides a method to determine conjugacy in free groups.

Theorem 14.2.5. *In a free group F two elements g_1, g_2 are conjugate if and only if a cyclically reduced word for g_1 is a cyclic permutation of a cyclically reduced word for g_2.*

The theory of free groups has a large and extensive literature. We close this section by stating several important properties. Proofs for these results can be found in [24], [23] or [15].

Theorem 14.2.6. *A free group is torsion-free.*

From Theorem 14.2.4 we can deduce:

Theorem 14.2.7. *An abelian subgroup of a free group must be cyclic.*

Finally a celebrated theorem of Nielsen and Schreier states that a subgroup of a free group must be free.

Theorem 14.2.8 (Nielsen–Schreier). *A subgroup of a free group is itself a free group.*

Combinatorially F is free on X if X is a set of generators for F and there are no nontrivial relations. In particular:

There are several different proofs of this result (see [24]) with the most straightforward being topological in nature. We give an outline of a simple topological proof in Section 14.4.

Nielsen, using a technique now called Nielsen transformations in his honor, first proved this theorem about 1920 for finitely generated subgroups. Schreier shortly after found a combinatorial method to extend this to arbitrary subgroups. A complete version of the original combinatorial proof appears in [24] and in the notes by Johnson [19].

Schreier's combinatorial proof also allows for a description of the free basis for the subgroup. In particular, let F be free on X, and $H \subset F$ a subgroup. Let $T = \{t_\alpha\}$ be a complete set of right coset representatives for F mod H with the property that if $t_\alpha = x_{v_1}^{e_1} x_{v_2}^{e_2} \dots x_{v_n}^{e_n} \in T$, with $\epsilon_i = \pm 1$ then all the initial segments $1, x_{v_1}^{e_1}, x_{v_1}^{e_1} x_{v_2}^{e_2}$, etc. are also in T. Such a system of coset representatives can always be found and is called a *Schreier system* or *Schreier transversal* for H. If $g \in F$ let $\overline{g}$ represent its coset representative in T and further define for $g \in F$ and $t \in T$, $S_{tg} = tg(\overline{tg})^{-1}$. Notice that $S_{tg} \in H$ for all t, g. We then have:

Theorem 14.2.9 (explicit form of Nielsen–Schreier). *Let F be free on X and H a subgroup of F. If T is a Schreier transversal for F* mod *H then H is free on the set* $\{S_{tx} : t \in T, x \in X, S_{tx} \neq 1\}$.

Example 14.2.10. Let F be free on $\{a, b\}$ and $H = F(X^2)$ the normal subgroup of F generated by all squares in F.

Then $F/F(X^2) = \langle a, b; a^2 = b^2 = (ab)^2 = 1\rangle = \mathbb{Z}_2 \times \mathbb{Z}_2$. It follows that a Schreier system for F mod H is $\{1, a, b, ab\}$ with $\overline{a} = a, \overline{b} = b$ and $\overline{ba} = ab$. From this it can be shown that H is free on the generating set

$$x_1 = a^2, \quad x_2 = bab^{-1}a^{-1}, \quad x_3 = b^2, \quad x_4 = abab^{-1}, \quad x_5 = ab^2a^{-1}.$$

The theorem also allows for a computation of the rank of H given the rank of F and the index. Specifically:

Corollary 14.2.11. *Suppose F is free of rank n and $|F : H| = k$. Then H is free of rank $nk - k + 1$.*

From the example we see that F is free of rank 2, H has index 4 so H is free of rank $2 \cdot 4 - 4 + 1 = 5$.

14.3 Group Presentations

The significance of free groups stems from the following result which is easily deduced from the definition and will lead us directly to a formal definition of a group presentation. Let G be any group and F the free group on the elements of G considered as a set. The identity map $f : G \to G$ can be extended to a homomorphism of F onto G, therefore:

Theorem 14.3.1. *Every group G is a homomorphic image of a free group. That is let G be any group. Then $G = F/N$ where F is a free group.*

In the above theorem instead of taking all the elements of G we can consider just a set X of generators for G. Then G is a factor group of $F(X)$, $G \cong F(X)/R$. The normal subgroup N is the kernel of the homomorphism from $F(X)$ onto G. We use Theorem 14.3.1 to formally define a group presentation.

If H is a subgroup of a group G then the *normal closure* of H denoted by $N(H)$ is the smallest normal subgroup of G containing H. This can be described alternatively in the following manner. The normal closure of H is the subgroup of G generated by all conjugates of elements of H.

Now suppose that G is a group with X a set of generators for G. We also call X a *generating system* for G. Now let $G = F(X)/N$ as in Theorem 14.3.1 and the comments after it. N is the kernel of the homomorphism $f : F(X) \to G$. It follows that if r is a free group word with $r \in N$ then $r = 1$ in G (under the homomorphism). We then call r a *relator* in G and the equation $r = 1$ a *relation* in G. Suppose that R is a subset of N such that $N = N(R)$, then R is called a set of *defining relators* for G. The equations $r = 1$, $r \in R$, are a set of *defining relations* for G. It follows that any relator in G is a product of conjugates of elements of R. Equivalently $r \in F(X)$ is a relator in G if and only if r can be reduced to the empty word by insertions and deletions of elements of R and trivial words.

Definition 14.3.2. Let G be a group. Then a *group presentation* for G consists of a set of generators X for G and a set R of defining relators. In this case we write $G = \langle X; R\rangle$. We could also write the presentation in terms of defining relations as $G = \langle X; r = 1, r \in R\rangle$.

From Theorem 14.3.1 it follows immediately that every group has a presentation. However in general there are many presentations for the same group. If $R \subset R_1$ then R_1 is also a set of defining relators.

Lemma 14.3.3. *Let G be a group. Then G has a presentation.*

If $G = \langle X; R\rangle$ and X is finite then G is said to be *finitely generated*. If R is finite G is *finitely related*. If both X and R are finite G is *finitely presented*.

Using group presentations we get another characterization of free groups.

Theorem 14.3.4. *F is a free group if and only if F has a presentation of the form $F = \langle X; \rangle$.*

Mimicking the construction of a free group from a set X we can show that to each presentation corresponds a group. Suppose that we are given a supposed presentation $\langle X; R\rangle$ where R is given as a set of words in X. Consider the free group $F(X)$ on X. Define two words w_1, w_2 on X to be equivalent if w_1 can be transformed into w_2 using insertions and deletions of elements of R and trivial words. As in the free group case this is an equivalence relation. Let G be the set of equivalence classes. If we define multiplication as before as concatenation followed by the appropriate equivalence class then G is a group. Further each $r \in R$ must equal the identity in G so that $G = \langle X; R\rangle$. Notice that here there may be no unique reduced word for an element of G.

Theorem 14.3.5. *Given $\langle X; R\rangle$ where X is a set and R is a set of words on X. Then there exists a group G with presentation $\langle X; R\rangle$.*

We now give some examples of group presentations.

Example 14.3.6. A free group of rank n has a presentation

$$F_n = \langle x_1, \ldots, x_n; \rangle.$$

Example 14.3.7. A free abelian group of rank n has a presentation

$$\mathbb{Z}^n = \langle x_1, \ldots, x_n; x_i x_j x_i^{-1} x_j^{-1}, i = 1, \ldots, n, j = 1, \ldots, n\rangle.$$

Example 14.3.8. A cyclic group of order n has a presentation

$$\mathbb{Z}_n = \langle x; x^n = 1\rangle.$$

Example 14.3.9. The dihedral groups of order $2n$ representing the symmetry group of a regular n-gon has a presentation

$$\langle r, f; r^n = 1, f^2 = 1, (rf)^2 = 1\rangle.$$

We look at this example in Section 14.3.1.

14.3.1 The Modular Group

In this section we give a more complicated example and then a nice application to number theory.

If R is any commutative ring with an identity, then the set of invertible $n \times n$ matrices with entries from R forms a group under matrix multiplication called the *n-dimensional general linear group over R* (see [25]). This group is denoted by $\mathrm{GL}_n(R)$. Since $\det(A)\det(B) = \det(AB)$ for square matrices A, B it follows that the subset of $\mathrm{GL}_n(R)$ consisting of those matrices of determinant 1 forms a subgroup. This subgroup is called the *special linear group over* R and is denoted by $\mathrm{SL}_n(R)$. In this section we concentrate on $\mathrm{SL}_2(\mathbb{Z})$, or more specifically a quotient of it, $\mathrm{PSL}_2(\mathbb{Z})$ and find presentations for them.

The group $\mathrm{SL}_2(\mathbb{Z})$ then consists of 2×2 integral matrices of determinant one:

$$\mathrm{SL}_2(\mathbb{Z}) = \left\{ \begin{pmatrix} a & b \\ c & d \end{pmatrix} : a, b, c, d \in \mathbb{Z}, ad - bc = 1 \right\}.$$

$\mathrm{SL}_2(\mathbb{Z})$ is called the *homogeneous modular group* and an element of $\mathrm{SL}_2(\mathbb{Z})$ is called a *unimodular matrix*.

If G is any group, recall that its *center* $Z(G)$ consists of those elements of G which commute with all elements of G:

$$Z(G) = \{g \in G : gh = hg, \forall h \in G\}.$$

$Z(G)$ is a normal subgroup of G and hence we can form the factor group $G/Z(G)$. For $G = \mathrm{SL}_2(\mathbb{Z})$ the only unimodular matrices that commute with all others are $\pm I = \pm\left(\begin{smallmatrix} 1 & 0 \\ 0 & 1 \end{smallmatrix}\right)$. Therefore $Z(\mathrm{SL}_2(\mathbb{Z})) = \{I, -I\}$. The quotient

$$\mathrm{SL}_2(\mathbb{Z})/Z(\mathrm{SL}_2(\mathbb{Z})) = \mathrm{SL}_2(\mathbb{Z})/\{I, -I\}$$

is denoted $\mathrm{PSL}_2(\mathbb{Z})$ and is called the *projective special linear group* or *inhomogeneous modular group*. More commonly $\mathrm{PSL}_2(\mathbb{Z})$ is just called the *Modular Group* and denoted by M.

M arises in many different areas of mathematics including number theory, complex analysis and Riemann surface theory and the theory of automorphic forms and functions. M is perhaps the most widely studied single finitely presented group. Complete discussions of M and its structure can be found in the books *Integral Matrices* by M. Newman [38] and *Algebraic Theory of the Bianchi Groups* by B. Fine [34].

Since $M = \mathrm{PSL}_2(\mathbb{Z}) = \mathrm{SL}_2(\mathbb{Z})/\{I, -I\}$ it follows that each element of M can be considered as $\pm A$ where A is a unimodular matrix. A *projective unimodular matrix* is then

$$\pm \begin{pmatrix} a & b \\ c & d \end{pmatrix}, \quad a, b, c, d \in \mathbb{Z},\ ad - bc = 1.$$

The elements of M can also be considered as linear fractional transformations over the complex numbers

$$z' = \frac{az + b}{cz + d}, \quad a, b, c, d \in \mathbb{Z},\ ad - bc = 1, \text{ where } z \in \mathbb{C}.$$

Thought of in this way, M forms a *Fuchsian group* which is a discrete group of isometries of the non-Euclidean hyperbolic plane. The book by Katok [20] gives a solid and clear introduction to such groups. This material can also be found in condensed form in [35].

We now determine presentations for both $\mathrm{SL}_2(\mathbb{Z})$ and $M = \mathrm{PSL}_2(\mathbb{Z})$.

Theorem 14.3.10. *The group* $\mathrm{SL}_2(\mathbb{Z})$ *is generated by the elements*

$$X = \begin{pmatrix} 0 & -1 \\ 1 & 0 \end{pmatrix} \quad \text{and} \quad Y = \begin{pmatrix} 0 & 1 \\ -1 & -1 \end{pmatrix}.$$

Further a complete set of defining relations for the group in terms of these generators is given by

$$X^4 = Y^3 = YX^2Y^{-1}X^{-2} = I.$$

It follows that $\mathrm{SL}_2(\mathbb{Z})$ *has the presentation*

$$\langle X, Y; X^4 = Y^3 = YX^2Y^{-1}X^{-2} = I \rangle.$$

Proof. We first show that $\mathrm{SL}_2(\mathbb{Z})$ is generated by X and Y, that is every matrix A in the group can be written as a product of powers of X and Y.

Let

$$U = \begin{pmatrix} 1 & 1 \\ 0 & 1 \end{pmatrix}.$$

Then a direct multiplication shows that $U = XY$ and we show that $\mathrm{SL}_2(\mathbb{Z})$ is generated by X and U which implies that it is also generated by X and Y. Further

$$U^n = \begin{pmatrix} 1 & n \\ 0 & 1 \end{pmatrix}$$

so that U has infinite order.

Let $A = \left(\begin{smallmatrix} a & b \\ c & d \end{smallmatrix}\right) \in \mathrm{SL}_2(\mathbb{Z})$. Then we have

$$XA = \begin{pmatrix} -c & -d \\ a & b \end{pmatrix} \quad \text{and} \quad U^k A = \begin{pmatrix} a + kc & b + kd \\ c & d \end{pmatrix}$$

for any $k \in \mathbb{Z}$. We may assume that $|c| \le |a|$ otherwise start with XA rather than A. If $c = 0$ then $A = \pm U^q$ for some q. If $A = U^q$ then certainly A is in the group generated by X and U. If $A = -U^q$ then $A = X^2U^q$ since $X^2 = -I$. It follows that here also A is in the group generated by X and U.

Now suppose $c \neq 0$. Apply the Euclidean algorithm to a and c in the following modified way:

$$\begin{aligned} a &= q_0 c + r_1 \\ -c &= q_1 r_1 + r_2 \\ r_1 &= q_2 r_2 + r_3 \\ &\;\;\vdots \\ (-1)^n r_{n-1} &= q_n r_n + 0 \end{aligned}$$

where $r_n = \pm 1$ since $(a, c) = 1$. Then

$$XU^{-q_n} \cdots XU^{-q_0}A = \pm U^{q_{n+1}} \quad \text{with } q_{n+1} \in \mathbb{Z}.$$

Therefore

$$A = X^m U^{q_0} X U^{q_1} \cdots X U^{q_n} X U^{q_{n+1}}$$

with $m = 0, 1, 2, 3$; $q_0, q_1, \ldots, q_{n+1} \in \mathbb{Z}$ and $q_0, \ldots, q_n \neq 0$. Therefore X and U and hence X and Y generate $\mathrm{SL}_2(\mathbb{Z})$.

We must now show that

$$X^4 = Y^3 = YX^2Y^{-1}X^{-2} = I$$

are a complete set of defining relations for $\mathrm{SL}_2(\mathbb{Z})$ or that every relation on these generators is derivable from these. It is straightforward to see that X and Y do satisfy these relations. Assume then that we have a relation

$$S = X^{\epsilon_1} Y^{\alpha_1} X^{\epsilon_2} Y^{\alpha_2} \cdots Y^{\alpha_n} X^{\epsilon_{n+1}} = I$$

with all $\epsilon_i, \alpha_j \in \mathbb{Z}$. Using the set of relations

$$X^4 = Y^3 = YX^2Y^{-1}X^{-2} = I$$

we may transform S so that

$$S = X^{\epsilon_1} Y^{\alpha_1} \cdots Y^{\alpha_m} X^{\epsilon_{m+1}}$$

with $\epsilon_1, \epsilon_{m+1} = 0, 1, 2$ or 3 and $\alpha_i = 1$ or 2 for $i = 1, \ldots, m$ and $m \geq 0$. Multiplying by a suitable power of X we obtain

$$Y^{\alpha_1} X \cdots Y^{\alpha_m} X = X^{\alpha} = S_1$$

with $m \geq 0$ and $\alpha = 0, 1, 2$ or 3. Assume that $m \geq 1$ and let

$$S_1 = \begin{pmatrix} a & -b \\ -c & d \end{pmatrix}.$$

We show by induction that

$$a, b, c, d \geq 0, \quad b + c > 0$$

or

$$a, b, c, d \leq 0, \quad b + c < 0.$$

This claim for the entries of S_1 is true for

$$YX = \begin{pmatrix} 1 & 0 \\ -1 & 1 \end{pmatrix} \quad \text{and} \quad Y^2X = \begin{pmatrix} -1 & 1 \\ 0 & -1 \end{pmatrix}.$$

Suppose it is correct for $S_2 = \left(\begin{smallmatrix} a_1 & -b_1 \\ -c_1 & d_1 \end{smallmatrix}\right)$. Then

$$YXS_2 = \begin{pmatrix} a_1 & -b_1 \\ -(a_1+c_1) & b_1+d_1 \end{pmatrix} \quad \text{and}$$

$$Y^2XS_2 = \begin{pmatrix} -a_1-c_1 & b_1+d_1 \\ c_1 & d_1 \end{pmatrix}.$$

Therefore the claim is correct for all S_1 with $m \geq 1$. This gives a contradiction, for the entries of X^α with $\alpha = 0, 1, 2$ or 3 do not satisfy the claim. Hence $m = 0$ and S can be reduced to a trivial relation by the given set of relations. Therefore they are a complete set of defining relations and the theorem is proved. □

Corollary 14.3.11. *The modular group* $M = \mathrm{PSL}_2(\mathbb{Z})$ *has the presentation*

$$M = \langle x, y; x^2 = y^3 = 1\rangle.$$

Further x, y *can be taken as the linear fractional transformations*

$$x : z' = -\frac{1}{z} \quad \textit{and} \quad y : z' = -\frac{1}{z+1}.$$

Proof. The center of $\mathrm{SL}_2(\mathbb{Z})$ is $\pm I$. Since $X^2 = -I$ setting $X^2 = I$ in the presentation for $\mathrm{SL}_2(\mathbb{Z})$ gives the presentation for M. Writing the projective matrices as linear fractional transformations gives the second statement. □

This corollary says that M is the *free product* of a cyclic group of order 2 and a cyclic group of order 3 a concept we will introduce in Section 14.7.

We note that there is an elementary alternative proof to Corollary 14.3.11 as far as showing that $X^2 = Y^3 = 1$ are a complete set of defining relations. As linear fractional transformations we have

$$X(z) = -\frac{1}{z}, \quad Y(z) = -\frac{1}{z+1}, \quad Y^2(z) = -\frac{z+1}{z}.$$

Now let

$$\mathbb{R}^+ = \{x \in \mathbb{R} : x > 0\} \quad \text{and} \quad \mathbb{R}^- = \{x \in \mathbb{R} : x < 0\}.$$

Then

$$X(\mathbb{R}^-) \subset \mathbb{R}^+ \quad \text{and} \quad Y^\alpha(\mathbb{R}^+) \subset \mathbb{R}^-, \quad \alpha = 1, 2.$$

Let $S \in \Gamma$. Using the relations $X^2 = Y^3 = 1$ and a suitable conjugation we may assume that either $S = 1$ is a consequence of these relations or that

$$S = Y^{\alpha_1} X Y^{\alpha_2} \cdots X Y^{\alpha_n}$$

with $1 \leq \alpha_i \leq 2$ and $\alpha_1 = \alpha_n$.

In this second case if $x \in \mathbb{R}^+$ then $S(x) \in \mathbb{R}^-$ and hence $S \neq 1$.

This type of *ping-pong argument* can be used in many examples (see [23], [15] and [19]). As another example consider the unimodular matrices

$$A = \begin{pmatrix} 0 & 1 \\ -1 & 2 \end{pmatrix}, \quad B = \begin{pmatrix} 0 & -1 \\ 1 & 2 \end{pmatrix}.$$

Let $\overline{A}, \overline{B}$ denote the corresponding linear fractional transformations in the modular group M. We have

$$A^n = \begin{pmatrix} -n+1 & n \\ -n & n+1 \end{pmatrix}, \quad B^n = \begin{pmatrix} -n+1 & -n \\ n & n+1 \end{pmatrix} \quad \text{for } n \in \mathbb{Z}.$$

In particular $\overline{A}$ and $\overline{B}$ have infinite order. Now

$$\overline{A}^n(\mathbb{R}^-) \subset \mathbb{R}^+ \quad \text{and} \quad \overline{B}^n(\mathbb{R}^+) \subset \mathbb{R}^-$$

for all $n \neq 0$. The ping-pong argument used for any element of the type

$$S = \overline{A}^{n_1}\overline{B}^{m_1} \cdots \overline{B}^{m_k}\overline{A}^{n_{k+1}}$$

with all $n_i, m_i \neq 0$ and $n_1 + n_{k+1} \neq 0$ shows that $S(x) \in \mathbb{R}^+$ if $x \in \mathbb{R}^-$. It follows that there are no nontrivial relations on $\overline{A}$ and $\overline{B}$ and therefore the subgroup of M generated by $\overline{A}, \overline{B}$ must be a free group of rank 2.

To close this section we give a nice number theoretical application of the modular group. First we need the following corollary to Corollary 14.3.11.

Corollary 14.3.12. *Let $M = \langle X, Y; X^2 = Y^3 = 1\rangle$ be the modular group. If A is an element of order 2 then A is conjugate to X. If B is an element of order 3 then B is conjugate to either Y or Y^2.*

Definition 14.3.13. Let a, n be relatively prime integers with $a \neq 0, n \geq 1$. Then a is a *quadratic residue* mod n if there exists an $x \in \mathbb{Z}$ with $x^2 = a$ mod n, that is $a = x^2 + kn$ for some $k \in \mathbb{Z}$.

The following is called Fermat's two-square theorem.

Theorem 14.3.14 (Fermat's two-square theorem). *Let $n > 0$ be a natural number. Then $n = a^2 + b^2$ with $(a, b) = 1$ if and only if -1 is a quadratic residue modulo n.*

Proof. Suppose -1 is a quadratic residue mod n. Then there exists an x with $x^2 \equiv -1$ mod n or $x^2 = -1 + mn$. This implies that $-x^2 - mn = 1$ so that there must exist a projective unimodular matrix

$$A = \pm\begin{pmatrix} x & n \\ m & -x \end{pmatrix}.$$

It is straightforward that $A^2 = 1$ so by Corollary 14.3.12 A is conjugate within M to X. Now consider conjugates of X within M. Let $T = \left(\begin{smallmatrix} a & b \\ c & d \end{smallmatrix}\right)$. Then

$$T^{-1} = \begin{pmatrix} d & -b \\ -c & a \end{pmatrix}$$

and

$$TXT^{-1} = \begin{pmatrix} a & b \\ c & d \end{pmatrix}\begin{pmatrix} 0 & 1 \\ -1 & 0 \end{pmatrix}\begin{pmatrix} d & -b \\ -c & a \end{pmatrix} = \pm\begin{pmatrix} -(bd+ac) & a^2+b^2 \\ -(c^2+d^2) & bd+ac \end{pmatrix}. \quad (*)$$

Therefore any conjugate of X must have form $(*)$ and therefore A must have form $(*)$. Therefore $n = a^2+b^2$. Further $(a,b) = 1$ since in finding form $(*)$ we had $ad-bc = 1$.

Conversely suppose $n = a^2 + b^2$ with $(a,b) = 1$. Then there exist $c, d \in \mathbb{Z}$ with $ad - bc = 1$ and hence there exists a projective unimodular matrix

$$T = \pm\begin{pmatrix} a & b \\ c & d \end{pmatrix}.$$

Then

$$TXT^{-1} = \pm\begin{pmatrix} \alpha & a^2+b^2 \\ \gamma & -\alpha \end{pmatrix} = \pm\begin{pmatrix} \alpha & n \\ \gamma & -\alpha \end{pmatrix}.$$

This has determinant one, so

$$-\alpha^2 - n\gamma = 1 \implies \alpha^2 = -1 - n\gamma \implies \alpha^2 \equiv -1 \mod n.$$

Therefore -1 is a quadratic residue mod n. □

This type of group theoretical proof can be extended in several directions. Kern-Isberner and Rosenberger [21] considered groups of matrices of the form

$$U = \begin{pmatrix} a & b\sqrt{N} \\ c\sqrt{N} & d \end{pmatrix}, \quad a,b,c,d,N \in \mathbb{Z},\ ad - Nbc = 1$$

or

$$U = \begin{pmatrix} a\sqrt{N} & b \\ c & d\sqrt{N} \end{pmatrix}, \quad a,b,c,d,N \in \mathbb{Z},\ Nad - bc = 1.$$

They then proved that if

$$N \in \{1,2,4,5,6,8,9,10,12,13,16,18,22,25,28,37,58\}$$

and $n \in \mathbb{N}$ with $(n,N) = 1$ then:

(1) If $-N$ is a quadratic residue mod n and n is a quadratic residue mod N then n can be written as $n = x^2 + Ny^2$ with $x, y \in \mathbb{Z}$.

(2) Conversely if $n = x^2 + Ny^2$ with $x, y \in \mathbb{Z}$ and $(x,y) = 1$ then $-N$ is a quadratic residue mod n and n is a quadratic residue mod N.

The proof of the above results depends on the class number of $\mathbb{Q}(\sqrt{-N})$ (see [21]).

In another direction Fine [33] and [32] showed that the Fermat two-square property is actually a property satisfied by many rings R. These are called *sum of squares rings*. For example if $p \equiv 3 \mod 4$ then $\mathbb{Z}_{p^n}$ for $n > 1$ is a sum of squares ring.

14.4 Presentations of Subgroups

Given a group presentation $G = \langle X; R\rangle$ it is possible to find a presentation for a subgroup H of G. The procedure to do this is called the *Reidemeister–Schreier process* and is a consequence of the explicit version of the Nielsen–Schreier theorem (Theorem 14.2.9). We give a brief description. A complete description and a verification of its correctness is found in [24] or in [15].

Let G be a group with the presentation $\langle a_1,\dots,a_n; R_1,\dots,R_k\rangle$. Let H be a subgroup of G and T a Schreier system for G mod H defined analogously as above.

Reidemeister–Schreier Process. Let G, H and T be as above. Then H is generated by the set

$$\{S_{ta_v} : t \in T, a_v \in \{a_1,\dots,a_n\}, S_{ta_v} \neq 1\}$$

with a complete set of defining relations given by conjugates of the original relators rewritten in terms of the subgroup generating set.

In order to actual rewrite the relators in terms of the new generators we use a mapping τ on words on the generators of G called the *Reidemeister rewriting process*. This map is defined as follows: If

$$W = a_{v_1}^{e_1} a_{v_2}^{e_2} \dots a_{v_j}^{e_j} \text{ with } e_i = \pm 1 \text{ defines an element of } H$$

then

$$\tau(W) = S_{t_1,a_{v_1}}^{e_1} S_{t_2,a_{v_2}}^{e_2} \cdots S_{t_j,a_{v_j}}^{e_j}$$

where t_i is the coset representative of the initial segment of W preceding a_{v_i} if $e_i = 1$ and t_i is the representative of the initial segment of W up to and including $a_{v_i}^{-1}$ if $e_i = -1$. The complete set of relators rewritten in terms of the subgroup generators is then given by

$$\{\tau(tR_it^{-1})\} \text{ with } t \in T \text{ and } R_i \text{ runs over all relators in } G.$$

We present two examples; one with a finite group and then an important example with a free group which shows that a countable free group contains free subgroups of arbitrary ranks.

Example 14.4.1. Let $G = A_4$ be the alternating group on 4 symbols. Then a presentation for G is

$$G = A_4 = \langle a, b; a^2 = b^3 = (ab)^3 = 1\rangle.$$

Let $H = A_4'$ the commutator subgroup. We use the above method to find a presentation for H. Now

$$G/H = A_4/A_4' = \langle a, b; a^2 = b^3 = (ab)^3 = [a,b] = 1\rangle = \langle b; b^3 = 1\rangle.$$

Therefore $|A_4 : A_4'| = 3$. A Schreier system is then $\{1, b, b^2\}$. The generators for A_4' are then

$$X_1 = S_{1a} = a, \quad X_2 = S_{ba} = bab^{-1}, \quad X_3 = S_{b^2a} = b^2ab$$

while the relations are

1. $\tau(aa) = S_{1a}S_{1a} = X_1^2$
2. $\tau(baab^{-1}) = X_2^2$
3. $\tau(b^2aab^{-2}) = X_3^2$
4. $\tau(bbb) = 1$
5. $\tau(bbbbb^{-1}) = 1$
6. $\tau(b^2bbbb^{-2}) = 1$
7. $\tau(ababab) = S_{1a}S_{ba}S_{b^2a} = X_1X_2X_3$
8. $\tau(bababab b^{-1}) = S_{ba}S_{b^2a}S_{1a} = X_2X_3X_1$
9. $\tau(b^2abababb^{-2}) = S_{b^2a}S_{1a}S_{ba} = X_3X_1X_2$.

Therefore after eliminating redundant relations and using that $X_3 = X_1X_2$ we get as a presentation for A_4',

$$\langle X_1, X_2; X_1^2 = X_2^2 = (X_1X_2)^2 = 1\rangle.$$

Example 14.4.2. Let $F = \langle x, y; \rangle$ be the free group of rank 2. Let H be the commutator subgroup. Then

$$F/H = \langle x, y; [x, y] = 1\rangle = \mathbb{Z} \times \mathbb{Z}$$

a free abelian group of rank 2. It follows that H has infinite index in F. As Schreier coset representatives we can take

$$t_{m,n} = x^m y^n, \quad m = 0, \pm 1, \pm 2, \ldots, \; n = 0, \pm 1, \pm 2, \ldots.$$

The corresponding Schreier generators for H are

$$x_{m,n} = x^m y^n x^{-m} y^{-n}, \quad m = 0, \pm 1, \pm 2, \ldots, \; n = 0, \pm 1, \pm 2, \ldots.$$

The relations are only trivial and therefore H is free on the countable infinitely many generators above. It follows that a free group of rank 2 contains as a subgroup a free group of countably infinite rank. Since a free group of countable infinite rank contains as subgroups free groups of all finite ranks it follows that a free group of rank 2 contains as a subgroup a free subgroup of any arbitrary finite rank.

Theorem 14.4.3. *Let F be free of rank 2. Then the commutator subgroup F' is free of countable infinite rank. In particular a free group of rank 2 contains as a subgroup a free group of any finite rank n.*

Corollary 14.4.4. *Let n, m be any pair of positive integers $n, m \geq 2$ and F_n, F_m free groups of ranks n, m respectively. Then F_n can be embedded into F_m and F_m can be embedded into F_n.*

14.5 Geometric Interpretation

Combinatorial group theory has its origins in topology and complex analysis. Especially important in the development is the theory of the fundamental group. This connection is so deep that many people consider combinatorial group theory as the study of the fundamental group – especially the fundamental group of a low-dimensional complex. This connection proceeds in both directions. The fundamental group provides methods and insights to study the topology. In the other direction the topology can be used to study the groups.

Recall that if X is a topological space then its *fundamental group* based at a point x_0, denoted $\pi(X, x_0)$, is the group of all homotopy classes of closed paths at x_0. If X is path connected then the fundamental groups at different points are all isomorphic and we can speak of the fundamental group of X which we will denote $\pi(X)$. Historically group presentations were developed to handle the fundamental groups of spaces which allowed simplicial or cellular decompositions. In these cases the presentation of the fundamental group can be read off from the combinatorial decomposition of the space.

An (abstract) *simplicial complex* or *cell complex* K is a topological space consisting of a set of points called the *vertices*, which we will denote $V(K)$, and collections of subsets of vertices called *simplexes* or *cells* which have the property that the intersection of any two simplices is again a simplex. If n is the number of vertices in a cell then $n-1$ is called its *dimension*. Hence the set of vertices are the 0-dimensional cells and a simplex $\{v_1, \dots, v_n\}$ is an $(n-1)$-dimensional cell. The 1-dimensional cells are called *edges*. These have the form $\{u, v\}$ where u and v are vertices. One should think of the cells in a geometric manner so that the edges are really edges, the 2-cells are filled triangles, that are equivalent to disks and so on. The maximum dimension of any cell in a complex K is called the dimension of K. From now on we will assume that our simplicial complexes are path connected.

A *graph* Γ is just a 1-dimensional simplicial complex. Hence Γ consists of just vertices and edges. If K is any complex then the set of vertices and edges is called the 1-*skeleton* of K. Similarly all the cells of dimension less than or equal to 2 comprise the 2-skeleton. A connected graph with no closed paths in it is called a *tree*. If K is any complex then a *maximal tree* in K is a tree that can be contained in no other tree within K.

From the viewpoint of combinatorial group theory what is relevant is that if K is a complex then a presentation of its fundamental group can be determined from its 2-skeleton and read off directly. In particular:

Theorem 14.5.1. *Suppose that K is a connected cell complex. Suppose that T is a maximal tree within the 1-skeleton of K. Then a presentation for $\pi(K)$ can be determined in the following manner:*

Generators: *all edges outside of the maximal tree T*

Relations: (a) $\{u, v\} = 1$ *if* $\{u, v\}$ *is an edge in* T

(b) $\{u, v\}\{v, w\} = \{u, w\}$ *if* u, v, w *lie in a simplex of* K.

From this the following is obvious:

Corollary 14.5.2. *The fundamental group of a connected graph is free. Further its rank is the number of edges outside a maximal tree.*

A connected graph is homotopic to a wedge or bouquet of circles. If there are n circles in a bouquet of circles then the fundamental group is free of rank n. The converse is also true. A free group can be realized as the fundamental group of a wedge of circles.

An important concept in applying combinatorial group theory is that of a *covering complex*.

Definition 14.5.3. Suppose that K is a complex. Then a complex K_1 is a covering complex for K if there exists a surjection $p : K_1 \to K$ called a *covering map* with the property that for any cell $s \in K$ the inverse image $p^{-1}(s)$ is a union of pairwise disjoint cells in K_1 and p restricted to any of the preimage cells is a homeomorphism.

That is for each simplex S in K we have

$$p^{-1}(S) = \bigcup \overline{S_i}$$

and $p : \overline{S_i} \to S$ is a bijection for each i.

The following then becomes clear.

Lemma 14.5.4. *If K_1 is a connected covering complex for K then K_1 and K have the same dimension.*

What is crucial in using covering complexes to study the fundamental group is that there is a *Galois theory* of covering complexes and maps. The covering map p induces a homomorphism of the fundamental group which we will also call p. Then:

Theorem 14.5.5. *Let K_1 be a covering complex of K with covering map p. Then $p(\pi(K_1))$ is a subgroup of $\pi(K)$. Conversely to each subgroup H of $\pi(K)$ there is a covering complex K_1 with $\pi(K_1) = H$. Hence there is a one-to-one correspondence between subgroups of the fundamental group of a complex K and covers of K.*

We will see the analog of this theorem in regard to algebraic field extensions in Chapter 15.

A topological space X is *simply connected* if $\pi(X) = \{1\}$. Hence the covering complex of K corresponding to the identity in $\pi(K)$ is simply connected. This is called the *universal cover* of K since it covers any other cover of K.

Based on Theorem 14.5.1 we get a very simple proof of the Nielsen–Schreier theorem.

Theorem 14.5.6 (Nielsen–Schreier). *Any subgroup of a free group is free.*

Proof. Let F be a free group. Then $F = \pi(K)$ where K is a connected graph. Let H be a subgroup of F. Then H corresponds to a cover K_1 of K. But a cover is also 1-dimensional and hence $H = \pi(K_1)$ where K_1 is a connected graph and hence H is also free. □

The fact that a presentation of a fundamental group of a simplicial complex is determined by its 2-skeleton goes in the other direction also. That is given an arbitrary presentation there exists a 2-dimensional complex whose fundamental group has that presentation. Essentially given a presentation $\langle X; R\rangle$ we consider a wedge of circles with cardinality $|X|$. We then paste on a 2-cell for each relator W in R bounded by the path corresponding to the word W.

Theorem 14.5.7. *Given an arbitrary presentation $\langle X; R\rangle$ there exists a connected 2-complex K with $\pi(K) = \langle X; R\rangle$.*

We note that the books by Rotman [26] and Camps, große Rebel and Rosenberger [15] have very nice detailed and accessible descriptions of groups and complexes.

Cayley and then Dehn introduced for each group G, a graph, now called its *Cayley graph*, as a tool to apply complexes to the study of G. The Cayley graph is actually tied to a presentation and not to the group itself. Gromov reversed the procedure and showed that by considering the geometry of the Cayley graph one could get information about the group. This led to the development of the theory of hyperbolic groups (see for instance [15]).

Definition 14.5.8. Let $G = \langle X; R\rangle$ be a presentation. We form a graph $\Gamma(G, X)$ in the following way. Let $A = X \cup X^{-1}$. For the vertex set of $\Gamma(G, X)$ we take the elements of G, that is $V(\Gamma) = \{g : g \in G\}$. The edges of Γ are given by the set $\{(g, x) : g \in G, x \in A\}$. We call g the initial point and gx is the terminal point. That is two points g, g_1 in the vertex set are connected by an edge if $g_1 = gx$ for some $x \in A$. We have $(g, x)^{-1} = (gx, x^{-1})$. This gives a directed graph called the *Cayley graph* of G on the generating set X.

Call x the *label* on the edge (g, x). Given a $g \in G$ then G is represented by at least one word W in A. This represents a path in the Cayley graph. The length of the word W is the length of the path. This is equivalent to making each edge have length one. If we take the distance between 2 points as the minimum path length we make the Cayley graph a metric space. This metric is called the *word metric*. If we extend this metric to all pairs of points in the Cayley graph in the obvious way (making each edge a unit real interval) then the Cayley graph becomes a geodesic metric space. Each closed path in the Cayley graph represents a relator.

By left multiplication the group G acts on the Cayley graph as a group of isometries. Further the action of G on the Cayley graph is without inversion, that is $ge \neq e^{-1}$, if e is an edge.

If we sew in a 2-cell for each closed path in the Cayley graph we get a simply connected 2-complex called the *Cayley complex*.

14.6 Presentations of Factor Groups

Let G be a group with a presentation $G = \langle X; R\rangle$. Suppose that H is a factor group of G, that is $H \cong G/N$ for some normal subgroup N of G. We show that a presentation for H is then $H = \langle X; R \cup R_1\rangle$ where R_1 is a, perhaps additional, system of relators.

Theorem 14.6.1 (Dyck's theorem). *Let $G = \langle X; R\rangle$ and suppose that $H \cong G/N$ where N is a normal subgroup of G. Then a presentation for H is $\langle X; R \cup R_1\rangle$ for some set of words R_1 on X. Conversely the presentation $\langle X; R \cup R_1\rangle$ defines a group that is a factor group of G.*

Proof. Since each element of H is a coset of N they have the form gN for $g \in G$. It is clear then that the images of X generate H. Further since H is a homomorphic image of G each relator in R is a relator in H. Let N_1 be a set of elements that generate N and let R_1 be the corresponding words in the free group on X. Then R_1 is an additional set of relators in H. Hence $R \cup R_1$ is a set of relators for H. Any relator in H is either a relator in G and hence a consequence of R or can be realized as an element of G that lies in N and hence is a consequence of R_1. Therefore $R \cup R_1$ is a complete set of defining relators for H and H has the presentation $H = \langle X; R \cup R_1\rangle$.

Conversely $G = \langle X; R\rangle, G_1 = \langle X; R \cup R_1\rangle$. Then $G = F(X)/N_1$ where $N_1 = N(R)$ and $G_1 = F(X)/N_2$ where $N_2 = N(R \cup R_1)$. Hence $N_1 \subset N_2$. The normal subgroup N_2/N_1 of $F(X)/N_1$ corresponds to a normal subgroup of H of G and therefore by the isomorphism theorem

$$G/H \cong (F(X)/N_1)/(N_2/N_1) \cong F(X)/N_2 \cong G_1. \qquad \square$$

14.7 Group Presentations and Decision Problems

We have seen that given any group G there exists a presentation for it, $G = \langle X; R \rangle$. In the other direction given any presentation $\langle X; R \rangle$ we have seen that there is a group with that presentation. In principle every question about a group can be answered via a presentation. However things are not that simple. Max Dehn in his pioneering work on combinatorial group theory about 1910 introduced the following three fundamental *group decision problems*.

(1) *Word Problem*: Suppose G is a group given by a finite presentation. Does there exist an algorithm to determine if an arbitrary word w in the generators of G defines the identity element of G?

(2) *Conjugacy Problem*: Suppose G is a group given by a finite presentation. Does there exist an algorithm to determine if an arbitrary pair of words u, v in the generators of G define conjugate elements of G?

(3) *Isomorphism Problem*: Does there exist an algorithm to determine given two arbitrary finite presentations whether the groups they present are isomorphic or not?

All three of these problems have negative answers in general. That is for each of these problems one can find a finite presentation for which these questions cannot be answered algorithmically (see [23]). Attempts for solutions and for solutions in restricted cases have been of central importance in combinatorial group theory. For this reason combinatorial group theory has always searched for and studied classes of groups in which these decision problems are solvable.

For finitely generated free groups there are simple and elegant solutions to all three problems. If F is a free group on $x_1, \dots, x_n$ and W is a freely reduced word in $x_1, \dots, x_n$ then $W \neq 1$ if and only if $L(W) \geq 1$. Since freely reducing any word to a freely reduced word is algorithmic this provides a solution to the word problem. Further a freely reduced word $W = x_{v_1}^{e_1} x_{v_2}^{e_2} \dots x_{v_n}^{e_n}$ is *cyclically reduced* if $v_1 \neq v_n$ or if $v_1 = v_n$ then $e_1 \neq -e_n$. Clearly then every element of a free group is conjugate to an element given by a cyclically reduced word called a cyclic reduction. This leads to a solution to the conjugacy problem. Suppose V and W are two words in the generators of F and $\overline{V}, \overline{W}$ are respective cyclic reductions. Then V is conjugate to W if and only if $\overline{V}$ is a cyclic permutation of $\overline{W}$. Finally two finitely generated free groups are isomorphic if and only if they have the same rank.

14.8 Group Amalgams: Free Products and Direct Products

Closely related to free groups in both form and properties are free products of groups. Let $A = \langle a_1, \dots; R_1, \dots \rangle$ and $B = \langle b_1, \dots; S_1, \dots \rangle$ be two groups. We consider A and B to be disjoint. Then:

Definition 14.8.1. The *free product* of A and B denoted $A * B$ is the group G with the presentation $\langle a_1, \dots, b_1, \dots; R_1, \dots, S_1, \dots \rangle$, that is the generators of G consist of the disjoint union of the generators of A and B with relators taken as the disjoint union of the relators R_i of A and S_j of B. A and B are called the *factors* of G.

In an analogous manner the concept of a free product can be extended to an arbitrary collection of groups.

Definition 14.8.2. If $A_\alpha = \langle \text{gens } A_\alpha; \text{rels } A_\alpha \rangle, \alpha \in \mathcal{I}$, is a collection of groups, then their free product $G = *A_\alpha$ is the group whose generators consist of the disjoint union of the generators of the A_α and whose relators are the disjoint union of the relators of the A_α.

Free products exist and are nontrivial. We have:

Theorem 14.8.3. *Let $G = A * B$. Then the maps $A \to G$ and $B \to G$ are injections. The subgroup of G generated by the generators of A has the presentation $\langle$generators of A; relators of $A\rangle$, that is, is isomorphic to A. Similarly for B. Thus A and B can be considered as subgroups of G. In particular $A * B$ is nontrivial if A and B are.*

Free products share many properties with free groups. First of all there is a categorical formulation of free products. Specifically we have:

Theorem 14.8.4. *A group G is the free product of its subgroups A and B if A and B generate G and given homomorphisms $f_1 : A \to H$, $f_2 : B \to H$ into a group H there exists a unique homomorphism $f : G \to H$ extending f_1 and f_2.*

Secondly each element of a free product has a *normal form* related to the reduced words of free groups. If $G = A * B$ then a *reduced sequence* or *reduced word* in G is a sequence $g_1 g_2 \dots g_n$, $n \geq 0$, with $g_i \neq 1$, each g_i in either A or B and g_i, g_{i+1} not both in the same factor. Then:

Theorem 14.8.5. *Each element $g \in G = A * B$ has a unique representation as a reduced sequence. The length n is unique and is called the syllable length. The case $n = 0$ is reserved for the identity.*

A reduced word $g_1 \dots g_n \in G = A * B$ is called *cyclically reduced* if either $n \leq 1$ or $n \geq 2$ and g_1 and g_n are from different factors. Certainly every element of G is conjugate to a cyclically reduced word.

From this we obtain several important properties of free products which are analogous to properties in free groups.

Theorem 14.8.6. *An element of finite order in a free product is conjugate to an element of finite order in a factor. In particular a finite subgroup of a free product is entirely contained in a conjugate of a factor.*

Theorem 14.8.7. *If two elements of a free product commute then they are both powers of a single element or are contained in a conjugate of an abelian subgroup of a factor.*

Finally a theorem of Kurosh extends the Nielsen–Schreier theorem to free products.

Theorem 14.8.8 (Kurosh). *A subgroup of a free product is also a free product. Explicitly if $G = A * B$ and $H \subset G$ then*

$$H = F * (*A_\alpha) * (*B_\beta)$$

where F is a free group and $(*A_\alpha)$ is a free product of conjugates of subgroups of A and $(*B_\beta)$ is a free product of conjugates of subgroups of B.

We note that the rank of F as well as the number of the other factors can be computed. A complete discussion of these is in [24], [23] and [15].

If A and B are disjoint groups then we now have two types of products forming new groups out of them, the free product and the direct product. In both these products the original factors inject. In the free product there are no relations between elements of A and elements of B while in a direct product each element of A commutes with each element of B. If $a \in A$ and $b \in B$ a *cross commutator* is $[a,b] = aba^{-1}b^{-1}$. The direct product is a factor group of the free product and the kernel is precisely the normal subgroup generated by all the cross commutators.

Theorem 14.8.9. *Suppose that A and B are disjoint groups. Then*

$$A \times B = (A \star B)/H$$

where H is the normal closure in $A \star B$ of all the cross commutators. In particular a presentation for $A \times B$ is given by

$$A \times B = \langle \text{gens } A, \text{gens } B; \text{rels } A, \text{rels } B, [a,b] \textit{ for all } a \in A, b \in B \rangle.$$

14.9 Exercises

1. Let X^{-1} be a set disjoint from X but bijective to X. A *word* in X is a finite sequence of letters from the alphabet. That is a word has the form
$$w = x_{i_1}^{\epsilon_{i_1}} x_{i_2}^{\epsilon_{i_2}} \dots x_{i_n}^{\epsilon_{i_n}},$$
where $x_{i_j} \in X$ and $\epsilon_{i_j} = \pm 1$. Let $W(X)$ be the set of all words on X.
If $w_1, w_2 \in W(X)$ we say that w_1 is *equivalent* to w_2, denoted $w_1 \sim w_2$, if w_1 can be converted to w_2 by a finite string of insertions and deletions of trivial words. Verify that this is an equivalence relation on $W(X)$.
2. In $F(X)$ let $N(X)$ be the subgroup generated by all squares in $F(X)$ that is
$$N(X) = \langle \{g^2 : g \in F(X)\} \rangle.$$
Show that $N(X)$ is a normal subgroup and the factor group $F(X)/N(X)$ is abelian where every nontrivial element has order 2.
3. Show that a free group F is torsion-free.
4. Let F be a free group and $a, b \in F$. Show: If $a^k = b^k$, $k \neq 0$, then $a = b$.
5. Let $F = \langle a, b; \rangle$ a free group with basis $\{a, b\}$. Let $c_i = a^{-i}ba^i$, $i \in \mathbb{Z}$. Then $G = \langle c_i, i \in \mathbb{Z} \rangle$ is free with basis $\{c_i \mid i \in \mathbb{Z}\}$.
6. Show that $\langle x, y; x^2y^3, x^3y^4 \rangle \cong \langle x; x \rangle = \{1\}$.
7. Let $G = \langle v_1, \dots, v_n; v_1^2 \cdots v_n^2 \rangle$, $n \geq 1$, and $\alpha : G \to \mathbb{Z}_2$ the epimorphism with $\alpha(v_i) = -1$ for all i. Let U be the kernel of α. Then U has a presentation $U = \langle x_1, \dots, x_{n-1}, y_1, \dots, y_{n-1}; y_1x_1 \cdots y_{n-1}x_{n-1}y_{n-1}^{-1}x_{n-1}^{-1} \cdots y_1^{-1}x_1^{-1} \rangle$.
8. Let $M = \langle x, y; x^2, y^3 \rangle \cong \mathrm{PSL}(2, \mathbb{Z})$ be the modular group. Let M' be the commutator subgroup. Show that M' is a free group of rank 2 with a basis $\{[x, y], [x, y^2]\}$.

Chapter 15
Finite Galois Extensions

15.1 Galois Theory and the Solvability of Polynomial Equations

As we mentioned in Chapter 1, one of the origins of abstract algebra was the problem of trying to determine a formula for finding the solutions in terms of radicals of a fifth degree polynomial. It was proved first by Ruffini in 1800 and then by Abel that it is impossible to find a formula in terms of radicals for such a solution. Galois in 1820 extended this and showed that such a formula is impossible for any degree five or greater. In proving this he laid the groundwork for much of the development of modern abstract algebra especially field theory and finite group theory. One of the goals of this book has been to present a comprehensive treatment of Galois theory and a proof of the results mentioned above. At this point we have covered enough general algebra and group theory to discuss Galois extensions and general Galois theory.

In modern terms, *Galois theory* is that branch of mathematics that deals with the interplay of the algebraic theory of fields, the theory of equations and finite group theory. This theory was introduced by Evariste Galois about 1830 in his study of the insolvability by radicals of quintic (degree 5) polynomials, a result proved somewhat earlier by Ruffini and independently by Abel. Galois was the first to see the close connection between field extensions and permutation groups. In doing so he initiated the study of finite groups. He was the first to use the term group, as an abstract concept, although his definition was really just for a closed set of permutations.

The method Galois developed not only facilitated the proof of the insolvability of the quintic and higher powers but led to other applications and to a much larger theory as well.

The main idea of Galois theory is to associate to certain special types of algebraic field extensions called *Galois extensions* a group called the *Galois group*. The properties of the field extension will be reflected in the properties of the group, which are somewhat easier to examine. Thus, for example, solvability by radicals can be translated into solvability of groups which was discussed in Chapter 12. Showing that for every degree five or greater there exists a field extension whose Galois group is not solvable proves that there cannot be a general formula for solvability by radicals.

The tie-in to the theory of equations is as follows: If $f(x) = 0$ is a polynomial equation over some field F, we can form the splitting field K. This is usually a Galois extension, and therefore has a Galois group called the *Galois group of the equation*. As before, properties of this group will reflect properties of this equation.

15.2 Automorphism Groups of Field Extensions

In order to define the Galois group we must first consider the automorphism group of a field extension. In this section K, L, M will always be (commutative) fields with additive identity 0 and multiplicative identity 1.

Definition 15.2.1. Let $L|K$ be a field extension. Then the set

$$\mathrm{Aut}(L|K) = \{\alpha \in \mathrm{Aut}(L) : \alpha_{|K} = \text{the identity on } K\}$$

is called the set of *automorphisms of L over K*. Notice that if $\alpha \in \mathrm{Aut}(L|K)$ then $\alpha(k) = k$ for all $k \in K$.

Lemma 15.2.2. *Let $L|K$ be a field extension. Then* $\mathrm{Aut}(L|K)$ *forms a group called the Galois group of $L|K$.*

Proof. $\mathrm{Aut}(L|K) \subset \mathrm{Aut}(L)$ and hence to show that $\mathrm{Aut}(L|K)$ is a group we only have to show that its a subgroup of $\mathrm{Aut}(L)$. Now the identity map on L is certainly the identity map on K so $1 \in \mathrm{Aut}(L|K)$ and hence $\mathrm{Aut}(L|K)$ is nonempty. If $\alpha, \beta \in \mathrm{Aut}(L|K)$ then consider $\alpha^{-1}\beta$. If $k \in K$ then $\beta(k) = k$ and $\alpha(k) = k$ so $\alpha^{-1}(k) = k$. Therefore $\alpha^{-1}\beta(k) = k$ for all $k \in K$ and hence $\alpha^{-1}\beta \in \mathrm{Aut}(L|K)$. It follows that $\mathrm{Aut}(L|K)$ is a subgroup of $\mathrm{Aut}(L)$ and therefore a group. □

If $f(x) \in K[x] \setminus K$ and L is the splitting field of $f(x)$ over K then $\mathrm{Aut}(L|K)$ is also called the *Galois group* of $f(x)$.

Theorem 15.2.3. *If P is the prime field of L then* $\mathrm{Aut}(L|P) = \mathrm{Aut}(L)$.

Proof. We must show that any automorphism of a prime field P is the identity. If $\alpha \in \mathrm{Aut}(L)$ then $\alpha(1) = 1$ and so $\alpha(n \cdot 1) = n \cdot 1$. Therefore in P, α fixes all integer multiples of the identity. However every element of P can be written as a quotient $\frac{m \cdot 1}{n \cdot 1}$ of integer multiples of the identity. Since α is a field homomorphism and α fixes both the top and the bottom it follows that α will fix every element of this form and hence fix each element of P. □

For splitting fields the Galois group is a permutation group on the roots of the defining polynomial.

Theorem 15.2.4. *Let $f(x) \in K[x]$ and L the splitting field of $f(x)$ over K. Suppose that $f(x)$ has roots $\alpha_1, \dots, \alpha_n \in L$.*

(a) *Then each* $\phi \in \mathrm{Aut}(L|K)$ *is a permutation on the roots. In particular* $\mathrm{Aut}(L|K)$ *is isomorphic to a subgroup of* S_n *and uniquely determined by the zeros of* $f(x)$.

(b) *If* $f(x)$ *is irreducible then* $\mathrm{Aut}(L|K)$ *operates transitively on* $\{\alpha_1, \dots, \alpha_n\}$. *Hence for each* i, j *there is a* $\phi \in \mathrm{Aut}(L|K)$ *such that* $\phi(\alpha_i) = \alpha_j$.

(c) *If* $f(x) = b(x-\alpha_1)\cdots(x-\alpha_n)$ *with* $\alpha_1, \dots, \alpha_n$ *distinct and* $\mathrm{Aut}(L|K)$ *operates transitively on* $\alpha_1, \dots, \alpha_n$ *then* $f(x)$ *is irreducible.*

Proof. For the proofs we use the results of Chapter 8.

(a) Let $\phi \in \mathrm{Aut}(K)$. Then from Theorem 8.1.5 we obtain that ϕ permutes the roots $\alpha_1, \dots, \alpha_n$. Hence $\phi_{|\{\alpha_1,\dots,\alpha_n\}} \in S_n$. This map then defines a homomorphism

$$\tau : \mathrm{Aut}(L|K) \to S_n \quad \text{by } \tau(\phi) = \phi_{|\{\alpha_1,\dots,\alpha_n\}}.$$

Further ϕ is uniquely determined by the images $\phi(\alpha_i)$. It follows that τ is a monomorphism.

(b) If $f(x)$ is irreducible then $\mathrm{Aut}(L|K)$ operates transitively on the set $\{\alpha_1, \dots, \alpha_n\}$ again following from Theorem 8.1.5.

(c) Suppose that $f(x) = b(x - \alpha_1)\cdots(x - \alpha_n)$ with $\alpha_1, \dots, \alpha_n$ distinct and $f \in \mathrm{Aut}(L|K)$ operates transitively on $\alpha_1, \dots, \alpha_n$. Assume that $f(x) = g(x)h(x)$ with $g(x), h(x) \in K[x] \setminus K$. Without loss of generality let α_1 be a zero of $g(x)$ and α_n be a zero of $h(x)$.

Let $\alpha \in \mathrm{Aut}(L|K)$ with $\alpha(\alpha_1) = \alpha_n$. However $\alpha(g(x)) = g(x)$, that is $\alpha(\alpha_1)$ is a zero of $\alpha(g(x)) = g(x)$ which gives a contradiction since α_n is not a zero of $g(x)$. Therefore $f(x)$ must be irreducible. □

Example 15.2.5. Let $f(x) = (x^2 - 2)(x^2 - 3) \in \mathbb{Q}[x]$. The field $L = \mathbb{Q}(\sqrt{2}, \sqrt{3})$ is the spitting field of $f(x)$.

Over L we have

$$f(x) = (x + \sqrt{2})(x - \sqrt{2})(x + \sqrt{3})(x - \sqrt{3}).$$

We want to determine the Galois group $\mathrm{Aut}(L|\mathbb{Q}) = \mathrm{Aut}(L) = G$.

Lemma 15.2.6. *The Galois group* G *above is the Klein* 4*-group.*

Proof. First we show that $|\mathrm{Aut}(L)| \le 4$. Let $\alpha \in \mathrm{Aut}(L)$. Then α is uniquely determined by $\alpha(\sqrt{2})$ and $\alpha(\sqrt{3})$ and

$$\alpha(2) = 2 = (\sqrt{2})^2 = \alpha(\sqrt{2}^2) = (\alpha(\sqrt{2}))^2.$$

Hence $\alpha(\sqrt{2}) = \pm\sqrt{2}$. Analogously $\alpha(\sqrt{3}) = \pm\sqrt{3}$. From this it follows that $|\mathrm{Aut}(L)| \le 4$. Further $\alpha^2 = 1$ for any $\alpha \in G$.

Next we show that the polynomial $f(x) = x^2 - 3$ is irreducible over $K = \mathbb{Q}(\sqrt{2})$. Assume that $x^2 - 3$ were reducible over K. Then $\sqrt{3} \in K$. This implies that $\sqrt{3} =$

$\frac{a}{b} + \frac{c}{d}\sqrt{2}$ with $a, b, c, d \in \mathbb{Z}$ and $b \neq 0 \neq d$ and $\gcd(c, d) = 1$. Then $bd\sqrt{3} = ad + bc\sqrt{2}$ hence $3b^2d^2 = a^2b^2 + 2b^2c^2 + 2\sqrt{2}adbc$. Since $bd \neq 0$ this implies that we must have $ac = 0$. If $c = 0$ then $\sqrt{3} = \frac{a}{b} \in \mathbb{Q}$ a contradiction. If $a = 0$ then $\sqrt{3} = \frac{c}{d}\sqrt{2}$ which implies $3d^2 = 2c^2$. It follows from this that $3 | \gcd(c, d) = 1$ again a contradiction. Hence $f(x) = x^2 - 3$ is irreducible over $K = \mathbb{Q}(\sqrt{2})$.

Since L is the splitting field of $f(x)$ and $f(x)$ is irreducible over K then there exists an automorphism $\alpha \in \mathrm{Aut}(L)$ with $\alpha(\sqrt{3}) = -\sqrt{3}$ and $\alpha_{|K} = I_K$, that is $\alpha(\sqrt{2}) = \sqrt{2}$. Analogously there is a $\beta \in \mathrm{Aut}(L)$ with $\beta(\sqrt{2}) = -\sqrt{2}$ and $\beta(\sqrt{3}) = \sqrt{3}$.

Clearly $\alpha \neq \beta$, $\alpha\beta = \beta\alpha$ and $\alpha \neq \alpha\beta \neq \beta$. It follows that $\mathrm{Aut}(L) = \{1, \alpha, \beta, \alpha\beta\}$. completing the proof. □

15.3 Finite Galois Extensions

We now define (finite) Galois extensions. First we introduce the concept of a *Fix field*. Let K be a field and G a subgroup of $\mathrm{Aut}(K)$. Define the set

$$\mathrm{Fix}(K, G) = \{k \in K : g(k) = k \ \forall g \in G\}.$$

Theorem 15.3.1. *For a $G \subset \mathrm{Aut}(K)$, the set $\mathrm{Fix}(K, G)$ is a subfield of K called the Fix field of G over K.*

Proof. $1 \in K$ is in $\mathrm{Fix}(K, G)$ so $\mathrm{Fix}(K, G)$ is not empty. Let $k_1, k_2 \in \mathrm{Fix}(K, G)$ and let $g \in G$. Then $g(k_1 \pm k_2) = g(k_1) \pm g(k_2)$ since g is an automorphism. Then $g(k_1) \pm g(k_2) = k_1 \pm k_2$ and it follows that $k_1 \pm k_2 \in \mathrm{Fix}(K, G)$. In an analogous manner $k_1k_2^{-1} \in \mathrm{Fix}(K, G)$ if $k_2 \neq 0$ and therefore $\mathrm{Fix}(K, G)$ is a subfield of K. □

Using the concept of a fix field we define a finite Galois extension.

Definition 15.3.2. $L|K$ is a *(finite) Galois extension* if there exists a finite subgroup $G \subset \mathrm{Aut}(L)$ such that $K = \mathrm{Fix}(L, G)$.

We now give some examples of finite Galois extensions.

Lemma 15.3.3. *Let $L = \mathbb{Q}(\sqrt{2}, \sqrt{3})$ and $K = \mathbb{Q}$. Then $L|K$ is a Galois extension.*

Proof. Let $G = \mathrm{Aut}(L|K)$. From the example in the previous section there are automorphisms $\alpha, \beta \in G$ with

$$\alpha(\sqrt{3}) = -\sqrt{3}, \quad \alpha(\sqrt{2}) = \sqrt{2} \quad \text{and} \quad \beta(\sqrt{2}) = -\sqrt{2}, \quad \beta(\sqrt{3}) = \sqrt{3}.$$

We have

$$\mathbb{Q}(\sqrt{2}, \sqrt{3}) = \{c + d\sqrt{3} : c, d \in \mathbb{Q}(\sqrt{2})\}.$$

Let $t = a_1 + b_1\sqrt{2} + (a_2 + b_2\sqrt{2})\sqrt{3} \in \mathrm{Fix}(L, G)$.

Then applying β we have

$$t = \beta(t) = a_1 - b_1\sqrt{2} + (a_2 - b_2\sqrt{2})\sqrt{3}.$$

It follows that $b_1 + b_2\sqrt{3} = 0$, that is, $b_1 = b_2 = 0$ since $\sqrt{3} \notin \mathbb{Q}$. Therefore $t = a_1 + a_2\sqrt{3}$. Applying α we have $\alpha(t) = a_1 - a_2\sqrt{3}$ and hence $a_2 = 0$. Therefore $t = a_1 \in \mathbb{Q}$. Hence $\mathbb{Q} = \operatorname{Fix}(L, G)$ and $L|K$ is a Galois extension. □

Lemma 15.3.4. *Let $L = \mathbb{Q}(2^{\frac{1}{4}})$ and $K = \mathbb{Q}$. Then $L|K$ is not a Galois extension.*

Proof. Suppose that $\alpha \in \operatorname{Aut}(L)$ and $a = 2^{\frac{1}{4}}$. Then a is a zero of $x^4 - 2$ and hence

$$\begin{aligned}
&\alpha(a) = 2^{\frac{1}{4}} \quad \text{or}\\
&\alpha(a) = i2^{\frac{1}{4}} \notin L \quad \text{since } i \notin L \quad \text{or}\\
&\alpha(a) = -2^{\frac{1}{4}} \quad \text{or}\\
&\alpha(a) = -i2^{\frac{1}{4}} \notin L \quad \text{since } i \notin L.
\end{aligned}$$

In particular $\alpha(\sqrt{2}) = \sqrt{2}$ and therefore

$$\operatorname{Fix}(L, \operatorname{Aut}(L)) = \mathbb{Q}(\sqrt{2}) \neq \mathbb{Q}.$$ □

15.4 The Fundamental Theorem of Galois Theory

We now state the fundamental theorem of Galois theory. This theorem describes the interplay between the Galois group and Galois extensions. In particular the result ties together subgroups of the Galois group and intermediate fields between L and K.

Theorem 15.4.1 (fundamental theorem of Galois theory). *Let $L|K$ be a Galois extension with Galois group $G = \operatorname{Aut}(L|K)$. For each intermediate field E let $\tau(E)$ be the subgroup of G fixing E. Then:*

(1) *τ is a bijection between intermediate fields containing K and subgroups of G.*

(2) *$L|K$ is a finite extension and if M is an intermediate field then*

$$\begin{aligned}
|L : M| &= |\operatorname{Aut}(L|M)|\\
|M : K| &= |\operatorname{Aut}(L|K) : \operatorname{Aut}(L|M)|.
\end{aligned}$$

(3) *If M is an intermediate field then*

  (a) *$L|M$ is always a Galois extension*

  (b) *$M|K$ is a Galois extension if and only if*

$$\operatorname{Aut}(L|M) \textit{ is a normal subgroup of } \operatorname{Aut}(L|K).$$

(4) *If M is an intermediate field and $M|K$ is a Galois extension then*

  (a) $\alpha(M) = M$ *for all* $\alpha \in \operatorname{Aut}(L|K)$,

  (b) *the map* $\phi : \operatorname{Aut}(L|K) \to \operatorname{Aut}(M|K)$ *with* $\phi(\alpha) = \alpha_{|M} = \beta$ *is an epimorphism,*

  (c) $\operatorname{Aut}(M|K) = \operatorname{Aut}(L|K)/\operatorname{Aut}(L|M)$.

(5) *The lattice of subfields of L containing K is the inverted lattice of subgroups of* $\operatorname{Aut}(K|L)$.

We will prove this main result via a series of theorems and then combine them all.

Theorem 15.4.2. *Let G be a group, K a field and $\alpha_1, \dots, \alpha_n$ pairwise distinct group homomorphisms from G to $K^\star$ the multiplicative group of K. Then $\alpha_1, \dots, \alpha_n$ are linearly independent elements of the K-vector space of all homomorphisms from G to K.*

Proof. The proof is by induction on n. If $n = 1$ and $k\alpha_1 = 0$ with $k \in K$ then $0 = k\alpha_1(1) = k \cdot 1$ and hence $k = 0$. Now suppose that $n \geq 2$ and suppose that each $n-1$ of the $\alpha_1, \dots, \alpha_n$ are linearly independent over K. If

$$\sum_{i=1}^{n} k_i \alpha_i = 0, \quad k_i \in K \tag{$*$}$$

then we must show that all $k_i = 0$. Since $\alpha_1 \neq \alpha_n$ there exists an $a \in G$ with $\alpha_1(a) \neq \alpha_n(a)$. Let $g \in G$ and apply the sum above to ag. We get

$$\sum_{i=1}^{n} k_i(\alpha_i(a))(\alpha_i(g)) = 0. \tag{$**$}$$

Now multiply equation $(*)$ by $\alpha_n(a) \in K$ to get

$$\sum_{i=1}^{n} k_i(\alpha_n(a))(\alpha_i(g)) = 0. \tag{$***$}$$

If we subtract equation $(***)$ from equation $(**)$ then the last term vanishes and we have an equation in the $n-1$ homomorphism $\alpha_1, \dots, \alpha_{n-1}$. Since these are linearly independent we obtain

$$k_1(\alpha_1(a)) - k_1(\alpha_n(a)) = 0$$

for the coefficient for α_1. Since $\alpha_1(a) \neq \alpha_n(a)$ we must have $k_1 = 0$. Now $\alpha_2, \dots, \alpha_{n-1}$ are by assumption linearly independent so $k_2 = \cdots = k_n = 0$ also. Hence all the coefficients must be zero and therefore the mappings are independent. □

Theorem 15.4.3. *Let $\alpha_1, \dots, \alpha_n$ be pairwise distinct monomorphisms from the field K into the field K'. Let*

$$L = \{k \in K : \alpha_1(k) = \alpha_2(k) = \cdots = \alpha_n(k)\}.$$

Then L is a subfield of K with $|L : K| \geq n$.

Proof. Certainly L is a field. Assume that $r = |K : L| < n$ and let $\{a_1, \dots, a_r\}$ be a basis of the L-vector space K. We consider the following system of linear equations with r equations and n unknowns.

$$(\alpha_1(a_1))x_1 + \cdots + (\alpha_n(a_1))x_n = 0$$
$$\vdots$$
$$(\alpha_1(a_r))x_1 + \cdots + (\alpha_n(a_r))x_n = 0.$$

Since $r < n$ there exists a nontrivial solution $(x_1, \dots, x_n) \in (K')^n$.

Let $a \in K$. Then

$$a = \sum_{j=1}^{r} l_j a_j \quad \text{with } l_j \in L.$$

From the definition of L we have

$$\alpha_1(l_j) = \alpha_i(l_j) \quad \text{for } i = 2, \dots, n.$$

Then with our nontrivial solution $(x_1, \dots, x_n)$ we have

$$\begin{aligned}\sum_{i=1}^{n} x_i(\alpha_i(a)) &= \sum_{i=1}^{n} x_i\Big(\sum_{j=1}^{r} \alpha_i(l_j)\alpha_i(a_j)\Big)\\ &= \sum_{j=1}^{r}(\alpha_1(l_j))\sum_{i=1}^{n} x_i(\alpha_i(a_j)) = 0\end{aligned}$$

since $\alpha_1(l_j) = \alpha_i(l_j)$ for $i = 2, \dots, n$. This holds for all $a \in K$ and hence $\sum_{i=1}^{n} x_i\alpha_i = 0$ contradicting Theorem 15.4.2. Therefore our assumption that $|K : L| < n$ must be false and hence $|K : L| \geq n$. □

Definition 15.4.4. Let K be a field and G a finite subgroup of $\mathrm{Aut}(K)$. The map $\mathrm{tr}_G : K \to K$ given by

$$\mathrm{tr}_G(k) = \sum_{\alpha \in G} \alpha(k)$$

is called the *G-trace of K*.

Theorem 15.4.5. *Let K be a field and G a finite subgroup of* Aut(K). *Then*

$$\{0\} \neq \mathrm{tr}_G(K) \subset \mathrm{Fix}(K, G).$$

Proof. Let $\beta \in G$. Then

$$\beta(\mathrm{tr}_G(k)) = \sum_{\alpha \in G} \beta\alpha(k) = \sum_{\alpha \in G} \alpha(k) = \mathrm{tr}_G(k).$$

Therefore $\mathrm{tr}_G(K) \subset \mathrm{Fix}(K, G)$.

Now assume that $\mathrm{tr}_G(k) = 0$ for all $k \in K$. Then $\sum_{\alpha \in G} \alpha(k) = 0$ for all $k \in K$. It follows that $\sum_{\alpha \in G} \alpha$ is the zero map and hence the set of all $\alpha \in G$ are linearly dependent as elements of the K-vector space of all maps from K to K. This contradicts Theorem 15.4.2 and hence the trace cannot be the zero map. □

Theorem 15.4.6. *Let K be a field and G a finite subgroup of* Aut(K). *Then*

$$|K : \mathrm{Fix}(K, G)| = |G|.$$

Proof. Let $L = \mathrm{Fix}(K, G)$ and suppose that $|G| = n$. From Theorem 15.4.3 we know that $|K : L| \geq n$. We must show that $|K : L| \leq n$.

Suppose that $G = \{\alpha_1, \dots, \alpha_n\}$. To prove the result we show that if $m > n$ and $a_1, \dots, a_m \in K$ then $a_1, \dots, a_m$ are linearly dependent.

We consider the system of equations

$$(\alpha_1^{-1}(a_1))x_1 + \cdots + (\alpha_1^{-1}(a_m))x_m = 0$$

$$\vdots$$

$$(\alpha_n^{-1}(a_1))x_1 + \cdots + (\alpha_n^{-1}(a_m))x_m = 0.$$

Since $m > n$ there exists a nontrivial solution $(y_1, \dots, y_m) \in K^m$. Suppose that $y_l \neq 0$. Using Theorem 15.4.5 we can choose $k \in K$ with $\mathrm{tr}_G(k) \neq 0$. Define

$$(x_1, \dots, x_m) = k y_l^{-1}(y_1, \dots, y_m).$$

This m-tuple $(x_1, \dots, x_m)$ is then also a nontrivial solution of the system of equations considered above.

Then we have

$$\mathrm{tr}_G(x_l) = \mathrm{tr}_G(k) \quad \text{since } x_l = k.$$

Now we apply α_i to the i-th equation to obtain

$$a_1(\alpha_1(x_1)) + \cdots + a_m(\alpha_1(x_m)) = 0$$

$$\vdots$$

$$a_1(\alpha_n(x_1)) + \cdots + a_m(\alpha_n(x_m)) = 0.$$

Summation leads to

$$0 = \sum_{j=1}^{m} a_j \sum_{i=1}^{n} (\alpha_i(x_j)) = \sum_{j=1}^{m} (\mathrm{tr}_G(x_j))a_j$$

by definition of the G-trace. Hence $a_1, \dots, a_m$ are linearly dependent over L since $\mathrm{tr}_G(x_l) \neq 0$. Therefore $|K : L| \leq n$. Combining this with Theorem 15.4.3 we get that $|K : L| = n = |G|$. □

Theorem 15.4.7. *Let K be a field and G a finite subgroup of* $\mathrm{Aut}(K)$. *Then*

$$\mathrm{Aut}(K|\,\mathrm{Fix}(K, G)) = G.$$

Proof. $G \subset \mathrm{Aut}(K|\,\mathrm{Fix}(K, G))$ since if $g \in G$ then $g \in \mathrm{Aut}(K)$ and g fixes $\mathrm{Fix}(K, G)$ by definition. Therefore we must show that $\mathrm{Aut}(K|\,\mathrm{Fix}(K, G)) \subset G$.

Assume then that there exists an $\alpha \in \mathrm{Aut}(K|\,\mathrm{Fix}(K, G))$ with $\alpha \notin G$. Suppose, as in the previous proof, $|G| = n$ and $G = \{\alpha_1, \dots, \alpha_n\}$ with $\alpha_1 = 1$. Now

$$\begin{aligned}\mathrm{Fix}(K, G) &= \{a \in K : a = \alpha_2(a) = \cdots = \alpha_n(a)\}\\ &= \{a \in K : \alpha(a) = a = \alpha_2(a) = \cdots = \alpha_n(a)\}.\end{aligned}$$

From Theorem 15.4.3 we have that $|K : \mathrm{Fix}(K, G)| \geq n + 1$. However from Theorem 15.4.6 $|K : \mathrm{Fix}(K, G)| = n$ getting a contradiction. □

Suppose that $L|K$ is a Galois extension. We now establish that the map τ between intermediate fields $K \subset E \subset L$ and subgroups of $\mathrm{Aut}(L|K)$ is a bijection.

Theorem 15.4.8. *Let $L|K$ be a Galois extension. Then:*

(1) $\mathrm{Aut}(L|K)$ *is finite and*

$$\mathrm{Fix}(L, \mathrm{Aut}(L|K)) = K.$$

(2) *If $H \subset$* $\mathrm{Aut}(L|K)$ *then*

$$\mathrm{Aut}(L|\,\mathrm{Fix}(L, H)) = H.$$

Proof. (1) If $(L|K)$ is a Galois extension there is a finite subgroup of $\mathrm{Aut}(L)$ with $K = \mathrm{Fix}(K, G)$. From Theorem 15.4.7 we have $G = \mathrm{Aut}(L|K)$. In particular $\mathrm{Aut}(L|K)$ is finite and $K = \mathrm{Fix}(L, \mathrm{Aut}(L|K))$.

(2) Let $H \subset \mathrm{Aut}(L|K)$. From part (1) H is finite and then $\mathrm{Aut}(L|\,\mathrm{Fix}(L, H)) = H$ from Theorem 15.4.7. □

Theorem 15.4.9. *Let $L|K$ be a field extension. Then the following are equivalent.*

(1) *$L|K$ is a Galois extension.*

(2) $|L : K| = |\mathrm{Aut}(L|K)| < \infty$.

(3) $|\mathrm{Aut}(L|K)| < \infty$ *and* $K = \mathrm{Fix}(L, \mathrm{Aut}(L|K))$.

Proof. (1) $\Rightarrow$ (2): Now $|\mathrm{Aut}(L|K)| < \infty$ and $\mathrm{Fix}(L, \mathrm{Aut}(L|K)) = K$ from Theorem 15.4.8. Therefore $|L : K| = |\mathrm{Aut}(L|K)|$ from Theorem 15.4.6.

(2) $\Rightarrow$ (3): Let $G = \mathrm{Aut}(L|K)$. Then $K \subset \mathrm{Fix}(L, G) \subset L$. From Theorem 15.4.6 we have

$$|L : \mathrm{Fix}(L, G)| = |G| = |L : K|.$$

(3) $\Rightarrow$ (1) follows directly from the definition completing the proof. □

We now show that if $L|K$ is a Galois extension then $L|M$ is also a Galois extension for any intermediate field M.

Theorem 15.4.10. *Let $L|K$ be a Galois extension and $K \subset M \subset L$ be an intermediate field. Then $L|M$ is always a Galois extension and*

$$|M : K| = |\mathrm{Aut}(L|K) : \mathrm{Aut}(L|M)|.$$

Proof. Let $G = \mathrm{Aut}(L|K)$. Then $|G| < \infty$ and further $K = \mathrm{Fix}(L, G)$ from Theorem 15.4.9. Define $H = \mathrm{Aut}(L|M)$ and $M' = \mathrm{Fix}(L, H)$. We must show that $M' = M$ for then $L|M$ is a Galois extension.

Since the elements of H fix M we have $M \subset M'$. Let $G = \bigcup_{i=1}^{r} \alpha_i H$ a disjoint union of the cosets of H. Let $\alpha_1 = 1$ and define $\beta_i = (\alpha_i)_{|M}$. The $\beta_1, \dots, \beta_r$ are pairwise distinct for if $\beta_i = \beta_j$, that is $(\alpha_i)_{|M} = (\alpha_j)_{|M}$, then $\alpha_j^{-1}\alpha_i \in H$ so α_i and α_j are in the same coset.

We claim that

$$\{a \in M : \beta_1(a) = \cdots = \beta_r(a)\} = M \cap \mathrm{Fix}(L, G).$$

Further we know that

$$M \cap \mathrm{Fix}(L, G) = M \cap K = K$$

from Theorem 15.4.9.

To establish the claim it is clear that

$$M \cap \mathrm{Fix}(L, G) \subset \{a \in M : \beta_1(a) = \cdots = \beta_r(a)\}$$

since

$$a = \beta_i(a) = \alpha_i(a) \quad \text{for } \alpha_i \in G,\ a \in K.$$

Hence we must show that

$$\{a \in M : \beta_1(a) = \cdots = \beta_r(a)\} \subset M \cap \text{Fix}(L, G).$$

To do this we must show that $\alpha(b) = b$ for all $\alpha \in G, b \in M$. We have $\alpha \in \alpha_i H$ for some i and hence $\alpha = \alpha_i \gamma$ for $\gamma \in H$. We obtain then

$$\alpha(b) = \alpha_i(\gamma(b)) = \alpha_i(b) = \beta_i(b) = b$$

proving the inclusion and establishing the claim.

Now $|M : K| \geq r$ from Theorem 15.4.3. From the degree formula we get

$$|L : M'||M' : M||M : K| = |L : K| = |G| = |G : H||H| = r|L : M'|$$

since $|L : K| = |G|$ from Theorem 15.4.9 and $|H| = |L : M'|$ from Theorem 15.4.6. Therefore $|M : M'| = 1$ and hence $M = M'$ since $|M : K| \geq r$. Now

$$|M : K| = |G : H| = |\text{Aut}(L|K) : \text{Aut}(L|M)|$$

completing the proof. □

Lemma 15.4.11. *Let $L|K$ be a field extension and $K \subset M \subset L$ be an intermediate field. If $\alpha \in \text{Aut}(L|K)$ then*

$$\text{Aut}(L|\alpha(M)) = \alpha \,\text{Aut}(L|M)\alpha^{-1}.$$

Proof. Now $\beta \in \text{Aut}(L|\alpha(M))$ if and only if $\beta(\alpha(a)) = \alpha(a)$ for all $a \in M$. This occurs if and only if $\alpha^{-1}\beta\alpha(a) = a$ for all $a \in M$ which is true if and only if $\beta \in \alpha \,\text{Aut}(L|M)\alpha^{-1}$. □

Lemma 15.4.12. *Let $L|K$ be a Galois extension and $K \subset M \subset L$ be an intermediate field. Suppose that $\alpha(M) = M$ for all $\alpha \in \text{Aut}(L|K)$. Then*

$$\phi : \text{Aut}(L|K) \to \text{Aut}(M|K) \quad \text{by } \phi(\alpha) = \alpha_{|M}$$

is an epimorphism with kernel $\ker(\phi) = \text{Aut}(L|M)$.

Proof. It is clear that ϕ is a homomorphism with $\ker(\phi) = \text{Aut}(L|M)$ (see exercises). We must show that it is an epimorphism.

Let $G = \text{im}(\phi)$. Since $L|K$ is a Galois extension we get that

$$\text{Fix}(M, G) = \text{Fix}(L, \text{Aut}(L|K)) \cap M = K \cap M = K.$$

Then from Theorem 15.4.8 we have

$$\text{Aut}(M|K) = \text{Aut}(M|\text{Fix}((M, G)) = G$$

and therefore ϕ is an epimorphism. □

Theorem 15.4.13. *Let $L|K$ be a Galois extension and $K \subset M \subset L$ be an intermediate field. Then the following are equivalent.*

(1) *$M|K$ is a Galois extension.*

(2) *If $\alpha \in \mathrm{Aut}(L|K)$ then $\alpha(M) = M$.*

(3) *$\mathrm{Aut}(L|M)$ is a normal subgroup of $\mathrm{Aut}(L|K)$.*

Proof. (1) $\Rightarrow$ (2): Suppose that $M|K$ is a Galois extension. Let $\mathrm{Aut}(M|K) = \{\alpha_1, \dots, \alpha_r\}$. Consider the α_i as monomorphisms from M into L. Let $\alpha_{r+1} : M \to L$ be a monomorphism with $\alpha_{r+1|K} = 1$. Then

$$\{a \in M : \alpha_1(a) = \alpha_2(a) \cdots = \alpha_r(a) = \alpha_{r+1}(a)\} = K$$

since $M|K$ is a Galois extension. Therefore from Theorem 15.4.3 we have that, if the $\alpha_1, \dots, \alpha_r, \alpha_{r+1}$ are distinct then

$$|M : K| \geq r + 1 > r = |\mathrm{Aut}(M|K)| = |M : K|$$

giving a contradiction. Hence if $\alpha_{r+1} \in \mathrm{Aut}(L|K)$ is arbitrary then $\alpha_{r+1|M} \in \{\alpha_1, \dots, \alpha_r\}$, that is α_{r+1} fixes M.

(2) $\Rightarrow$ (1): Suppose that if $\alpha \in \mathrm{Aut}(L|K)$ then $\alpha(M) = M$. The map $\phi : \mathrm{Aut}(L|K) \to \mathrm{Aut}(M|K)$ with $\phi(\alpha) = \alpha_{|M}$ is surjective. Since $L|K$ is a Galois extension then $\mathrm{Aut}(L|K)$ is finite. Therefore also $H = \mathrm{Aut}(M|K)$ is finite. To prove (1) then it is sufficient to show that $K = \mathrm{Fix}(M, H)$.

The field $K \subset \mathrm{Fix}(M, H)$ from the definition of the Fix field. Hence we must show that $\mathrm{Fix}(M, H) \subset K$. Assume that there exists an $\alpha \in \mathrm{Aut}(L|K)$ with $\alpha(a) \neq a$. Recall that $L|K$ is a Galois extension and therefore $\mathrm{Fix}(L, \mathrm{Aut}(L|K)) = K$. Define $\beta = \alpha_{|M}$. Then $\beta \in H$ since $\alpha(M) = M$ and our original assumption. Then $\beta(a) \neq a$ contradicting $a \in \mathrm{Fix}(M, H)$. Therefore $K = \mathrm{Fix}(M, H)$ and $M|K$ is a Galois extension.

(2) $\Rightarrow$ (3): Suppose that if $\alpha \in \mathrm{Aut}(L|K)$ then $\alpha(M) = M$. Then $\mathrm{Aut}(L|M)$ is a normal subgroup of $\mathrm{Aut}(L|K)$ follows from Lemma 15.4.12 since $\mathrm{Aut}(L|M)$ is the kernel of ϕ.

(3) $\Rightarrow$ (2): Suppose that $\mathrm{Aut}(L|M)$ is a normal subgroup of $\mathrm{Aut}(L|K)$. Let $\alpha \in \mathrm{Aut}(L|K)$ then from our assumption and Lemma 15.4.11 we get that

$$\mathrm{Aut}(L|\alpha(M)) = \mathrm{Aut}(L|M).$$

Now $L|M$ and $L|\alpha(M)$ are Galois extensions by Theorem 15.4.10. Therefore

$$\alpha(M) = \mathrm{Fix}(L, \mathrm{Aut}(L|\alpha(M))) = \mathrm{Fix}(L, \mathrm{Aut}(L|M)) = M$$

completing the proof. □

We now combine all of these results to give the proof of Theorem 15.4.1, the fundamental theorem of Galois theory.

Proof of Theorem 15.4.1. Let $L|K$ be a Galois extension.

(1) Let $G \subset \mathrm{Aut}(L|K)$. Both G and $\mathrm{Aut}(L|K)$ are finite from Theorem 15.4.8. Further $G = \mathrm{Aut}(L|\,\mathrm{Fix}(L, G))$ from Theorem 15.4.7.

Now let M be an intermediate field of $L|K$. Then $L|M$ is a Galois extension from Theorem 15.4.10 and then $\mathrm{Fix}(L, \mathrm{Aut}(L|M)) = M$ from Theorem 15.4.8.

(2) Let M be an intermediate field of $L|K$. From Theorem 15.4.10 $L|M$ is a Galois extension. From Theorem 15.4.9 we have $|L : M| = |\mathrm{Aut}(L|M)|$. Applying Theorem 15.4.10 we get the result on indices

$$|M : K| = |\mathrm{Aut}(L|K) : \mathrm{Aut}(L|M)|.$$

(3) Let M be an intermediate field of $L|K$.

(a) From Theorem 15.4.10 we have that $L|M$ is a Galois extension.

(b) From Theorem 15.4.13 $M|K$ is a Galois extension if and only if

$$\mathrm{Aut}(L|M) \text{ is a normal subgroup of } \mathrm{Aut}(L|K).$$

(4) Let $M|K$ be a Galois extension.

(a) $\alpha(M) = M$ for all $\alpha \in \mathrm{Aut}(L|K)$ from Theorem 15.4.13.

(b) The map $\phi : \mathrm{Aut}(L|K) \to \mathrm{Aut}(M|K)$ with $\phi(\alpha) = \alpha_{|M} = \beta$ is an epimorphism follows from Lemma 15.4.12 and Theorem 15.4.13.

(c) $\mathrm{Aut}(M|K) = \mathrm{Aut}(L|K)/\,\mathrm{Aut}(L|M)$ follows directly from the group isomorphism theorem.

(5) That the lattice of subfields of L containing K is the inverted lattice of subgroups of $\mathrm{Aut}(L|K)$ follows directly from the previous results. □

In Chapter 8 we looked at the following example (Example 8.1.7). Here we analyze it further using the Galois theory.

Example 15.4.14. Let $f(x) = x^3 - 7 \in \mathbb{Q}[x]$. This has no zeros in $\mathbb{Q}$ and since it is of degree 3 it follows that it must be irreducible in $\mathbb{Q}[x]$.

Let $\omega = -\frac{1}{2} + \frac{\sqrt{3}}{2}i \in \mathbb{C}$. Then it is easy to show by computation that

$$\omega^2 = -\frac{1}{2} - \frac{\sqrt{3}}{2}i \quad \text{and} \quad \omega^3 = 1.$$

Therefore the three zeros of $f(x)$ in $\mathbb{C}$ are

$$a_1 = 7^{1/3}, \quad a_2 = \omega(7^{1/3}), \quad a_3 = \omega^2(7^{1/3}).$$

Hence $L = \mathbb{Q}(a_1, a_2, a_3)$ is the splitting field of $f(x)$. Since the minimal polynomial of all three zeros over $\mathbb{Q}$ is the same $(f(x))$ it follows that

$$\mathbb{Q}(a_1) \cong \mathbb{Q}(a_2) \cong \mathbb{Q}(a_3).$$

Since $\mathbb{Q}(a_1) \subset \mathbb{R}$ and a_2, a_3 are nonreal it is clear that $a_2, a_3 \notin \mathbb{Q}(a_1)$.

Suppose that $\mathbb{Q}(a_2) = \mathbb{Q}(a_3)$. Then $\omega = a_3 a_2^{-1} \in \mathbb{Q}(a_2)$ and so $7^{1/3} = \omega^{-1} a_2 \in \mathbb{Q}(a_2)$. Hence $Q(a_1) \subset \mathbb{Q}(a_2)$ and therefore $\mathbb{Q}(a_1) = \mathbb{Q}(a_2)$ since they are the same degree over $\mathbb{Q}$. This contradiction shows that $\mathbb{Q}(a_2)$ and $\mathbb{Q}(a_3)$ are distinct.

By computation we have $a_3 = a_1^{-1} a_2^2$ and hence

$$L = \mathbb{Q}(a_1, a_2, a_3) = \mathbb{Q}(a_1, a_2) = \mathbb{Q}(7^{1/3}, \omega).$$

Now the degree of L over $\mathbb{Q}$ is

$$|L : \mathbb{Q}| = |Q(7^{1/3}, \omega) : \mathbb{Q}(\omega)||\mathbb{Q}(\omega) : \mathbb{Q}|.$$

Now $|\mathbb{Q}(\omega) : \mathbb{Q}| = 2$ since the minimal polynomial of ω over $\mathbb{Q}$ is $x^2 + x + 1$. Since no zero of $f(x)$ lies in $\mathbb{Q}(\omega)$ and the degree of $f(x)$ is 3 it follows that $f(x)$ is irreducible over $\mathbb{Q}(\omega)$. Therefore we have that the degree of L over $\mathbb{Q}(\omega)$ is 3. Hence $|L : \mathbb{Q}| = (2)(3) = 6$.

Clearly then we have the following lattice of intermediate fields:

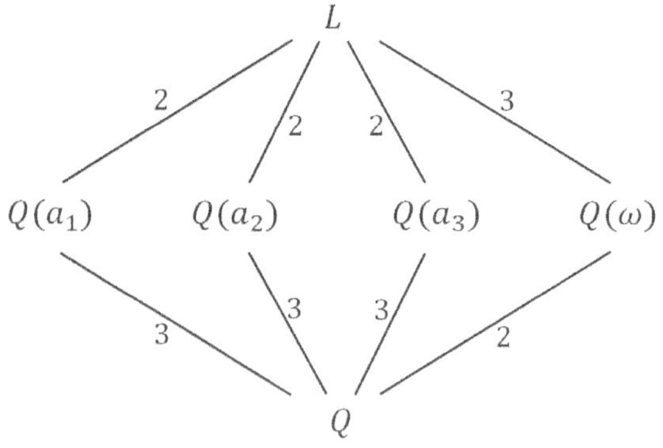

The question then arises as to whether these are all the intermediate fields. The answer is yes which we now prove.

Let $G = \text{Aut}(L|\mathbb{Q}) = \text{Aut}(L)$. $(\text{Aut}(L|\mathbb{Q}) = \text{Aut}(L)$ since $\mathbb{Q}$ is a prime field). Now $G \cong S_3$. G acts transitively on $\{a_1, a_2, a_3\}$ since f is irreducible. Let $\delta : \mathbb{C} \to \mathbb{C}$ be the automorphism of $\mathbb{C}$ taking each element to its complex conjugate, that is $\delta(z) = \overline{z}$.

Then $\delta(f) = f$ and $\delta_{|L} \in G$ (see Theorem 8.2.2). Since $a_1 \in \mathbb{R}$ we get that $\delta_{|\{a_1, a_2, a_3\}} = (a_2, a_3)$ the 2-cycle that maps a_2 to a_3 and a_3 to a_2. Since G is transitive on $\{a_1, a_2, a_3\}$ there is a $\tau \in G$ with $\tau(a_1) = a_2$.

Case 1: $\tau(a_3) = a_3$. Then $\tau = (a_1, a_2)$ and $(a_1, a_2)(a_2, a_3) = (a_1, a_2, a_3) \in G$.

Case 2: $\tau(a_3) \neq a_3$. Then τ is a 3-cycle. In either case G is generated by a transposition and a 3-cycle. Hence G is all of S_3. Then $L|\mathbb{Q}$ is a Galois extension from Theorem 15.4.9 since $|G| = |L : \mathbb{Q}|$.

The subgroups of S_3 are as follows:

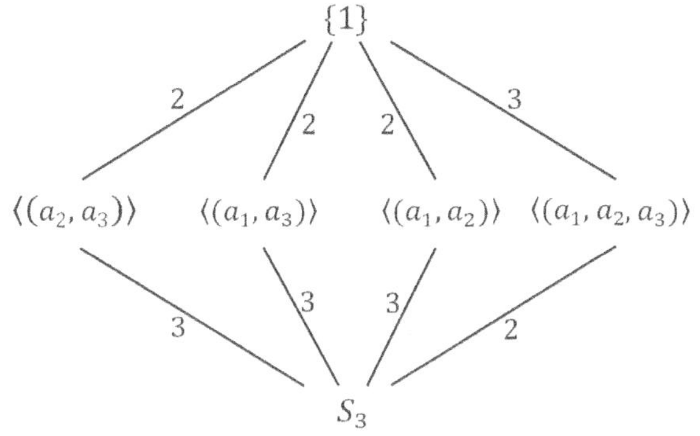

Hence the above lattice of fields is complete. $L|\mathbb{Q}, \mathbb{Q}|\mathbb{Q}, \mathbb{Q}(\omega)|\mathbb{Q}$ and $L|\mathbb{Q}(a_i)$ are Galois extensions while $\mathbb{Q}(a_i)|\mathbb{Q}$ with $i = 1, 2, 3$ are not Galois extensions.

15.5 Exercises

1. Let $K \subset M \subset L$ be a chain of fields and let $\phi : \operatorname{Aut}(L|K) \to \operatorname{Aut}(M|K)$ be defined by $\phi(\alpha) = \alpha_{|M}$. Show that ϕ is an epimorphism with kernel $\ker(\phi) = \operatorname{Aut}(L|M)$.

2. Show that $\mathbb{Q}(5^{\frac{1}{4}})|\mathbb{Q}(\sqrt{5})$ and $\mathbb{Q}(\sqrt{5})|\mathbb{Q}$ are Galois extensions and $\mathbb{Q}(5^{\frac{1}{4}})|\mathbb{Q}$ is not a Galois extension.

3. Let $L|K$ be a field extension and $u, v \in L$ algebraic over K with $|K(u) : K| = m$ and $|K(v) : K| = n$. If m and n are coprime, then $|K(u, v) : K| = n \cdot m$.

4. Let p, q be prime numbers with $p \neq q$. Let $L = \mathbb{Q}(\sqrt{p}, q^{\frac{1}{3}})$. Show that $L = \mathbb{Q}(\sqrt{p} \cdot q^{\frac{1}{3}})$. Determine a basis of L over $\mathbb{Q}$ and the minimal polynomial of $\sqrt{p} \cdot q^{\frac{1}{3}}$.

5. Let $K = \mathbb{Q}(2^{\frac{1}{n}})$ with $n \geq 2$.

   (i) Determine the number of $\mathbb{Q}$-embeddings $\sigma : K \to \mathbb{R}$. Show that for each such embedding we have $\sigma(K) = K$.

   (ii) Determine $\operatorname{Aut}(K|\mathbb{Q})$.

6. Let $\alpha = \sqrt{5 + 2\sqrt{5}}$.
   (i) Determine the minimal polynomial of α over $\mathbb{Q}$.
   (ii) Show that $\mathbb{Q}(a)|\mathbb{Q}$ is a Galois extension.
   (iii) Determine $\mathrm{Aut}(\mathbb{Q}(a)|\mathbb{Q})$.
7. Let K be a field of prime characteristic p and let $f(x) = x^p - x + a \in K$ be an irreducible polynomial. Let $L = K(v)$, where v is a zero of $f(x)$.
   (i) If α is a zero of $f(x)$ then also $\alpha + 1$.
   (ii) $L|K$ is a Galois extension.
   (iii) There is exactly one K-automorphism σ of L with $\sigma(v) = v + 1$.
   (iv) The Galois group $\mathrm{Aut}(L|K)$ is cyclic with generating element σ.

Chapter 16
Separable Field Extensions

16.1 Separability of Fields and Polynomials

In the previous chapter we introduced and examined Galois extensions. Recall that $L|K$ is a *Galois extension* if there exists a finite subgroup $G \subset \mathrm{Aut}(L)$ such that $K = \mathrm{Fix}(L, G)$. The following questions immediately arise.

(1) Under what conditions is a field extension $L|K$ a Galois extension?

(2) When is $L|K$ a Galois extension when L is the splitting field of a polynomial $f(x) \in K[x]$?

In this chapter we consider these questions and completely characterize Galois extensions. In order to do this we must introduce separable extensions.

Definition 16.1.1. Let K be a field. Then a nonconstant polynomial $f(x) \in K[x]$ is called *separable over* K if each irreducible factor of $f(x)$ has only simple zeros in its splitting field.

We now extend this definition to field extensions.

Definition 16.1.2. Let $L|K$ be a field extension and $a \in L$. Then a is *separable over* K if a is a zero of a separable polynomial. The field extension $L|K$ is a *separable field extension* or just *separable* if all $a \in L$ are separable over K. In particular a separable extension is an algebraic extension.

Finally we consider fields where every nonconstant polynomial is separable.

Definition 16.1.3. A field K is *perfect* if each nonconstant polynomial in $K[x]$ is separable over K.

The following is straightforward from the definitions. An element a is separable over K if and only if its minimal polynomial $m_a(x)$ is separable.

If $f(x) \in K[x]$ then $f(x) = \sum_{i=0}^{n} k_i x^i$ with $k_i \in K$. The *formal derivative* of $f(x)$ is then $f'(x) = \sum_{i=1}^{n} i k_i x^{i-1}$. As in ordinary Calculus we have the usual differentiation rules

$$(f(x) + g(x))' = f'(x) + g'(x)$$

and

$$(f(x)g(x))' = f'(x)g(x) + f(x)g'(x)$$

for $f(x), g(x) \in K[x]$.

Lemma 16.1.4. *Let K be a field and $f(x)$ an irreducible nonconstant polynomial in $K[x]$. Then $f(x)$ is separable if and only if its formal derivative is nonzero.*

Proof. Let L be the splitting field of $f(x)$ over K. Let $f(x) = (x-a)^r g(x)$ where $(x-a)$ does not divide $g(x)$. Then

$$f'(x) = (x-a)^{r-1}(rg(x) + (x-a)g'(x)).$$

If $f'(x) \neq 0$ then a is a zero of $f(x)$ in L over K of multiplicity $m \geq 2$ if and only if $(x-a) | f(x)$ and also $(x-a) | f'(x)$.

Let $f(x)$ be a separable polynomial over $K[x]$ and let a be a zero of $f(x)$ in L. Then if $f(x) = (x-a)^r g(x)$ with $(x-a)$ not dividing $g(x)$ we must have $r = 1$. Then

$$f'(x) = g(x) + (x-a)g'(x).$$

If $g'(x) = 0$ then $f'(x) = g(x) \neq 0$. Now suppose that $g'(x) \neq 0$. Assume that $f'(x) = 0$. Then necessarily $(x-a)|g(x)$ giving a contradiction. Therefore $f'(x) \neq 0$.

Conversely suppose that $f'(x) \neq 0$. Assume that $f(x)$ is not separable. Then both $f(x)$ and $f'(x)$ have a common zero $a \in L$. Let $m_a(x)$ be the minimal polynomial of a in $K[x]$. Then $m_a(x)|f(x)$ and $m_a(x)|f'(x)$. Since $f(x)$ is irreducible then the degree of $m_a(x)$ must equal the degree of $f(x)$. But $m_a(x)$ must also have the same degree as $f'(x)$ which is less than that of $f(x)$ giving a contradiction. Therefore $f(x)$ must be separable. □

We now consider the following example of a nonseparable polynomial over the finite field $\mathbb{Z}_p$ of p elements. We will denote this field now as GF(p), the Galois field of p elements.

Example 16.1.5. Let $K = \text{GF}(p)$ and $L = K(t)$ the field of rational functions in t over K. Consider the polynomial $f(x) = x^p - t \in L[x]$.

Now $K[t]/tK[t] \cong K$. Since K is a field this implies that $tK[t]$ is a maximal ideal and hence a prime ideal in $K[t]$ with prime element $t \in K[t]$ (see Theorem 3.2.7). By the Eisenstein criteria $f(x)$ is an irreducible polynomial in $L[x]$ (see Theorem 4.4.8). However $f'(x) = px^{p-1} = 0$ since $\text{char}(K) = p$. Therefore $f(x)$ is not separable.

16.2 Perfect Fields

We now consider when a field K is perfect. First we show that in general any field of characteristic 0 is perfect. In particular the rationals $\mathbb{Q}$ are perfect and hence any extension of the rationals is separable.

Theorem 16.2.1. *Each field K of characteristic zero is perfect.*

Proof. Suppose that K is a field with $\operatorname{char}(K) = 0$. Suppose that $f(x)$ is a nonconstant polynomial in $K[x]$. Then $f'(x) \neq 0$. If $f(x)$ is irreducible then $f(x)$ is separable from Lemma 16.1.4. Therefore by definition each nonconstant polynomial $f(x) \in K[x]$ is separable. □

We remark that in the original motivation for Galois theory the ground field was the rationals $\mathbb{Q}$. Since this has characteristic zero it is perfect and all extensions are separable. Hence the question of separability didn't arise until the question of extensions of fields of prime characteristic arose.

Corollary 16.2.2. *Any finite extension of the rationals $\mathbb{Q}$ is separable.*

We now consider the case of prime characteristic.

Theorem 16.2.3. *Let K be a field with* $\operatorname{char}(K) = p \neq 0$. *If $f(x)$ is a nonconstant polynomial in $K[x]$ then the following are equivalent:*

(1) $f'(x) = 0$.

(2) *$f(x)$ is a polynomial in x^p, that is, there is a $g(x) \in K[x]$ with $f(x) = g(x^p)$.*

If in (1) *and* (2) *$f(x)$ is irreducible then $f(x)$ is not separable over K if and only if $f(x)$ is a polynomial in x^p.*

Proof. Let $f(x) = \sum_{i=1}^{n} a_i x^i$. Then $f'(x) = 0$ if and only if $p|i$ for all i with $a_i \neq 0$. But this is equivalent to

$$f(x) = a_0 + a_p x^p + \cdots + a_m x^{mp}.$$

If $f(x)$ is irreducible then we have that $f(x)$ is not separable if and only if $f'(x) = 0$ from Lemma 16.1.4. □

Theorem 16.2.4. *Let K be a field with* $\operatorname{char}(K) = p \neq 0$. *Then the following are equivalent:*

(1) *K is perfect.*

(2) *Each element in K has a p-th root in K.*

(3) *The Frobenius homomorphism $x \mapsto x^p$ is an automorphism of K.*

Proof. First we show that (1) implies (2). Suppose that K is perfect and $a \in K$. Then $x^p - a$ is separable over K. Let $g(x) \in K[x]$ be an irreducible factor of $x^p - a$. Let L be the splitting field of $g(x)$ over K and b a zero of $g(x)$ in L. Then $b^p = a$. Further $x^p - b^p = (x-b)^p \in L[x]$ since the characteristic of K is p. Hence $g(x) = (x-b)^s$ and then s must equal 1 since $g(x)$ is irreducible. Therefore $b \in K$ and b is a p-th root of a.

Now we show that (2) implies (3). Recall that the Frobenius homomorphism $\tau : x \mapsto x^p$ is injective (see Theorem 1.8.8). We must show that it is also surjective. Let

$a \in K$ and let b be a p-th root of a so that $a = b^p$. Then $\tau(b) = b^p = a$ and τ is surjective.

Finally we show that (3) implies (1). Let $\tau : x \mapsto x^p$ be surjective. It follows that each $a \in K$ has a p-th root in K. Now let $f(x) \in K[x]$ be irreducible. Assume that $f(x)$ is not separable. From Theorem 16.2.3 there is a $g(x) \in K[x]$ with $f(x) = g(x^p)$, that is

$$f(x) = a_0 + a_1 x^p + \cdots + a_m x^{mp}.$$

Let $b_i \in K$ with $a_i = b_i^p$. Then

$$f(x) = b_o^p + b_1^p x^p + \cdots + b_m^p x^{mp} = (b_0 + b_1 x + \cdots + b_m x^m)^p.$$

However this is a contradiction since $f(x)$ is irreducible. Therefore $f(x)$ is separable completing the proof. □

Theorem 16.2.5. *Let K be a field with $\operatorname{char}(K) = p \neq 0$. Then each element of K has at most one p-th power in K.*

Proof. Suppose that $b_1, b_2 \in K$ with $b_1^p = b_2^p = a$. Then

$$0 = b_1^p - b_2^p = (b_1 - b_2)^p.$$

Since K has no zero divisors it follows that $b_1 = b_2$. □

16.3 Finite Fields

In this section we consider finite fields. In particular we show that if K is a finite field then $|K| = p^m$ for some prime p and natural number $m > 0$. Further we show that if K_1, K_2 are finite fields with $|K_1| = |K_2|$ then $K_1 \cong K_2$. Hence there is a unique finite field for each possible order.

Notice that if K is a finite field then by necessity $\operatorname{char} K = p \neq 0$. We first show that in this case K is always perfect.

Theorem 16.3.1. *A finite field is perfect.*

Proof. Let K be a finite field of characteristic $p > 0$. Then the Frobenius map $\tau : x \mapsto x^p$ is surjective since its injective and K is finite. Therefore K is perfect from Theorem 16.2.4. □

Next we show that each finite field has order p^m for some prime p and natural number $m > 0$.

Lemma 16.3.2. *Let K be a finite field. Then $|K| = p^m$ for some prime p and natural number $m > 0$.*

Proof. Let K be a finite field with characteristic $p > 0$. Then K can be considered as a vector space over $K = \mathrm{GF}(p)$ and hence of finite dimension since $|K| < \infty$. If $\alpha_1, \dots, \alpha_m$ is a basis then each $f \in K$ can be written as $f = c_1\alpha_1 + \dots + c_n\alpha_m$ with each $c_i \in \mathrm{GF}(p)$. Hence there are p choices for each c_i and therefore p^m choices for each f. □

In Theorem 9.5.16 we proved that any finite subgroup of the multiplicative group of a field is cyclic. If K is a finite field then its multiplicative subgroup $K^\star$ is finite and hence is cyclic.

Lemma 16.3.3. *Let K be a finite field. Then its multiplicative subgroup $K^\star$ is cyclic.*

If K is a finite field with order p^m then its multiplicative subgroup $K^\star$ has order $p^m - 1$. Then from Lagrange's theorem each nonzero element to the power p^m is the identity. Therefore we have the result.

Lemma 16.3.4. *Let K be a field of order p^m. Then each $\alpha \in K$ is a zero of the polynomial $x^{p^m} - x$. In particular if $\alpha \neq 0$ then α is a zero of $x^{p^m-1} - 1$.*

If K is a finite field of order p^m, it is a finite extension of $\mathrm{GF}(p)$. Since the multiplicative group is cyclic we must have $K = \mathrm{GF}(p)(\alpha)$ for some $\alpha \in K$. From this we obtain that for a given possible finite order there is only one finite field up to isomorphism.

Theorem 16.3.5. *Let K_1, K_2 be finite fields with $|K_1| = |K_2|$. Then $K_1 \cong K_2$.*

Proof. Let $|K_1| = |K_2| = p^m$. From the remarks above $K_1 = \mathrm{GF}(p)(\alpha)$ where α has order $p^m - 1$ in $K_1^\star$. Similarly $K_2 = \mathrm{GF}(p)(\beta)$ where β also has order $p^m - 1$ in $K_2^\star$. Hence $\mathrm{GF}(p)(\alpha) \cong \mathrm{GF}(p)(\beta)$ and therefore $K_1 \cong K_2$. □

In Lemma 16.3.2 we saw that if K is a finite field then $|K| = p^n$ for some prime p and positive integer n. We now show that given a prime power p^n there does exist a finite field of that order.

Theorem 16.3.6. *Let p be a prime and $n > 0$ a natural number. Then there exists a field K of order p^n.*

Proof. Given a prime p consider the polynomial $g(x) = x^{p^n} - x \in \mathrm{GF}(p)[x]$. Let K be the splitting field of this polynomial over $\mathrm{GF}(p)$. Since a finite field is perfect K is a separable extension and hence all the zeros of $g(x)$ are distinct in K.

Let F be the set of p^n distinct zeros of $g(x)$ within K. Let $a, b \in F$. Since

$$(a \pm b)^{p^n} = a^{p^n} \pm b^{p^n} \quad \text{and} \quad (ab)^{p^n} = a^{p^n} b^{p^n}$$

it follows that F forms a subfield of K. However F contains all the zeros of $g(x)$ and since K is the smallest extension of GF(p) containing all the zeros of $g(x)$ we must have $K = F$. Since F has p^n elements it follows that the order of K is p^n. □

Combining Theorems 16.3.5 and 16.3.6 we get the following summary result indicating that up to isomorphism there exists one and only one finite field of order p^n.

Theorem 16.3.7. *Let p be a prime and $n > 0$ a natural number. Then up to isomorphism there exists a unique finite field of order p^n.*

16.4 Separable Extensions

In this section we consider some properties of separable extensions.

Theorem 16.4.1. *Let K be a field with $K \subset L$ and L algebraically closed. Let $\alpha : K \to L$ be a monomorphism. Then the number of monomorphisms $\beta : K(a) \to L$ with $\beta_{|K} = \alpha$ is equal to the number of pairwise distinct zeros in L of the minimal polynomial m_a of a over K.*

Proof. Let β be as in the statement of the theorem. Then β is uniquely determined by $\beta(a)$ and $\beta(a)$ is a zero of the polynomial $\beta(m_a(x)) = \alpha(m_a(x))$. Now let a' be a zero of $\alpha(m_a(x))$ in L. Then there exists a $\beta : K(a) \to L$ with $\beta(a) = a'$ from Theorem 7.1.4. Therefore α has exactly as many extensions β as $\alpha(m_a(x))$ has pairwise distinct zeros in L. The number of pairwise distinct zeros of $\alpha(m_a(x))$ is equal to the number of pairwise distinct zeros of $m_a(x)$. This can be seen as follows. Let L_0 be a splitting field of $m_a(x)$ and $L_1 \subset L$ a splitting field of $\alpha(m_a(x))$. From Theorems 8.1.5 and 8.1.6 there is an isomorphism $\psi : L_0 \to L_1$ which maps the zeros of $m_a(x)$ onto the zeros of $\alpha(m_a(x))$. □

Lemma 16.4.2. *Let $L|K$ be a finite extension with $L \subset \overline{L}$ and $\overline{L}$ algebraically closed. In particular $L = K(a_1, \dots, a_n)$ where the a_i are algebraic over K. Let k_i be the number of pairwise distinct zeros of the minimal polynomial m_{a_i} of a_i over $K(a_1, \dots, a_{n-1})$ in $\overline{L}$. Then there are exactly $k_1, \dots, k_n$ monomorphisms $\beta : L \to \overline{L}$ with $\beta_{|K} = 1_K$.*

Proof. From Theorem 16.4.1 there are exactly p_1 monomorphisms $\alpha : K(a_1) \to \overline{L}$ with $\alpha_{|K}$ equal to the identity on K. Each such α has exactly p_2 extensions of the identity on K to $K(a_1, a_2)$. We now continue in this manner. □

Theorem 16.4.3. *Let $L|K$ be a field extension with M an intermediate field. If $a \in L$ is separable over K then it is also separable over M.*

Proof. This follows directly from the fact that the minimal polynomial of a over M divides the minimal polynomial of a over K. □

Theorem 16.4.4. *Let $L|K$ be a field extension. Then the following are equivalent.*

(1) *$L|K$ is finite and separable.*

(2) *There are finitely many separable elements $a_1, \dots, a_n$ over K with $K = K(a_1, \dots, a_n)$.*

(3) *$L|K$ is finite and if $L \subset \overline{L}$ with $\overline{L}$ algebraically closed then there are exactly $[L : K]$ monomorphisms $\alpha : L \to \overline{L}$ with $\alpha_{|K} = 1_K$.*

Proof. That (1) implies (2) follows directly from the definitions. We show then that (2) implies (3). Let $L = K(a_1, \dots, a_n)$ where $a_1, \dots, a_n$ are separable elements over K. The extension $L|K$ is finite (see Theorem 5.3.4). Let p_i be the number of pairwise distinct zeros in $\overline{L}$ of the minimal polynomial $m_{a_i}(x) = f_i(x)$ of a_i over $K(a_1, \dots, a_{i-1})$. Then $p_i \le \deg(f_i) = |K(a_1, \dots, a_i) : K(a_1, \dots, a_{i-1})|$. Hence $p_i = \deg(f_i(x))$ since a_i is separable over $K(a_1, \dots, a_{i-1})$ from Theorem 16.4.3. Therefore

$$[L : K] = p_1 \cdots p_n$$

is equal to the number of monomorphisms $\alpha : L \to \overline{L}$ with $\alpha_{|K}$ the identity on K.

Finally we show that (3) implies (1). Suppose then the conditions of (3). Since $L|K$ is finite there are finitely many $a_1, \dots, a_n \in L$ with $L = K(a_1, \dots, a_n)$. Let p_i and $f_i(x)$ be as in the proof above and hence $p_i \le \deg(f_i(x))$. By assumption we have

$$[L : K] = p_1 \cdots p_n$$

equal to the number of monomorphisms $\alpha : L \to \overline{L}$ with $\alpha_{|K}$ the identity on K. Also

$$[L : K] = p_1 \cdots p_n \le \deg(f_1(x)) \cdots \deg(f_n(x)) = [L : K].$$

Hence $p_i = \deg(f_i(x))$. Therefore by definition each a_i is separable over K.

To complete the proof we must show that $L|K$ is separable. Inductively it suffices to prove that $K(a_1)|K$ is separable over K whenever a_1 is separable over K and not in K.

This is clear if $\operatorname{char}(K) = 0$ because K is perfect. Suppose then that $\operatorname{char}(K) = p > 0$. First we show that $K(a_1^p) = K(a_1)$. Certainly $K(a_1^p) \subset K(a_1)$. Assume that $a_1 \notin K(a_1^p)$. Then $g(x) = x^p - a_1^p$ is the minimal polynomial of a_1 over K. This follows from the fact that $x^p - a_1^p = (x - a_1)^p$ and hence there can be no irreducible factor of $x^p - a_1^p$ of the form $(x - a_1)^m$ with $m < p$ and $m|p$.

However it follows then in this case that $g'(x) = 0$ contradicting the separability of a_1 over K. Therefore $K(a_1) = K(a_1^p)$.

Let $E = K(a_1)$ then also $E = K(E^p)$ where E^p is the field generated by the p-th powers of E. Now let $b \in E = K(a_1)$. We must show that the minimal polynomial of b, say $m_b(x)$, is separable over K.

Assume that $m_b(x)$ is not separable over K. Then

$$m_b(x) = \sum_{i=0}^{k} b_i x^{pi}, \quad b_i \in K, \; b_k = 1$$

from Theorem 16.2.3. We have

$$b_0 + b_1 b^p + \cdots + b_k b^{pk} = 0.$$

Therefore the elements $1, b^p, \dots, b^{pk}$ are linearly dependent over K. Since $K(a_1) = E = K(E^p)$ we find that $1, b, \dots, b^k$ are linearly dependent also since if they were independent the p-th powers would also be independent. However this is not possible since $k < \deg(m_b(x))$. Therefore $m_b(x)$ is separable over K and hence $K(a_1)|K$ is separable. Altogether $L|K$ is then finite and separable completing the proof. □

Theorem 16.4.5. *Let $L|K$ be a field extension and let M be an intermediate field. Then the following are equivalent.*

(1) *$L|K$ is separable.*

(2) *$L|M$ and $M|K$ are separable.*

Proof. We first show that (1) $\Rightarrow$ (2): If $L|K$ is separable then $L|M$ is separable by Theorem 16.4.3 and $M|K$ is separable.

Now suppose (2) and let $M|K$ and $L|M$ be separable. Let $a \in L$ and let

$$m_a(x) = f(x) = b_0 + \cdots + b_{n-1}x^{n-1} + x^n$$

be the minimal polynomial of a over M. Then $f(x)$ is separable. Let

$$M' = K(b_1, \dots, b_{n-1}).$$

We have $K \subset M' \subset M$ and hence $M'|K$ is separable since $M|K$ is separable. Further a is separable over M' since $f(x)$ is separable and $f(x) \in M'[x]$. From Theorem 16.4.1 there are $m = \deg(f(x)) = [M'(a) : M']$ extensions of $\alpha : M' \to \overline{M}$ with $\overline{M}$ the algebraic closure of M'.

Since $M'|K$ is separable and finite there are $[M' : K]$ monomorphisms $\alpha : M' \to \overline{M}$ from Theorem 16.4.4. Altogether there are $[M'(a) : K]$ monomorphisms $\alpha : M' \to \overline{M}$ with $\alpha_{|K}$ the identity on K. Therefore $M'(a)|K$ is separable from Theorem 16.4.4. Hence a is separable over K and then $L|K$ is separable. Therefore (2) implies (1). □

Theorem 16.4.6. *Let $L|K$ be a field extension and let $S \subset L$ such that all elements of S are separable over K. Then $K(S)|K$ is separable and $K[S] = K(S)$.*

Proof. Let W be the set of finite subsets of S. Let $T \in W$. From Theorem 16.4.4 we obtain that $K(T)|K$ is separable. Since each element of $K(S)$ is contained in some $K(T)$ we have that $K(S)|K$ is separable. Since all elements of S are algebraic we have that $K[S] = K(S)$, □

Theorem 16.4.7. *Let $L|K$ be a field extension. Then there exists in L a uniquely determined maximal field M with the property that $M|K$ is separable. If $a \in L$ is separable over M then $a \in M$. M is called the separable hull of K in L.*

Proof. Let S be the set of all elements in L which are separable over K. Define $M = K(S)$. Then $M|K$ is separable from Theorem 16.4.6. Now, let $a \in L$ be separable over M. Then $M(a)|M$ is separable from Theorem 16.4.4. Further $M(a)|K$ is separable from Theorem 16.4.5. It follows that $a \in M$. □

16.5 Separability and Galois Extensions

We now completely characterize Galois extensions $L|K$ as finite, normal, separable extensions.

Theorem 16.5.1. *Let $L|K$ be a field extension. Then the following are equivalent.*

(1) *$L|K$ is a Galois extension.*

(2) *L is the splitting field of a separable polynomial in $K[x]$.*

(3) *$L|K$ is finite, normal and separable.*

Therefore we may characterize Galois extensions of a field K as finite, normal and separable extensions of K.

Proof. Recall from Theorem 8.2.2 that an extension $L|K$ is *normal* if

(1) $L|k$ is algebraic and

(2) each irreducible polynomial $f(x) \in K[x]$ that has a zero in L splits into linear factors in $L[x]$.

Now suppose that $L|K$ is a Galois extension. Then $L|K$ is finite from Theorem 15.4.1. Let $L = K(b_1, \dots, b_m)$ and $m_{b_i}(x) = f_i(x)$ be the minimal polynomial of b_i over K. Let $a_{i_1}, \dots, a_{i_n}$ be the pairwise distinct elements from

$$H_i = \{\alpha(b_i) : \alpha \in \mathrm{Aut}(L|K)\}.$$

Define

$$g_i(x) = (x - a_{i_1}) \cdots (x - a_{i_n}) \in L[x].$$

If $\alpha \in \mathrm{Aut}(L|K)$ then $\alpha(g_i) = g_i$ since α permutes the elements of H_i. This means that the coefficients of $g_i(x)$ are in $\mathrm{Fix}(L, \mathrm{Aut}(L|K)) = K$. Further $g_i(x) \in K[x]$

because b_i is one of the a_{i_j} and $f_i(x)|g_i(x)$. The group $\mathrm{Aut}(L|K)$ acts transitively on $\{a_{i_1}, \dots, a_{i_n}\}$ by the choice of $a_{i_1}, \dots, a_{i_n}$. Therefore each $g_i(x)$ is irreducible (see Theorem 15.2.4). It follows that $f_i(x) = g_i(x)$. Now $f_i(x)$ has only simple zeros in L, that is no zero has multiplicity ≥ 2 and hence $f_i(x)$ splits over L. Therefore L is a splitting field of $f(x) = f_1(x) \cdots f_m(x)$ and $f(x)$ is separable by definition. Hence (1) implies (2).

Now suppose that L is a splitting field of the separable polynomial $f(x) \in K[x]$ and $L|K$ is finite. From Theorem 16.4.4 we get that $L|K$ is separable since $L = K(a_1, \dots, a_n)$ with each a_i separable over K. Therefore $L|K$ is normal from Definition 8.2.1. Hence (2) implies (3).

Finally suppose that $L|K$ is finite, normal and separable. Since $L|K$ is finite and separable from Theorem 16.4.4 there exist exactly $[L : K]$ monomorphisms $\alpha : L \to \overline{L}$, $\overline{L}$ the algebraic closure of L, with $\alpha_{|K}$ the identity on K. Since $L|K$ is normal these monomorphisms are already automorphisms of L from Theorem 8.2.2. Hence $[L : K] \leq |\mathrm{Aut}(L|K)|$. Further $|L : K| \geq |\mathrm{Aut}(L|K)|$ from Theorem 15.4.3. Combining these we have $[L : K] = \mathrm{Aut}(L|K)$ and hence $L|K$ is a Galois extension from Theorem 15.4.9. Therefore (3) implies (1) completing the proof. □

Recall that any field of characteristic 0 is perfect and therefore any finite extension is separable. Applying this to $\mathbb{Q}$ implies that the Galois extensions of the rationals are precisely the splitting fields of polynomials.

Corollary 16.5.2. *The Galois extensions of the rationals are precisely the splitting fields of polynomials in* $\mathbb{Q}[x]$.

Theorem 16.5.3. *Let $L|K$ be a finite, separable field extension. Then there exists an extension field M of L such that $M|K$ is a Galois extension.*

Proof. Let $L = K(a_1, \dots, a_n)$ with all a_i separable over K. Let $f_i(x)$ be the minimal polynomial of a_i over K. Then each $f_i(x)$ and hence also $f(x) = f_1(x) \cdots f_n(x)$ is separable over K. Let M be the splitting field of $f(x)$ over K. Then $M|K$ is a Galois extension from Theorem 16.5.1. □

Example 16.5.4. Let $K = \mathbb{Q}$ be the rationals and let $f(x) = x^4 - 2 \in \mathbb{Q}[x]$. From Chapter 8 we know that $L = \mathbb{Q}(\sqrt[4]{2}, i)$ is a splitting field of $f(x)$. By the Eisenstein criteria $f(x)$ is irreducible and $[L : \mathbb{Q}] = 8$. Moreover

$$\sqrt[4]{2}, \quad i\sqrt[4]{2}, \quad -\sqrt[4]{2}, \quad -i\sqrt[4]{2}$$

are the zeros of $f(x)$. Since the rationals are perfect, $f(x)$ is separable. $L|K$ is a Galois extension by Theorem 16.5.1. From the calculations in Chapter 15 we have

$$|\mathrm{Aut}(L|K)| = |\mathrm{Aut}(L)| = [L : K] = 8.$$

Let

$$G = \mathrm{Aut}(L|K) = \mathrm{Aut}(L|\mathbb{Q}) = \mathrm{Aut}(L).$$

We want to determine the subgroup lattice of the Galois group G. We show $G \cong D_4$ the dihedral group of order 8. Since there are 4 zeros of $f(x)$ and G permutes these G must be a subgroup of S_4 and since the order is 8, G is a 2-Sylow subgroup of S_4. From this we have that

$$G = \langle (2,4), (1,2,3,4) \rangle.$$

If we let $\tau = (2,4)$ and $\sigma = (1,2,3,4)$ we get the isomorphism between G and D_4. From Theorem 14.1.1 we know that $D_4 = \langle r, f; r^4 = f^2 = (rf)^2 = 1 \rangle$.

This can also be seen in the following manner. Let

$$a_1 = \sqrt[4]{2}, \quad a_2 = i\sqrt[4]{2}, \quad a_3 = -\sqrt[4]{2}, \quad a_4 = -i\sqrt[4]{2}.$$

Let $\alpha \in G$. α is determined if we know $\alpha(\sqrt[4]{2})$ and $\alpha(i)$. The possibilities for $\alpha(i)$ are i or $-i$ that is the zeros of $x^2 + 1$.

The possibilities for $\sqrt[4]{2}$ are the 4 zeros of $f(x) = x^4 - 2$. Hence we have 8 possibilities for α. These are exactly the elements of the group G. We have $\delta, \tau \in G$ with

$$\delta(\sqrt[4]{2}) = i\sqrt[4]{2}, \quad \delta(i) = i$$

and

$$\tau(\sqrt[4]{2}) = \sqrt[4]{2}, \quad \tau(i) = -i.$$

It is straightforward to show that δ has order 4, τ has order 2 and $\delta\tau$ has order 2. These define a group of order 8 isomorphic to D_4 and since G has 8 elements this must be all of G.

We now look at the subgroup lattice of G and then the corresponding field lattice. Let δ and τ be as above. Then G has 5 subgroups of order 2

$$\{1, \delta^2\}, \{1, \tau\}, \{1, \delta\tau\}, \{1, \delta^2\tau\}, \{1, \delta^3\tau\}.$$

Of these only $\{1, \delta^2\}$ is normal in G.

G has 3 subgroups of order 4

$$\{1, \delta, \delta^2, \delta^3\}, \{1, \delta^2, \tau, \tau\delta^2\}, \{1, \delta^2, \delta\tau, \delta^3\tau\}$$

and all are normal since they all have index 2.

Hence we have the following subgroup lattice:

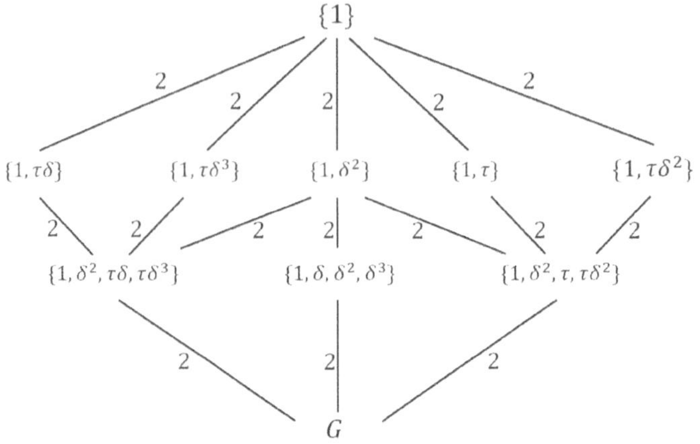

From this we construct the lattice of fields and intermediate fields. Since there are 10 proper subgroups of G, from the fundamental theorem of Galois theory there are 10 intermediate fields in $L|\mathbb{Q}$ namely the fix fields $\mathrm{Fix}(L, H)$ where H is a proper subgroup of G. In the identification the extension field corresponding to the whole group G is the ground field $\mathbb{Q}$ (recall that the lattice of fields is the inverted lattice of the subgroups), while the extension field corresponding to the identity is the whole field L. We now consider the other proper subgroups. Let δ, τ be as before.

(1) Consider $M_1 = \mathrm{Fix}(L, \{1, \tau\})$. Now $\{1, \tau\}$ fixes $\mathbb{Q}(\sqrt[4]{2})$ elementwise so that $\mathbb{Q}(\sqrt[4]{2}) \subset M_1$. Further $[L : M_1] = |\{1, \tau\}| = 2$ and hence $[L : \mathbb{Q}(\sqrt[4]{2})] = 2$. Therefore $M_1 = \mathbb{Q}(\sqrt[4]{2})$.

(2) Consider $M_2 = \mathrm{Fix}(L, \{1, \tau\delta\})$. We have

$$\begin{aligned}
\tau\delta(\sqrt[4]{2}) &= \tau(i\sqrt[4]{2}) = -i\sqrt[4]{2} \\
\tau\delta(i\sqrt[4]{2}) &= \tau(-\sqrt[4]{2}) = -\sqrt[4]{2} \\
\tau\delta(-\sqrt[4]{2}) &= \tau(-i\sqrt[4]{2}) = i\sqrt[4]{2} \\
\tau\delta(-i\sqrt[4]{2}) &= \tau(\sqrt[4]{2}) = \sqrt[4]{2}.
\end{aligned}$$

It follows that $\tau\delta$ fixes $(1-i)\sqrt[4]{2}$ and hence $M_2 = \mathbb{Q}((1-i)\sqrt[4]{2})$.

(3) Consider $M_3 = \mathrm{Fix}(L, \{1, \tau\delta^2\}$. The map $\tau\delta^2$ interchanges a_1 and a_3 and fixes a_2 and a_4. Therefore $M_3 = \mathbb{Q}(i\sqrt[4]{2})$.

In an analogous manner we can then consider the other 5 proper subgroups and corresponding intermediate fields. If we let $b_1 = (1-i)\sqrt[4]{2}$ and $b_2 = (1+i)\sqrt[4]{2}$ we get the following lattice of fields and subfields.

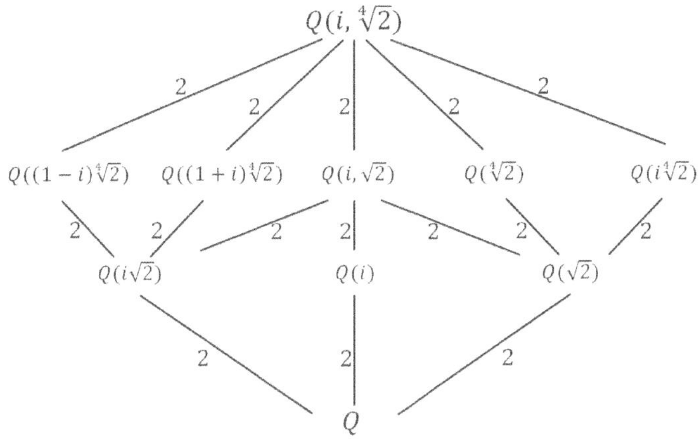

16.6 The Primitive Element Theorem

In this section we describe finite separable field extensions as simple extensions. If follows that a Galois extension is always a simple extension.

Theorem 16.6.1 (primitive element theorem). *Let $L = K(\gamma_1, \dots, \gamma_n)$ and suppose that each γ_i is separable over K. Then there exists a $\gamma_0 \in L$ such that $L = K(\gamma_0)$. The element γ_0 is called a primitive element.*

Proof. Suppose first that K is a finite field. Then L is also a finite field and therefore $L^\star = \langle \gamma_0 \rangle$ is cyclic. Therefore $L = K(\gamma_0)$ and the theorem is proved if K is a finite field.

Now suppose that K is infinite. Inductively it suffices to prove the theorem for $n = 2$. Hence let $\alpha, \beta \in L$ be separable over K. We must show that there exists a $\gamma \in L$ with $K(\alpha, \beta) = K(\gamma)$.

Let $\overline{L}$ be the splitting field of the polynomial $m_\alpha(x) m_\beta(x)$ over L where $m_\alpha(x)$, $m_\beta(x)$ are respectively the minimal polynomials of α, β over K. In $\overline{L}[x]$ we have

$$m_\alpha(x) = (x-\alpha_1)(x-\alpha_2)\cdots(x-\alpha_s) \quad \text{with } \alpha = \alpha_1$$
$$m_\beta(x) = (x-\beta_1)(x-\beta_2)\cdots(x-\beta_t) \quad \text{with } \beta = \beta_1.$$

By assumption the α_i and the β_j are respectively pairwise distinct.

For each pair (i, j) with $1 \le i \le s$, $2 \le j \le t$ the equation

$$\alpha_1 + z\beta_1 = \alpha_i + z\beta_j$$

has exactly one solution $z \in \overline{L}$ since $\beta_j - \beta_1 \neq 0$ if $j \geq 2$. Since K is infinite there exists a $c \in K$ with

$$\alpha_1 + c\beta_1 \neq \alpha_i + c\beta_j$$

for all i, j with $1 \leq i \leq s, 2 \leq j \leq t$. With such a value $c \in K$ we define

$$\gamma = \alpha + c\beta = \alpha_1 + c\beta_1.$$

We claim that $K(\alpha, \beta) = K(\gamma)$. It suffices to show that $\beta \in K(\gamma)$ for then $\alpha = \gamma - c\beta \in K(\gamma)$. This implies that $K(\alpha, \beta) \subset K(\gamma)$ and since $\gamma \in K(\alpha, \beta)$ it follows that $K(\alpha, \beta) = K(\gamma)$.

To show that $\beta \in K(\gamma)$ we first define $f(x) = m_\alpha(\gamma - cx)$ and let $d(x) = \gcd(f(x), m_\beta(x))$. We may assume that $d(x)$ is monic. We show that $d(x) = x - \beta$. Then $\beta \in K(\gamma)$ since $d(x) \in K(\gamma)[x]$.

Assume first that $d(x) = 1$. Then $\gcd(f(x), m_\beta(x)) = 1$ and $f(x)$ and $m_\beta(x)$ are also relatively prime in $\overline{L}[x]$. This is a contradiction since $f(x)$ and $m_\beta(x)$ have the common zero $\beta \in \overline{L}$ and hence the common divisor $x - \beta$.

Therefore $d(x) \neq 1$ so $\deg(d(x)) \geq 1$.

The polynomial $d(x)$ is a divisor of $m_\beta(x)$ and hence $d(x)$ splits into linear factors of the form $x - \beta_j$, $1 \leq j \leq t$ in $\overline{L}[x]$. The proof is completed if we can show that no linear factor of the form $x - \beta_j$ with $2 \leq j \leq t$ is a divisor of $f(x)$. That is, we must show that $f(\beta_j) \neq 0$ in $\overline{L}$ if $j \geq 2$.

Now $f(\beta_j) = m_\alpha(\gamma - c\beta_j) = m_\alpha(\alpha_1 + c\beta_1 - c\beta_j)$. Suppose that $f(\beta_j) = 0$ for some $j \geq 2$. This would imply that $\alpha_i = \alpha_1 + c\beta_1 - c\beta_j$, that is, $\alpha_1 + c\beta_1 = \alpha_j + c\beta_j$ for $j \geq 2$. This contradicts the choice of the value c. Therefore $f(\beta_j) \neq 0$ if $j \geq 2$ completing the proof. □

In the above theorem it is sufficient to assume that $n - 1$ of $\gamma_1, \ldots, \gamma_n$ are separable over K. The proof is similar. We only need that the $\beta_1, \ldots, \beta_t$ are pairwise distinct if β is separable over K to show that $K(\alpha, \beta) = K(\gamma)$ for some $\gamma \in L$.

If K is a perfect field then every finite extension is separable. Therefore we get the following corollary.

Corollary 16.6.2. *Let $L|K$ be a finite extension with K a perfect field. Then $L = K(\gamma)$ for some $\gamma \in L$.*

Corollary 16.6.3. *Let $L|K$ be a finite extension with K a perfect field. Then there exist only finitely many intermediate fields E with $K \subset E \subset L$.*

Proof. Since K is a perfect field we have $L = K(\gamma)$ for some $\gamma \in L$. Let $m_\gamma(x) \in K[x]$ be the minimal polynomial of γ over K and let $\overline{L}$ be the splitting field of $m_\gamma(x)$ over K. Then $\overline{L}|K$ is a Galois extension and hence there are only finitely many intermediate fields between K and $\overline{L}$. Therefore also only finitely many between K and L. □

Suppose that $L|K$ is algebraic. Then in general $L = K(\gamma)$ for some $\gamma \in L$ if and only if there exist only finitely many intermediate fields E with $K \subset E \subset L$.

This condition on intermediate fields implies that $L|K$ is finite if $L|K$ is algebraic. Hence we have proved this result in the case that K is perfect. The general case is discussed in the book of S. Lang [8].

16.7 Exercises

1. Let $f(x) = x^4 - 8x^3 + 24x^2 - 32x + 14 \in \mathbb{Q}[x]$ and let $v \in \mathbb{C}$ be a zero of f. Let $\alpha := v(4-v)$ and K a splitting field of f over $\mathbb{Q}$. Show:
   (i) f is irreducible over $\mathbb{Q}$ and $f(x) = f(4-x)$.
   (ii) There is exactly one automorphism σ of $\mathbb{Q}(v)$ with $\sigma(v) = 4 - v$.
   (iii) $L := \mathbb{Q}(\alpha)$ is the Fix field of σ and $|L : \mathbb{Q}| = 2$.
   (iv) Determine the minimal polynomial of α over $\mathbb{Q}$ and determine α.
   (v) $|\mathbb{Q}(v) : L| = 2$ and determine the minimal polynomial of v over L and determine v and all other zeros of $f(x)$.
   (vi) Determine the degree of $|K : \mathbb{Q}|$.
   (vii) Determine the structure of $\mathrm{Aut}(K|\mathbb{Q})$.
2. Let $L|K$ be a field extension and $f \in K[x]$ a separable polynomial. Let Z be a splitting field of f over L and Z_0 a splitting field of f over K. Show that $\mathrm{Aut}(Z|L)$ is isomorphic to a subgroup of $\mathrm{Aut}(Z_0|K)$.
3. Let $L|K$ be a field extension and $v \in L$. For each element $c \in K$ it is $K(v+c) = K(v)$. For $c \neq 0$ it is $K(cv) = K(v)$.
4. Let $v = \sqrt{2} + \sqrt{3}$ and let $K = \mathbb{Q}(v)$. Show that $\sqrt{2}$ and $\sqrt{3}$ are presentable as a $\mathbb{Q}$-linear combination of $1, v, v^2, v^3$. Conclude that $K = \mathbb{Q}(\sqrt{2}, \sqrt{3})$.
5. Let L be the splitting field of $x^3 - 5$ over $\mathbb{Q}$ in $\mathbb{C}$. Determine a primitive element t of L over $\mathbb{Q}$.

Chapter 17
Applications of Galois Theory

17.1 Applications of Galois Theory

As we mentioned in Chapter 1 Galois theory was originally developed as part of the proof that polynomial equations of degree 5 or higher over the rationals cannot be solved by formulas in terms of radicals. In this chapter we do this first and prove the insolvability of the quintic by radicals. To do this we must examine in detail what we call radical extensions.

We then return to some geometric material we started in Chapter 6. There using general field extensions we proved the impossibility of certain geometric compass and straightedge constructions. Here we use Galois theory to consider constructible n-gons.

Finally we will use Galois theory to present a proof of the fundamental theorem of algebra which says essentially that the complex number field $\mathbb{C}$ is algebraically closed.

17.2 Field Extensions by Radicals

We would like to use Galois theory to prove the insolvability by radicals of polynomial equations of degree 5 or higher. To do this we must introduce *extensions by radicals* and *solvability by radicals*.

Definition 17.2.1. Let $L|K$ be a field extension.

(1) Each zero of a polynomial $x^n - a \in K[x]$ in L is called a *radical* (over K). We denote it by $\sqrt[n]{a}$ (if a more detailed identification is not necessary).

(2) L is called a *simple extension of K by a radical* if $L = K(\sqrt[n]{a})$ for some $a \in K$.

(3) L is called an *extension of K by radicals* if there is a chain of fields

$$K = L_0 \subset L_1 \subset \cdots \subset L_m = L$$

such that each L_i is a simple extension of L_{i-1} by a radical for each $i = 1, \ldots, m$.

(4) Let $f(x) \in K[x]$. Then the equation $f(x) = 0$ is *solvable by radicals* or just *solvable* if the splitting field of $f(x)$ over K is contained in an extension of K by radicals.

In proving the insolvability of the quintic we will look for necessary and sufficient conditions for the solvability of polynomial equations. Our main result will be that if $f(x) \in K[x]$ then $f(x) = 0$ is solvable over K if the Galois group of the splitting field of $f(x)$ over K is a solvable group (see Chapter 11).

In the remainder of this section we assume that *all fields have characteristic zero*. The next theorem gives a characterization of simple extensions by radicals.

Theorem 17.2.2. *Let $L|K$ be a field extension and $n \in \mathbb{N}$. Assume that the polynomial $x^n - 1$ splits into linear factors in $K[x]$ so that K contains all the n-th roots of unity. Then $L = K(\sqrt[n]{a})$ for some $a \in K$ if and only if L is a Galois extension over K and* $\operatorname{Aut}(L|K) = \mathbb{Z}/m\mathbb{Z}$ *for some $m \in \mathbb{N}$ with $m|n$.*

Proof. The n-th roots of unity, that is the zeros of the polynomial $x^n - 1 \in K[x]$, form a cyclic multiplicative group $\mathcal{F} \subset K^\star$ of order n since each finite subgroup of the multiplicative group $K^\star$ of K is cyclic and $|\mathcal{F}| = n$. We call an n-th root of unity ω *primitive* if $\mathcal{F} = \langle\omega\rangle$.

Now let $L = K(\sqrt[n]{a})$ with $a \in K$, that is, $L = K(\beta)$ with $\beta^n = a \in K$. Let ω be a primitive n-th root of unity. With this β the elements $\omega\beta, \omega^2\beta, \dots, \omega^n\beta = \beta$ are zeros of $x^n - a$. Hence the polynomial $x^n - a$ splits into linear factors over L and hence $L = K(\beta)$ is a splitting field of $x^n - a$ over K. It follows that $L|K$ is a Galois extension.

Let $\sigma \in \operatorname{Aut}(L|K)$. Then $\sigma(\beta) = \omega^\nu\beta$ for some $0 < \nu \leq n$. The element ω^ν is uniquely determined by σ and we may write $\omega^\nu = \omega_\sigma$.

Consider the map $\phi : \operatorname{Aut}(L|K) \to \mathcal{F}$ given by $\sigma \to \omega_\sigma$ where ω_σ is defined as above by $\sigma(\beta) = \omega_\sigma\beta$. If $\tau, \sigma \in \operatorname{Aut}(L|K)$ then

$$\sigma\tau(\beta) = \sigma(\omega_\tau)\sigma(\beta) = \omega_\tau\omega_\sigma\beta$$

because $\omega_\tau \in K$.

Therefore $\phi(\sigma\tau) = \phi(\sigma)\phi(\tau)$ and hence ϕ is a homomorphism. The kernel $\ker(\phi)$ contains all the K-automorphisms of L for which $\sigma(\beta) = \beta$. However since $K = K(\beta)$ it follows that $\ker(\phi)$ contains only the identity. The Galois group $\operatorname{Aut}(L|K)$ is therefore isomorphic to a subgroup of $\mathcal{F}$. Since $\mathcal{F}$ is cyclic of order n we have that $\operatorname{Aut}(L|K)$ is cyclic of order m for some $m|n$ completing one way in the theorem.

Conversely first suppose that $L|K$ is a Galois extension with $\operatorname{Aut}(L|K) = \mathbb{Z}_n$ a cyclic group of order n. Let σ be a generator of $\operatorname{Aut}(L|K)$. This is equivalent to

$$\operatorname{Aut}(L|K) = \{\sigma, \sigma^2, \dots, \sigma^n = 1\}.$$

Let ω be a primitive n-th root of unity. Then by assumption $\omega \in K$, $\sigma(\omega) = \omega$ and $\mathcal{F} = \{\omega, \omega^2, \dots, \omega^n = 1\}$. Further the pairwise distinct automorphism σ^ν, $\nu = 1, 2, \dots, n$, of L are linearly independent, that is there exists an $\eta \in L$ such that

$$\omega \star \eta = \sum_{\nu=1}^{n} \omega^\nu\sigma^\nu(\eta) \neq 0.$$

The element $\omega \star \eta$ is called the *Lagrange resolvent* of ω by η. We fix such an element $\eta \in L$. Then we get, since $\sigma(\omega) = \omega$,

$$\sigma(\omega \star \eta) = \sum_{\nu=1}^{n} \omega^{\nu} \sigma^{\nu+1}(\eta) = \omega^{-1} \sum_{\nu=1}^{n} \omega^{\nu+1} \sigma^{\nu+1}(\eta) = \omega^{-1} \sum_{\nu=2}^{n+1} \omega^{\nu} \sigma^{\nu}(\eta)$$
$$= \omega^{-1} \sum_{\nu=1}^{n} \omega^{\nu} \sigma^{\nu}(\eta) = \omega^{-1}(\omega \star \eta).$$

Further $\sigma^{\mu}(\omega \star \eta) = \omega^{-\mu}(\omega \star \eta)$, $\mu = 1, 2, \ldots, n$. Hence the only K-automorphism of L which fixes $\omega \star \eta$ is the identity. Therefore $\mathrm{Aut}(L|K(\omega \star \eta)) = \{1\}$ and hence $L = K(\omega \star \eta)$ by the fundamental theorem of Galois theory.

Further

$$\sigma((\omega \star \eta)^n) = (\sigma(\omega \star \eta))^n = (\omega^{-1}(\omega \star \eta))^n = \omega^{-n}(\omega \star \eta)^n = (\omega \star \eta)^n.$$

Therefore $(\omega \star \eta)^n \in \mathrm{Fix}(L, \mathrm{Aut}(L|K)) = K$ again from the fundamental theorem of Galois theory. If $a = (\omega \star \eta)^n \in K$ then first $a \in K$ and second $L = K(\sqrt[n]{a}) = K(\omega \star \eta)$. This proves the result in the case where $m = n$. We now use this to prove it in general.

Suppose that $L|K$ is a Galois extension with $\mathrm{Aut}(L|K) = \mathbb{Z}_m$ a cyclic group of order m where $n = qm$ for some $q \geq 1$. If $n = qm$ then $L = K(\sqrt[m]{b})$ for some $b \in K$ by the above argument. Hence $L = K(\beta)$ with $\beta^m \in K$. Then certainly $a = \beta^n = (\beta^m)^q \in K$ and therefore $L = K(\beta) = K(\sqrt[n]{a})$ for some $a \in K$ completing the general case. □

We next show that every extension by radicals is contained in a Galois extension by radicals.

Theorem 17.2.3. *Each extension L of K by radicals is contained in a Galois extension $\tilde{L}$ of K by radicals. This means that there is an extension $\tilde{L}$ of K by radicals with $L \subset \tilde{L}$ and $\tilde{L}|K$ is a Galois extension.*

Proof. We use induction on the degree $m = [L : K]$. Suppose that $m = 1$. If $L = K(\sqrt[n]{a})$ then if ω is a primitive n-th root of unity define $\tilde{K} = K(\omega)$ and $\tilde{L} = \tilde{K}(\sqrt[n]{a})$. We then get the chain $K \subset \tilde{K} \subset \tilde{L}$ with $L \subset \tilde{L}$ and $\tilde{L}|K$ is a Galois extension. This last statement is due to the fact that $\tilde{L}$ is the splitting field of the polynomial $x^n - a \in K[x]$ over K. Hence the theorem is true if $m = 1$.

Now suppose that $m \geq 2$ and suppose that the theorem is true for all extensions F of K by radicals with $[F : K] < m$.

Since $m \geq 2$ by the definition of extension by radicals there exists a simple extension $L|E$ by a radical. That is there exists a field E with

$$K \subset E \subset L, \quad [L : E] \geq 2$$

and $L = E(\sqrt[n]{a})$ for some $a \in E, n \in \mathbb{N}$. Now $[E : K] < m$ so be the inductive hypothesis there exists a Galois extension by radicals $\tilde{E}$ of K with $E \subset \tilde{E}$. Let $G = \operatorname{Aut}(\tilde{E}|K)$ and let $\tilde{L}$ be the splitting field of the polynomial $f(x) = m_a(x^n) \in K[x]$ over $\tilde{E}$ where $m_a(x)$ is the minimal polynomial of a over K. We show that $\tilde{L}$ has the desired properties.

Now $\sqrt[n]{a} \in L$ is a zero of the polynomial $f(x)$ and $E \subset \tilde{E} \subset \tilde{L}$. Therefore $\tilde{L}$ contains an E-isomorphic image of $L = K(\sqrt[n]{a})$ and hence we may consider $\tilde{L}$ as an extension of L.

Since $\tilde{E}$ is a Galois extension of K the polynomial $f(x)$ may be factored as

$$f(x) = (x^n - \alpha_1) \cdots (x^n - \alpha_s)$$

with $\alpha_i \in \tilde{E}$ for $i = 1, \dots, s$. All zeros of $f(x)$ in $\tilde{L}$ are radicals over $\tilde{E}$. Therefore $\tilde{L}$ is an extension by radicals of $\tilde{E}$. Since $\tilde{E}$ is also an extension by radicals of K we obtain that $\tilde{L}$ is an extension by radicals of K.

Since $\tilde{E}$ is a Galois extension of K we have that $\tilde{E}$ is a splitting field of a polynomial $g(x) \in K[x]$. Further $\tilde{L}$ is a splitting field of $f(x) \in K[x]$ over $\tilde{E}$. Altogether then we have that $\tilde{L}$ is a splitting field of $f(x)g(x) \in K[x]$ over K. Therefore $\tilde{L}$ is a Galois extension of K completing the proof. □

We will eventually show that a polynomial equation is solvable by radicals if and only if the corresponding Galois group is a solvable group. We now begin to find conditions where the Galois group is solvable.

Lemma 17.2.4. *Let $K = L_0 \subset L_1 \subset \cdots \subset L_r = L$ be a chain of fields such that the following hold:*

(i) *L is a Galois extension of K.*

(ii) *L_j is a Galois extension of L_{j-1} for $j = 1, \dots, r$.*

(iii) *$G_j = \operatorname{Aut}(L_j|L_{j-1})$ is abelian for $j = 1, \dots, r$.*

Then $G = \operatorname{Aut}(L|K)$ is solvable.

Proof. We prove the lemma by induction on r. If $r = 0$ then $G = \{1\}$ and there is nothing to prove. Suppose then that $r \geq 1$ and assume that the lemma holds for all such chains of fields with a length $r' < r$. Since $L_1|K$ is a Galois extension then $\operatorname{Aut}(L_1|K)$ is a normal subgroup of G by the fundamental theorem of Galois theory and further

$$G_1 = \operatorname{Aut}(L_1|K) = G/\operatorname{Aut}(L|L_1).$$

Since G_1 is an abelian group it is solvable and by assumption $\operatorname{Aut}(L|L_1)$ is solvable. Therefore G is solvable (see Theorem 12.2.4). □

Lemma 17.2.5. *Let $L|K$ be a field extension. Let $\tilde{K}$ and $\tilde{L}$ be the splitting fields of the polynomial $x^n - 1 \in K[x]$ over K and L respectively. Since $K \subset L$ we have $\tilde{K} \subset \tilde{L}$. Then the following hold:*

(1) *If* $\sigma \in \mathrm{Aut}(\tilde{L}|L)$ *then* $\sigma_{|\tilde{K}} \in \mathrm{Aut}(\tilde{K}|K)$ *and the map*

$$\mathrm{Aut}(\tilde{L}|L) \to \mathrm{Aut}(\tilde{K}|K) \quad \textit{given by } \sigma \mapsto \sigma_{|\tilde{K}}$$

is an injective homomorphism.

(2) *Suppose that in addition* $L|K$ *is a Galois extension. Then* $\tilde{L}|K$ *is also a Galois extension. If further* $\sigma \in \mathrm{Aut}(\tilde{L}|\tilde{K})$ *then* $\sigma_{|L} \in \mathrm{Aut}(L|K)$ *and*

$$\mathrm{Aut}(\tilde{L}|\tilde{K}) \to \mathrm{Aut}(L|K) \quad \textit{given by } \sigma \mapsto \sigma_{|L}$$

is an injective homomorphism.

Proof. (1) Let ω be a primitive n-th root of unity. Then $\tilde{K} = K(\omega)$ and $\tilde{L} = L(\omega)$. Each $\sigma \in \mathrm{Aut}(\tilde{L}|L)$ maps ω onto a primitive n-th root of unity and fixes $K \subset L$ elementwise. Hence from $\sigma \in \mathrm{Aut}(\tilde{L}|L)$ we get that $\sigma_{|\tilde{K}} \in \mathrm{Aut}(\tilde{K}|K)$. Certainly the map $\sigma \mapsto \sigma_{|\tilde{K}}$ defines a homomorphism $\mathrm{Aut}(\tilde{L}|L) \to \mathrm{Aut}(\tilde{K}|K)$. Let $\sigma_{|\tilde{K}} = 1$ with $\sigma \in \mathrm{Aut}(\tilde{L}|L)$. Then $\sigma(\omega) = \omega$ and therefore we have already that $\sigma = 1$ since $\tilde{L} = L(\omega)$.

(2) If L is the splitting field of a polynomial $g(x)$ over K then $\tilde{L}$ is the splitting field of $g(x)(x^n - 1)$ over K. Hence $\tilde{L}|K$ is a Galois extension. Therefore $K \subset L \subset \tilde{L}$ and $L|K, \tilde{L}|L$ and $\tilde{L}|K$ are all Galois extensions. Therefore from the fundamental theorem of Galois theory

$$\mathrm{Aut}(L|K) = \{\sigma_{|L}; \sigma \in \mathrm{Aut}(\tilde{L}|K)\}.$$

In particular $\sigma_{|L} \in \mathrm{Aut}(L|K)$ if $\sigma \in \mathrm{Aut}(\tilde{L}|\tilde{K})$. Certainly the map $\mathrm{Aut}(\tilde{L}|\tilde{K}) \to \mathrm{Aut}(L|K)$ given by $\sigma \mapsto \sigma_{|L}$ is a homomorphism. From $\sigma \in \mathrm{Aut}(\tilde{L}|\tilde{K})$ we get that $\sigma(\omega) = \omega$ where as above ω is a primitive n-th root of unity. Therefore if $\sigma_{|L} = 1$ then already $\sigma = 1$ since $\tilde{L} = L(\omega)$. Hence the map is injective. □

17.3 Cyclotomic Extensions

Very important in the solvability by radicals problem are the splitting fields of the polynomials $x^n - 1$ over $\mathbb{Q}$. These are called *cyclotomic fields*.

Definition 17.3.1. The splitting field of the polynomial $x^n - 1 \in \mathbb{Q}[x]$ with $n \geq 2$ is called the n-th *cyclotomic field* denoted k_n.

We have $k_n = \mathbb{Q}(\omega)$ where ω is a primitive n-th root of unity, for example $\omega = e^{\frac{2\pi i}{n}}$ over $\mathbb{Q}$. $k_n|\mathbb{Q}$ is a Galois extension and the Galois group $\mathrm{Aut}(k_n|\mathbb{Q})$ is the set of automorphisms $\sigma_m : \omega \to \omega^m$ with $1 \leq m \leq n$ and $\gcd(m, n) = 1$.

To understand this group G we need the following concept. A *prime residue class mod n* is a residue class $a + n\mathbb{Z}$ with $\gcd(a, n) = 1$. The set of the prime residue classes mod n is just the set of invertible elements with respect to multiplication of the $\mathbb{Z}/n\mathbb{Z}$. This forms a multiplicative group that we denote by $(\mathbb{Z}/n\mathbb{Z})^\star = P_n$. We have $|P_n| = \phi(n)$ where $\phi(n)$ is the Euler phi-function.

If $G = \operatorname{Aut}(k_n|\mathbb{Q})$ then clearly $G \cong P_n$ under the map $\sigma_m \mapsto m + n\mathbb{Z}$.

If $n = p$ is a prime number then $G = \operatorname{Aut}(k_n|\mathbb{Q})$ is cyclic with $|G| = p - 1$.

If $n = p^2$ then $|G| = |\operatorname{Aut}(k_{p^2}|\mathbb{Q})| = p(p-1)$ since

$$\frac{x^{p^2-1}}{x-1}\frac{x-1}{x^p-1} = x^{p(p-1)} + x^{p(p-1)-1} + \cdots + 1$$

and each primitive p-th root of unity is a zero of this polynomial.

Lemma 17.3.2. *Let K be a field and $\tilde{K}$ be the splitting field of $x^n - 1$ over K. Then* $\operatorname{Aut}(\tilde{K}|K)$ *is abelian.*

Proof. We apply Lemma 17.2.5 for the field extension $K|\mathbb{Q}$. This can be done since the characteristic of K is zero and $\mathbb{Q}$ is the prime field of K. It follows that $\operatorname{Aut}(\tilde{K}|K)$ is isomorphic to a subgroup of $\operatorname{Aut}(\tilde{\mathbb{Q}}|\mathbb{Q})$ from part (1) of Lemma 17.2.5. But $\tilde{\mathbb{Q}} = k_n$ and hence $\operatorname{Aut}(\tilde{\mathbb{Q}}|\mathbb{Q})$ is abelian. Therefore $\operatorname{Aut}(\tilde{K}|K)$ is abelian. □

17.4 Solvability and Galois Extensions

In this section we prove that solvability by radicals is equivalent to the solvability of the Galois group.

Theorem 17.4.1. *Let $L|K$ be a Galois extension of K by radicals. Then $G =$* $\operatorname{Aut}(L|K)$ *is a solvable group.*

Proof. Suppose that $L|K$ is a Galois extension. Then we have a chain of fields

$$K = L_0 \subset L_1 \subset \cdots \subset L_r = L$$

such that $L_j = L_{j-1}(\sqrt[n_j]{a_j})$ for some $a_j \in L_j$. Let $n = n_1 \cdots n_r$ and let $\tilde{L}_j$ be the splitting field of the polynomial $x^n - 1 \in K[x]$ over L_j for each $j = 0, 1, \ldots, r$. Then $\tilde{L}_j = \tilde{L}_{j-1}(\sqrt[n_j]{a_j})$ and we get the chain

$$K \subset \tilde{K} = \tilde{L}_0 \subset \tilde{L}_1 \subset \cdots \subset \tilde{L}_r = \tilde{L}.$$

From part (2) of Lemma 17.2.5 we get that $\tilde{L}|K$ is a Galois extension. Further $\tilde{L}_j|\tilde{L}_{j-1}$ is a Galois extension with $\operatorname{Aut}(\tilde{L}_j|\tilde{L}_{j-1})$ cyclic from Theorem 17.2.2. Especially $\operatorname{Aut}(\tilde{L}_j|\tilde{L}_{j-1})$ is abelian. The group $\operatorname{Aut}(\tilde{K}|K)$ is abelian from Lemma 17.3.2. Therefore we may apply Lemma 17.2.4 to the chain

$$K \subset \tilde{K} = \tilde{L}_0 \subset \cdots \subset \tilde{L}_r = \tilde{L}.$$

Therefore $\tilde{G} = \mathrm{Aut}(\tilde{L}|K)$ is solvable. The group $G = \mathrm{Aut}(L|K)$ is a homomorphic image of $\tilde{G}$ from the fundamental theorem of Galois theory. Since homomorphic images of solvable groups are still solvable (see Theorem 12.2.3) it follows that G is solvable. □

Lemma 17.4.2. *Let $L|K$ be a Galois extension and suppose that $G = \mathrm{Aut}(L|K)$ is solvable. Assume further that K contains all q-th roots of unity for each prime divisor q of $m = [L : K]$. Then L is an extension of K by radicals.*

Proof. Let $L|K$ be a Galois extension and suppose that $G = \mathrm{Aut}(L|K)$ is solvable and assume that K contains all the q-th roots of unity for each prime divisor q of $m = [L : K]$. We prove the result by induction on m.

If $m = 1$ then $L = K$ and the result is clear. Now suppose that $m \geq 2$ and assume that the result holds for all Galois extensions $L'|K'$ with $[L' : K'] < m$. Now $G = \mathrm{Aut}(L|K)$ is solvable and G is nontrivial since $m \geq 2$. Let q be a prime divisor of m. From Lemma 12.2.2 and Theorem 13.3.5 it follows that there is a normal subgroup H of G with G/H cyclic of order q. Let $E = \mathrm{Fix}(L, H)$. From the fundamental theorem of Galois theory $E|K$ is a Galois extension with $\mathrm{Aut}(E|K) \cong G/H$ and hence $\mathrm{Aut}(E|K)$ is cyclic of order q. From Theorem 17.2.2 $E|K$ is a simple extension of K by a radical. The proof is completed if we can show that L is an extension of E by radicals.

The extension $L|E$ is a Galois extension and the group $\mathrm{Aut}(L|E)$ is solvable since it is a subgroup of $G = \mathrm{Aut}(L|K)$. Each prime divisor p of $[L : E]$ is also a prime divisor of $m = [L : K]$ by the degree formula. Hence as an extension of K the field E contains all the p-th roots of unity. Finally

$$[L : E] = \frac{[L : K]}{[E : K]} = \frac{m}{q} < m.$$

Therefore $L|E$ is an extension of E by radicals from the inductive assumption completing the proof. □

17.5 The Insolvability of the Quintic

We are now able to prove the insolvability of the quintic. This is one of the most important applications of Galois theory. As we mentioned we do this by equating the solvability of a polynomial equation by radicals to the solvability of the Galois group of the splitting field of this polynomial.

Theorem 17.5.1. *Let K be a field of characteristic 0 and let $f(x) \in K[x]$. Suppose that L is the splitting field of $f(x)$ over K. Then the polynomial equation $f(x) = 0$ is solvable by radicals if and only if $\mathrm{Aut}(L|K)$ is solvable.*

Proof. Suppose first that $f(x) = 0$ is solvable by radicals. Then L is contained in an extension L' of K by radicals. Hence L is contained in a Galois extension $\tilde{L}$ of K by radicals from Theorem 17.2.3. The group $\tilde{G} = \mathrm{Aut}(\tilde{L}|K)$ is solvable from Theorem 17.4.1. Further $L|K$ is a Galois extension. Therefore the Galois group $\mathrm{Aut}(L|K)$ is solvable as a subgroup of $\tilde{G}$.

Conversely suppose that the group $\mathrm{Aut}(L|K)$ is solvable. Let $q_1, \dots, q_r$ be the prime divisors of $m = [K : K]$ and let $n = q_1 \cdots q_r$. Let $\tilde{K}$ and $\tilde{L}$ be the splitting fields of the polynomial $x^n - 1 \in K[x]$ over K and L respectively. We have $\tilde{K} \subset \tilde{L}$. From part (2) of Lemma 17.2.5 we have that $\tilde{L}|K$ is a Galois extension and $\mathrm{Aut}(\tilde{L}|\tilde{K})$ is isomorphic to a subgroup of $\mathrm{Aut}(L|K)$. From this we first obtain that $[\tilde{L} : \tilde{K}] = |\mathrm{Aut}(\tilde{L}|\tilde{K})|$ is a divisor of $[L : K] = |\mathrm{Aut}(L|K)|$. Hence each prime divisor q of $[\tilde{L} : \tilde{K}]$ is also a prime divisor of $[L : K]$. Therefore $\tilde{L}$ is an extension by radicals of $\tilde{K}$ by Lemma 17.4.2. Since $\tilde{K} = K(\omega)$ where ω is a primitive n-th root of unity we obtain that $\tilde{L}$ is also an extension of K by radicals. Therefore L is contained in an extension $\tilde{L}$ of K by radicals and therefore $f(x) = 0$ is solvable by radicals. □

Corollary 17.5.2. *Let K be a field of characteristic 0 and let $f(x) \in K[x]$ be a polynomial of degree m with $1 \le m \le 4$. Then the equation $f(x) = 0$ is solvable by radicals.*

Proof. Let L be the splitting field of $f(x)$ over K. The Galois group $\mathrm{Aut}(L|K)$ is isomorphic to the subgroup of the symmetric group S_m. Now the group S_4 is solvable via the chain

$$\{1\} \subset \mathbb{Z}_2 \subset D_2 \subset A_4 \subset S_4$$

where $\mathbb{Z}_2$ is the cyclic group of order 2 and D_2 is the Klein 4-group which is isomorphic to $\mathbb{Z}_2 \times \mathbb{Z}_2$. Because $S_m \subset S_4$ for $1 \le m \le 4$ it follows that $\mathrm{Aut}(L|K)$ is solvable. From Theorem 17.5.1 the equation $f(x) = 0$ is solvable by radicals. □

Corollary 17.5.2 uses the general theory to show that any polynomial equation of degree less than or equal to 4 is solvable by radicals. This however does not provide explicit formulas for the solutions. We present these below.

Let K be a field of characteristic 0 and let $f(x) \in K[x]$ be a polynomial of degree m with $1 \le m \le 4$.

Case (1): If $\deg(f(x)) = 1$ then $f(x) = ax + b$ with $a, b \in K$ and $a \ne 0$. A zero is then given by $k = -\frac{b}{a}$.

Case (2): If $\deg(f(x)) = 2$ then $f(x) = ax^2 + bx + c$ with $a, b, c \in K$ and $a \ne 0$. The zeros are then given by the quadratic formula

$$k = \frac{-b \pm \sqrt{b^2 - 4ac}}{2a}.$$

We note that the quadratic formula holds over any field of characteristic not equal to 2. Whether there is a solution within the field K then depends on whether $b^2 - 4ac$ has a square root within K.

For the cases of degrees 3 and 4 we have the general forms of what are known as *Cardano's formulas*.

Case (3): If $\deg(f(x)) = 3$ then $f(x) = ax^3 + bx^2 + cx + d$ with $a, b, c, d \in K$ and $a \neq 0$. Dividing through by a we may assume without loss of generality that $a = 1$.

By a substitution $x = y - \frac{b}{3}$ the polynomial is transformed into

$$g(y) = y^3 + py + q \in K[y].$$

Let L be the splitting field of $g(y)$ over K and let $\alpha \in L$ be a zero of $g(y)$ so that

$$\alpha^3 + p\alpha + q = 0.$$

If $p = 0$ then $\alpha = \sqrt[3]{-q}$ so that $g(y)$ has the three zeros

$$\sqrt[3]{-q}, \quad \omega \sqrt[3]{-q}, \quad \omega^2 \sqrt[3]{-q}$$

where ω is a primitive third root of unity, $\omega^3 = 1$ with $\omega \neq \omega^2$.

Now let $p \neq 0$ and let β be a zero of $x^2 - \alpha x - \frac{p}{3}$ in a suitable extension L' of L. We have $\beta \neq 0$ since $p \neq 0$. Hence $\alpha = \beta - \frac{p}{3\beta}$. Putting this into the transformed cubic equation

$$\alpha^3 + p\alpha + q = 0$$

we get

$$\beta^3 - \frac{p^3}{27\beta^3} + q = 0.$$

Define $\gamma = \beta^3$ and $\delta = (\frac{-p}{3\beta})^3$ so that

$$\gamma + \delta + q = 0.$$

Then

$$\gamma^2 + q\gamma - \left(\frac{p}{3}\right)^3 = 0 \quad \text{and} \quad -\frac{p^3}{27\delta} + \delta + q = 0 \quad \text{and} \quad \delta^2 + q\delta - \left(\frac{p}{3}\right)^3 = 0.$$

Hence the zeros of the polynomial

$$x^2 + qx - \left(\frac{p}{3}\right)^3$$

are

$$\gamma, \delta = -\frac{q}{2} \pm \sqrt{\left(\frac{q}{2}\right)^2 + \left(\frac{p}{3}\right)^3}.$$

If we have $\gamma = \delta$ then both are equal to $-\frac{q}{2}$ and

$$\sqrt{\left(\frac{q}{2}\right)^2 + \left(\frac{p}{3}\right)^3} = 0.$$

Then from the definitions of γ, δ we have $\gamma = \beta^3$ and $\delta = (\frac{-p}{3\beta})^3$. From above $\alpha = \beta - \frac{p}{3\beta}$. Therefore we get α by finding the cube roots of γ and δ.

There are certain possibilities and combinations with these cube roots but because of the conditions the cube roots of γ and δ are not independent. We must satisfy the condition

$$\sqrt[3]{\gamma}\,\sqrt[3]{\delta} = \beta\frac{-p}{3\beta} = -\frac{p}{3}.$$

Therefore we get the final result:

The zeros of $g(y) = y^3 + py + q$ with $p \neq 0$ are

$$u + v, \quad \omega u + \omega^2 v, \quad \omega^2 u + \omega v$$

where ω is a primitive third root of unity and

$$u = \sqrt[3]{-\frac{q}{2} + \sqrt{\left(\frac{q}{2}\right)^2 + \left(\frac{p}{3}\right)^3}} \quad \text{and} \quad v = \sqrt[3]{-\frac{q}{2} - \sqrt{\left(\frac{q}{2}\right)^2 + \left(\frac{p}{3}\right)^3}}.$$

The above is known as the *cubic formula* or *Cardano's formula.*

Case (4): If $\deg(f(x)) = 4$ then $f(x) = ax^4 + bx^3 + cx^2 + dx + e$ with $a, b, c, d, e \in K$ and $a \neq 0$. Dividing through by a we may assume without loss of generality that $a = 1$.

By a substitution $x = y - \frac{b}{4a}$ the polynomial $f(x)$ is transformed into

$$g(y) = y^4 + py^2 + qy + r.$$

The zeros of $g(y)$ are

$$\begin{aligned}
y_1 &= \frac{1}{2}(\sqrt{\alpha_1} + \sqrt{\alpha_2} + \sqrt{\alpha_3})\\
y_2 &= \frac{1}{2}(\sqrt{\alpha_1} - \sqrt{\alpha_2} - \sqrt{\alpha_3})\\
y_3 &= \frac{1}{2}(-\sqrt{\alpha_1} + \sqrt{\alpha_2} - \sqrt{\alpha_3})\\
y_4 &= \frac{1}{2}(-\sqrt{\alpha_1} - \sqrt{\alpha_2} + \sqrt{\alpha_3})
\end{aligned}$$

where $\alpha_1, \alpha_2, \alpha_3$ are the zeros of the cubic polynomial

$$h(z) = z^3 + 2pz + (p^2 - 4r)z - q^2 \in \mathbb{Q}[z].$$

The polynomial $h(z)$ is called the *cubic resolvent* of $g(y)$. For a detailed proof of the case where $m = 4$ see [8].

The following theorem is due to Abel and shows the insolvability of the general degree 5 polynomial over the rationals $\mathbb{Q}$.

Theorem 17.5.3. *Let L be the splitting field of the polynomial $f(x) = x^5 - 2x^4 + 2 \in \mathbb{Q}[x]$ over $\mathbb{Q}$. Then* $\operatorname{Aut}(L|K) = S_5$ *the symmetric group on 5 letters. Since S_5 is not solvable the equation $f(x) = 0$ is not solvable by radicals.*

Proof. The polynomial $f(x)$ is irreducible over $\mathbb{Q}$ by the Eisenstein criterion. Further $f(x)$ has five zeros in the complex numbers $\mathbb{C}$ by the fundamental theorem of algebra (see Section 17.7). We claim that $f(x)$ has exactly 3 real zeros and 2 nonreal zeros which then necessarily are complex conjugates. In particular the 5 zeros are pairwise distinct.

To see the claim notice first that $f(x)$ has at least 3 real zeros from the intermediate value theorem. As a real function $f(x)$ is continuous and $f(-1) = -1 < 0$ and $f(0) = 2 > 0$ so it must have a real zero between -1 and 0. Further $f(\frac{3}{2}) = -\frac{81}{3} < 0$ and $f(2) = 2 > 0$. Hence there must be distinct real zeros between 0 and $\frac{3}{2}$ and between $\frac{3}{2}$ and 2. Suppose that $f(x)$ has more than 3 real zeros. Then $f'(x) = x^3(5x - 8)$ has at least 3 pairwise distinct real zeros from Rolle's theorem. But $f'(x)$ clearly has only 2 real zeros so this is not the case. Therefore $f(x)$ has exactly 3 real zeros and hence 2 nonreal zeros that are complex conjugates.

Let L be the splitting field of $f(x)$. The field L lies in $\mathbb{C}$ and the restriction of the map $\delta : z \mapsto \overline{z}$ of $\mathbb{C}$ to L maps the set of zeros of $f(x)$ onto themselves. Therefore δ is an automorphism of L. The map δ fixes the 3 real zeros and transposes the 2 nonreal zeros. From this we now show that $\operatorname{Aut}(L|\mathbb{Q}) = \operatorname{Aut} L = G = S_5$ the full symmetric group on 5 symbols. Clearly $G \subset S_5$ since G acts as a permutation group on the 5 zeros of $f(x)$.

Since δ transposes the 2 nonreal roots, G (as a permutation group) contains at least one transposition. Since $f(x)$ is irreducible G acts transitively on the zeros of $f(x)$. Let x_0 be one of the zeros of $f(x)$ and let G_{x_0} be the stabilizer of x_0. Since G acts transitively x_0 has five images under G and therefore the index of the stabilizer must be 5 (see Chapter 10).

$$5 = [G : G_{x_0}]$$

which by Lagrange's theorem must divide the order of G. Therefore from the Sylow theorems G contains an element of order 5. Hence G contains a 5-cycle and a transposition and therefore by Theorem 11.4.3 it follows that $G = S_5$. Since S_5 is not solvable it follows that $f(x)$ cannot be solved by radicals. □

Since Abel's theorem shows that there exists a degree 5 polynomial that cannot be solved by radicals it follows that there can be no formula like Cardano's formula in terms of radicals for degree 5.

Corollary 17.5.4. *There is no general formula for solving by radicals a fifth degree polynomial over the rationals.*

We now show that this result can be further extended to any degree greater than 5.

Theorem 17.5.5. *For each $n \geq 5$ there exist polynomials $f(x) \in \mathbb{Q}[x]$ of degree n for which the equation $f(x) = 0$ is not solvable by radicals.*

Proof. Let $f(x) = x^{n-5}(x^5 - 2x^4 + 2)$ and let L be the splitting field of $f(x)$ over $\mathbb{Q}$. Then $\text{Aut}(L|\mathbb{Q}) = \text{Aut}(L)$ contains a subgroup that is isomorphic to S_5. It follows that $\text{Aut}(L)$ is not solvable and therefore the equation $f(x) = 0$ is not solvable by radicals. □

This immediately implies the following.

Corollary 17.5.6. *There is no general formula for solving by radicals polynomial equations over the rationals of degree 5 or greater.*

17.6 Constructibility of Regular n-Gons

In Chapter 6 we considered certain geometric material related to field extensions. There, using general field extensions, we proved the impossibility of certain geometric compass and straightedge constructions. In particular there were four famous insolvable (to the Greeks) construction problems. The first is the *squaring of the circle.* This problem is, given a circle, to construct using straightedge and compass a square having area equal to that of the given circle. The second is the *doubling of the cube.* This problem is given a cube of given side length, to construct, using a straightedge and compass, a side of a cube having double the volume of the original cube. The third problem is the *trisection of an angle*. This problem is to trisect a given angle using only a straightedge and compass. The final problem is the *construction of a regular n-gon*. This problems asks which regular n-gons could be constructed using only straightedge and compass. In Chapter 6 we proved the impossibility of the first 3 problems. Here we use Galois theory to consider constructible n-gons.

Recall that a *Fermat number* is a positive integer of the form

$$F_n = 2^{2^n} + 1, \quad n = 0, 1, 2, 3, \ldots.$$

If a particular F_m is prime it is called a *Fermat prime*.

Fermat believed that all the numbers in this sequence were primes. In fact F_0, F_1, F_2, F_3, F_4 are all prime but F_5 is composite and divisible by 641 (see exercises). It is still an open question whether or not there are infinitely many Fermat primes. It has been conjectured that there are only finitely many. On the other hand if a number of the form $2^n + 1$ is a prime for some integer n then it must be a Fermat prime that is n must be a power of 2.

We first need the following.

Theorem 17.6.1. *Let $p = 2^n + 1$, $n = 2^s$ with $s \geq 0$ be a Fermat prime. Then there exists a chain of fields*

$$\mathbb{Q} = L_0 \subset L_1 \subset \cdots \subset L_n = k_p$$

where k_p is the p-th cyclotomic field such that

$$[L_j : L_{j-1}] = 2$$

for $j = 1, \dots, n$.

Proof. The extension $k_p|\mathbb{Q}$ is a Galois extension and $[k_p : \mathbb{Q}] = p - 1$. Further $\mathrm{Aut}(k_p)$ is cyclic of order $p - 1 = 2^n$. Hence there is a chain of subgroups

$$\{1\} = U_n \subset U_{n-1} \subset \cdots \subset U_0 = \mathrm{Aut}(k_p)$$

with $[U_{j-1} : U_j] = 2$ for $j = 1, \dots, n$. From the fundamental theorem of Galois theory the fields $L_j = \mathrm{Fix}(k_p, U_j)$ with $j = 0, \dots, n$ have the desired properties. □

The following corollaries describe completely the constructible n-gons tying them to Fermat primes.

Corollary 17.6.2. *Consider the numbers 0, 1, that is a unit line segment or a unit circle. A regular p-gon with $p \geq 3$ prime is constructible from $\{0, 1\}$ using a straightedge and compass if and only if $p = 2^{2^s} + 1$, $s \geq 0$ is a Fermat prime.*

Proof. From Theorem 6.3.13 we have that if a regular p-gon is constructible with a straightedge and compass then p must be a Fermat prime. The sufficiency follows from Theorem 17.6.1. □

We now extend this to general n-gons. Let $m, n \in \mathbb{N}$. Assume that we may construct from $\{0, 1\}$ a regular n-gon and a regular m-gon. In particular this means that we may construct the real numbers $\cos(\frac{2\pi}{n})$, $\sin(\frac{2\pi}{n})$, $\cos(\frac{2\pi}{m})$ and $\sin(\frac{2\pi}{m})$. If the $\gcd(m, n) = 1$ then we may construct from $\{0, 1\}$ a regular mn-gon.

To see this notice that

$$\cos\left(\frac{2\pi}{n} + \frac{2\pi}{m}\right) = \cos\left(\frac{2(n+m)\pi}{nm}\right) = \cos\left(\frac{2\pi}{n}\right)\cos\left(\frac{2\pi}{m}\right) - \sin\left(\frac{2\pi}{n}\right)\sin\left(\frac{2\pi}{m}\right)$$

and

$$\sin\left(\frac{2\pi}{n} + \frac{2\pi}{m}\right) = \sin\left(\frac{2(n+m)\pi}{nm}\right) = \sin\left(\frac{2\pi}{n}\right)\cos\left(\frac{2\pi}{m}\right) + \cos\left(\frac{2\pi}{n}\right)\sin\left(\frac{2\pi}{m}\right).$$

Therefore we may construct from $\{0, 1\}$ the numbers $\cos(\frac{2\pi}{mn})$ and $\sin(\frac{2\pi}{mn})$ because $\gcd(n+m, mn) = 1$. Therefore we may construct from $\{0, 1\}$ a regular mn-gon.

Now let $p \geq 3$ be a prime. Then $[k_{p^2} : \mathbb{Q}] = p(p-1)$ which is not a power of 2. Therefore from $\{0, 1\}$ it is not possible to construct a regular p^2-gon. Hence altogether we have the following.

Corollary 17.6.3. *Consider the numbers* $0, 1$, *that is a unit line segment or a unit circle. A regular* n*-gon with* $n \in \mathbb{N}$ *is constructible from* $\{0, 1\}$ *using a straightedge and compass if and only if*

(i) $n = 2^m$, $m \geq 0$ *or*

(ii) $p = 2^m p_1 p_2 \cdots p_r$, $m \geq 0$ *and the* p_i *are pairwise distinct Fermat primes.*

Proof. Certainly we may construct a 2^m-gon. Further if $r, s \in \mathbb{N}$ with $\gcd(r, s) = 1$ and if we can construct a regular rs-gon then clearly we may construct a regular r-gon and a regular s-gon. □

17.7 The Fundamental Theorem of Algebra

The *fundamental theorem of algebra* is one of the most important algebraic results. This says that any nonconstant complex polynomial must have a complex zero. In the language of field extensions this says that the field of complex numbers $\mathbb{C}$ is algebraically closed. There are many distinct and completely different proofs of this result. In [3] twelve proofs were given covering a wide area of mathematics. In this section we use Galois theory to present a proof. Before doing this we briefly mention some of the history surrounding this theorem.

The first mention of the fundamental theorem of algebra, in the form that every polynomial equation of degree n has exactly n roots, was given by Peter Roth of Nurnberg in 1608. However its conjecture is generally credited to Girard who also stated the result in 1629. It was then more clearly stated by Descartes in 1637 who also distinguished between real and imaginary roots. The first published proof of the fundamental theorem of algebra was then given by D'Alembert in 1746. However there were gaps in D'Alembert's proof and the first fully accepted proof was that given by Gauss in 1797 in his Ph.D. thesis. This was published in 1799. Interestingly enough, in reviewing Gauss' original proof, modern scholars tend to agree that there are as many holes in this proof as in D'Alembert's proof. Gauss, however, published three other proofs with no such holes. He published second and third proofs in 1816 while his final proof, which was essentially another version of the first, was presented in 1849.

Theorem 17.7.1. *Each nonconstant polynomial* $f(x) \in \mathbb{C}[x]$, *where* $\mathbb{C}$ *is the field of complex numbers, has a zero in* $\mathbb{C}$. *Therefore* $\mathbb{C}$ *is an algebraically closed field.*

Proof. Let $f(x) \in \mathbb{C}[x]$ be a nonconstant polynomial and let K be the splitting field of $f(x)$ over $\mathbb{C}$. Since the characteristic of the complex numbers $\mathbb{C}$ is zero this will be a Galois extension of $\mathbb{C}$. Since $\mathbb{C}$ is a finite extension of $\mathbb{R}$ this field K would also be a Galois extension of $\mathbb{R}$. The fundamental theorem of algebra asserts that K must be $\mathbb{C}$ itself, and hence the fundamental theorem of algebra is equivalent to the fact that any nontrivial Galois extension of $\mathbb{C}$ must be $\mathbb{C}$.

Let K be any finite extension of $\mathbb{R}$ with $|K : \mathbb{R}| = 2^m q, (2, q) = 1$. If $m = 0$, then K is an odd-degree extension of $\mathbb{R}$. Since K is separable over $\mathbb{R}$, from the primitive element theorem it is a simple extension, and hence $K = \mathbb{R}(\alpha)$, where the minimal polynomial $m_\alpha(x)$ over $\mathbb{R}$ has odd degree. However, odd-degree real polynomials always have a real root, and therefore $m_\alpha(x)$ is irreducible only if its degree is one. But then $\alpha \in \mathbb{R}$ and $K = \mathbb{R}$. Therefore, if K is a nontrivial finite extension of $\mathbb{R}$ of degree $2^m q$ we must have $m > 0$. This shows more generally that there are no odd-degree finite extensions of $\mathbb{R}$.

Suppose that K is a degree 2 extension of $\mathbb{C}$. Then $K = \mathbb{C}(\alpha)$ with $\deg m_\alpha(x) = 2$ where $m_\alpha(x)$ is the minimal polynomial of α over $\mathbb{C}$. But from the quadratic formula complex quadratic polynomials always have roots in $\mathbb{C}$ so a contradiction. Therefore, $\mathbb{C}$ has no degree 2 extensions.

Now, let K be a Galois extension of $\mathbb{C}$. Then K is also Galois over $\mathbb{R}$. Suppose $|K : \mathbb{R}| = 2^m q$, $(2, q) = 1$. From the argument above we must have $m > 0$. Let $G = \mathrm{Gal}(K/\mathbb{R})$ be the Galois group. Then $|G| = 2^m q, m > 0, (2, q) = 1$. Thus G has a 2-Sylow subgroup of order 2^m and index q (see Theorem 13.3.4). This would correspond to an intermediate field E with $|K : E| = 2^m$ and $|E : \mathbb{R}| = q$. However, then E is an odd-degree finite extension of $\mathbb{R}$. It follows that $q = 1$ and $E = \mathbb{R}$. Therefore, $|K : \mathbb{R}| = 2^m$ and $|G| = 2^m$.

Now, $|K : \mathbb{C}| = 2^{m-1}$ and suppose $G_1 = \mathrm{Gal}(K/\mathbb{C})$. This is a 2-group. If it were not trivial, then from Theorem 13.4.1 there would exist a subgroup of order 2^{m-2} and index 2. This would correspond to an intermediate field E of degree 2 over $\mathbb{C}$. However from the argument above $\mathbb{C}$ has no degree 2 extensions. It follows then that G_1 is trivial, that is, $|G_1| = 1$, so $|K : \mathbb{C}| = 1$ and $K = \mathbb{C}$ completing the proof. □

The fact that $\mathbb{C}$ is algebraically closed limits the possible algebraic extensions of the reals.

Corollary 17.7.2. *Let K be a finite field extension of the real numbers $\mathbb{R}$. Then $K = \mathbb{R}$ or $K = \mathbb{C}$.*

Proof. Since $|K : \mathbb{R}| < \infty$ by the primitive element theorem $K = \mathbb{R}(\alpha)$ for some $\alpha \in K$. Then the minimal polynomial $m_\alpha(x)$ of α over $\mathbb{R}$ is in $\mathbb{R}[x]$ and hence in $\mathbb{C}[x]$. Therefore form the fundamental theorem of algebra it has a root in $\mathbb{C}$. Hence $\alpha \in \mathbb{C}$. If $\alpha \in \mathbb{R}$ then $K = \mathbb{R}$, if not then $K = \mathbb{C}$. □

17.8 Exercises

1. For $f(x) \in \mathbb{Q}[x]$ with

$$f(x) = x^6 - 12x^4 + 36x^2 - 50$$
$$(f(x) = 4x^4 - 12x^2 + 20x - 3)$$

determine for each complex zero α of $f(x)$ a finite number of radicals $\gamma_i = \beta_i^{\frac{1}{m_i}}$, $i = 1, \dots, r$, and a presentation of α as a rational function in $\gamma_1, \dots, \gamma_r$ over $\mathbb{Q}$ such that γ_{i+1} is irreducible over $\mathbb{Q}(\gamma_1, \dots, \gamma_i)$ and $\beta_{i+1} \in \mathbb{Q}(\gamma_1, \dots, \gamma_i)$ for $i = 0, \dots, r-1$.

2. Let K be a field of prime characteristic p. Let $n \in \mathbb{N}$ and K_n the splitting field of $x^n - 1$ over K. Show that $\operatorname{Aut}(K_n|K)$ is cyclic.

3. Let $f(x) = x^4 - x + 1 \in \mathbb{Z}[x]$. Show:

   (i) f has a real zero.

   (ii) f is irreducible over $\mathbb{Q}$.

   (iii) If $u + iv$ $(u, v \in \mathbb{R})$ is a zero of f in $\mathbb{C}$, then $g = x^3 - 4x - 1$ is the minimal polynomial of $4u^2$ over $\mathbb{Q}$.

   (iv) The Galois group of f over $\mathbb{Q}$ has an element of order 3.

   (v) No zero $a \in \mathbb{C}$ of f is constructible from the points 0 and 1 with straightedge and compass.

4. Show that each polynomial $f(x)$ over $\mathbb{R}$ decomposes in linear factors and quadratic factors ($f(x) = d(x - a_1)\cdot(x - a_2)\cdots(x^2 + b_1x + c_1)\cdot(x^2 + b_2x + c_2)\cdots$, $d \in \mathbb{R}$).

5. Let E be a finite (commutative) field extension of $\mathbb{R}$. Then $E \cong \mathbb{R}$ or $E \cong \mathbb{C}$.

6. Let $n \geq 1$ be a natural number and x an indeterminate over $\mathbb{C}$. Consider the polynomial $x^n - 1 \in \mathbb{Z}[x]$. In $\mathbb{C}[x]$ it decomposes in linear factors:

$$x^n - 1 = (x - \xi_1)(x - \xi_2)\cdots(x - \xi_n),$$

where the complex numbers

$$\xi_\nu = e^{2\pi i \frac{\nu}{n}} = \cos\frac{2\pi\nu}{n} + i \cdot \sin\frac{2\pi\nu}{n}, \qquad 1 \leq \nu \leq n,$$

are all (different) n-th roots of unity, that is especially $\xi_n = 1$. These ξ_ν form a from ξ_1 generated multiplicative cyclic group $G = \{\xi_1, \xi_2, \dots, \xi_n\}$. It is $\xi_\nu = \xi_1^\nu$. An n-th root of unity ξ_ν is called a *primitive n-th root of unity*, if ξ_ν is not an m-th root of unity for any $m < n$.

Show that the following are equivalent:

(i) ξ_ν is a primitive n-th root of unity.

(ii) ξ_ν is a generating element of G.

(iii) $\gcd(\nu, n)=1$.

7. The polynomial $\phi_n(x) \in \mathbb{C}[x]$, whose zeros are exactly the primitive n-th roots of unity, is called the *n-th cyclotomic polynomial.* With Exercise 6 it is:

$$\phi_n(x) = \prod_{\substack{1 \le \nu \le n \\ \gcd(\nu,n)=1}} (x - \xi_\nu) = \prod_{\substack{1 \le \nu \le n \\ \gcd(\nu,n)=1}} (x - e^{2\pi i \frac{\nu}{n}}).$$

The degree of $\phi_n(x)$ is the number of the integers $\{1, \ldots, n\}$, which are coprime to n. Show:

(i) $x^n - 1 = \prod_{\substack{d \ge 1 \\ d \mid n}} \phi_d(x)$.

(ii) $\phi_n(x) \in \mathbb{Z}[x]$ for all $n \ge 1$.

(iii) $\phi_n(x)$ is irreducible over $\mathbb{Q}$ (and therefore also over $\mathbb{Z}$) for all $n \ge 1$.

8. Show that the Fermat numbers F_0, F_1, F_2, F_3, F_4 are all prime but F_5 is composite and divisible by 641.

Chapter 18
The Theory of Modules

18.1 Modules Over Rings

Recall that a vector space V over a field F is an abelian group V with a scalar multiplication $\cdot : F \times V \to V$ satisfying

(1) $f(v_1 + v_2) = fv_1 + fv_2$ for $f \in F$ and $v_1, v_2 \in V$.

(2) $(f_1 + f_2)v = f_1 v + f_2 v$ for $f_1, f_2 \in F$ and $v \in V$.

(3) $(f_1 f_2)v = f_1(f_2 v)$ for $f_1, f_2 \in F$ and $v \in V$.

(4) $1v = v$ for $v \in V$.

Vector spaces are the fundamental algebraic structures in linear algebra and the study of linear equations. Vector spaces have been crucial in our study of fields and Galois theory since any field extension is a vector space over any subfield. In this context the degree of a field extension is just the dimension of the extension field as a vector space over the base field.

If we modify the definition of a vector space to allow scalar multiplication from an arbitrary ring we obtain a more general structure called a *module*. We will formally define this below. Modules generalize vector spaces but the fact that the scalars do not necessarily have inverses makes the study of modules much more complicated. Modules will play an important role in both the study of rings and the study of abelian groups. In fact any abelian group is a module over the integers $\mathbb{Z}$ so that modules, besides being generalizations of vector spaces can also be considered as generalizations of abelian groups.

In this chapter we will introduce the theory of modules. In particular we will extend to modules the basic algebraic properties such as the isomorphism theorems that have been introduced earlier for groups, rings and fields.

In this chapter we restrict ourselves to commutative rings so that throughout R is always a commutative ring. If R has an identity 1 then we always consider only the case that $1 \neq 0$. Throughout this chapter we use letters $\mathfrak{a}, \mathfrak{b}, \mathfrak{c}, \mathfrak{m}, \ldots$ for ideals in R. For principal ideals we write $\langle a \rangle$ or aR for the ideal generated by $a \in R$. We note however that the definition can be extended to include modules over noncommutative rings. In this case we would speak of left modules and right modules.

Definition 18.1.1. Let $R = (R, +, \cdot)$ a commutative ring and $M = (M, +)$ an abelian group. M together with a scalar multiplication $\cdot : R \times M \to M, (\alpha, x) \mapsto \alpha x$, is called a *R-module* or module over R if the following axioms hold:

(M1) $(\alpha + \beta)x = \alpha x + \beta x$,

(M2) $\alpha(x + y) = \alpha x + \alpha y$ and

(M3) $(\alpha\beta)x = \alpha(\beta x)$ for all $\alpha, \beta \in R$ and $x, y \in M$.

If R has an identity 1 then M is called an *unitary R-module* if in addition

(M4) $1 \cdot x = x$ for all $x \in M$ holds.

In the following, R always is a commutative ring. If R contains an identity 1 then M always is an unitary R-module. If R has an identity 1 then we always assume $1 \neq 0$.

As usual we have the rules:

$$0 \cdot x = 0, \quad \alpha \cdot 0 = 0, \quad -(\alpha x) = (-\alpha)x = \alpha(-x)$$

for all $\alpha \in R$ and for all $x \in M$.

We next present a series of examples of modules.

Example 18.1.2. (1) If $R = K$ is a field then a K-module is a K-vector space.

(2) Let $G = (G, +)$ be an abelian group. If $n \in \mathbb{Z}$ and $x \in G$ then nx is defined as usual:

$$\begin{aligned} 0 \cdot x &= 0, \\ nx &= \underbrace{x + \cdots + x}_{n\text{-times}} \quad \text{if } n > 0 \quad \text{and} \\ nx &= (-n)(-x) \quad \text{if } n < 0. \end{aligned}$$

Then G is an unitary $\mathbb{Z}$-module via the scalar multiplication

$$\cdot : \mathbb{Z} \times G \to G, \quad (n, x) \mapsto nx.$$

(3) Let S be a subring of R. Then via $(s, r) \mapsto sr$ the ring R itself becomes an S-module.

(4) Let K be a field, V a K-vector space and $f : V \to V$ a linear map of V. Let $p = \sum_i \alpha_i t^i \in K[t]$. Then $p(f) := \sum_i \alpha_i f^i$ defines a linear map of V and V is an unitary $K[t]$-module via the scalar multiplication

$$K[t] \times V \to V, \quad (p, v) \mapsto pv := p(f)(v).$$

(5) If R is a commutative ring and $\mathfrak{a}$ is an ideal in R then $\mathfrak{a}$ is a module over R.

Basic to all algebraic theory is the concept of substructures. Next we define *submodules*.

Definition 18.1.3. Let M be an R-module. $\emptyset \neq U \subset M$ is called a *submodule* of M if

(UMI) $(U, +) < (M, +)$ and

(UMII) $\alpha \in R,\ u \in U \Rightarrow \alpha u \in U$, that is, $RU \subset U$.

Example 18.1.4. (1) In an abelian group G, considered as a $\mathbb{Z}$-module, the subgroups are precisely the submodules.

(2) The submodules of R, considered as a R-module, are precisely the ideals.

(3) $Rx := \{\alpha x : \alpha \in R\}$ is a submodule of M for each $x \in M$.

(4) Let K be a field, V a K-vector space and $f : V \to V$ a linear map of V. Let U be a submodule of V, considered as a $K[t]$-module as above. Then the following holds:

   (a) $U < V$.

   (b) $pU = p(f)U \subset U$ for all $p \in K[t]$. Especially $\alpha U \subset U$ for $p = \alpha \in K$ and $tU = f(U) \subset U$ for $p = t$, that is, U is an f-invariant subspace. On the other side also, $p(f)U \subset U$ for all $p \in K[t]$ if U is an f-invariant subspace.

We next extend to modules the concept of a generating system. For a single generator, as with groups, this is called *cyclic*.

Definition 18.1.5. A submodule U of the R-module M is called *cyclic* if there exists an $x \in M$ with $U = Rx$.

As in vector spaces, groups and rings the following constructions are standard leading us to generating systems.

(1) Let M be a R-module and $\{U_i : i \in I\}$ a family of submodules. Then $\bigcap_{i \in I} U_i$ is a submodule of M.

(2) Let M be a R-module. If $A \subset M$ then we define

$$\langle A \rangle := \bigcap \{U : U \text{ submodule of M with } A \subset U\}.$$

$\langle A \rangle$ is the smallest submodule of M which contains A. If R has an identity 1 then $\langle A \rangle$ is the set of all linear combinations $\sum_i \alpha_i a_i$ with all $\alpha_i \in R$, all $a_i \in A$. This holds because M is unitary and $na = n(1 \cdot a) = (n \cdot 1)a$ for $n \in \mathbb{Z}$ and $a \in A$, that is, we may consider the pseudoproduct na as a real product in the module. Especially, if R has a unit 1 then $aR = \langle \{a\} \rangle =: \langle a \rangle$.

Definition 18.1.6. Let R have an identity 1. If $M = \langle A \rangle$ then A is called a generating system of M. M is called finitely generated if there are $a_1, \dots, a_n \in M$ with $M = \langle \{a_1, \dots, a_n\} \rangle =: \langle a_1, \dots, a_n \rangle$.

The following is clear.

Lemma 18.1.7. *Let U_i be submodules of M, $i \in I$, I an index set. Then*

$$\left\langle \bigcup_{i \in I} U_i \right\rangle = \left\{ \sum_{i \in L} a_i : a_i \in U_i, L \subset I \text{ finite} \right\}.$$

We write $\langle \bigcup_{i \in I} U_i \rangle =: \sum_{i \in I} U_i$ and call this submodule the sum of the U_i. A sum $\sum_{i \in I} U_i$ is called a direct sum if for each representation of 0 as $0 = \sum a_i, a_i \in U_i$, it follows that all $a_i = 0$. This is equivalent to $U_i \cap \sum_{i \neq j} U_j = 0$ for all $i \in I$. Notation: $\bigoplus_{i \in I} U_i$; and if $I = \{1, \dots, n\}$ then we write $U_1 \oplus \cdots \oplus U_n$, too.

In analogy with our previously defined algebraic structure we extend to modules the concepts of quotient modules and module homomorphisms.

Definition 18.1.8. Let U be a submodule of the R-module M. Let M/U be the factor group. We define a (well-defined) scalar multiplication

$$R \times M/U \to M/U, \quad \alpha(x + U) := \alpha x + U.$$

With this M/U is a R-module, the *factor module* or *quotient module* of M by U. In M/U we have the operations

$$(x + U) + (y + U) = (x + y) + U$$

and

$$\alpha(x + U) = \alpha x + U.$$

A module M over a ring R can also be considered as a module over a quotient ring of R. The following is straightforward to verify (see exercises).

Lemma 18.1.9. *Let $\mathfrak{a} \lhd R$ an ideal in R and M a R-module. The set of all finite sums of the form $\sum \alpha_i x_i$, $\alpha_i \in \mathfrak{a}$, $x_i \in M$, is a submodule of M which we denote by $\mathfrak{a}M$. The factor group $M/\mathfrak{a}M$ becomes a $R/\mathfrak{a}$-module via the well-defined scalar multiplication*

$$(\alpha + \mathfrak{a})(m + \mathfrak{a}M) = \alpha m + \mathfrak{a}M.$$

If here R has an identity 1 and $\mathfrak{a}$ is a maximal ideal then $M/\mathfrak{a}M$ becomes a vector space over the field $K = R/\mathfrak{a}$.

We next define module homomorphisms

Definition 18.1.10. Let R be a ring and M, N be R-modules. A map $f : M \to N$ is called a *R-module homomorphism* (or R-linear) if

$$f(x + y) = f(x) + f(y)$$

and

$$f(\alpha x) = \alpha f(x)$$

for all $\alpha \in R$ and all $x, y \in M$. Endo-, epi-, mono-, iso- and automorphisms are defined analogously via the corresponding properties of the maps. If $f : M \to N$ and $g : N \to P$ are module homomorphisms then $g \circ f : M \to P$ is also a module homomorphism. If $f : M \to N$ is an isomorphism then also $f^{-1} : N \to M$.

We define kernel and image in the usual way:

$$\ker(f) := \{x \in M : f(x) = 0\}$$

and

$$\operatorname{im}(f) := f(M) = \{f(x) : x \in M\}.$$

$\ker(f)$ is a submodule of M and $\operatorname{im}(f)$ is a submodule of N. As usual:

$$f \text{ is injective} \iff \ker(f) = \{0\}.$$

If U is a submodule of M then the map $x \mapsto x + U$ defines a module epimorphism (the canonical epimorphism) from M onto M/U with kernel U.

There are module isomorphism theorems. The proofs are straightforward extensions of the corresponding proofs for groups and rings.

Theorem 18.1.11 (module isomorphism theorems). *Let M, N be R-modules.*

(1) *If $f : M \to N$ is a module homomorphism then*

$$f(M) \cong M/\ker(f).$$

(2) *If U, V are submodules of the R-module M then*

$$U/(U \cap V) \cong (U + V)/V.$$

(3) *If U and V are submodules of the R-module M with $U \subset V \subset M$ then*

$$(M/U)/(V/U) \cong M/V.$$

For the proofs, as for groups, just consider the map $f : U + V \to U/(U \cap V)$, $u + v \mapsto u + (U \cap V)$ which is well-defined because $U \cap V$ is a submodule of U; we have $\ker(f) = V$.

Note that $\alpha \mapsto \alpha\rho$, $\rho \in R$ fixed, defines a module homomorphism $R \to R$ if we consider R itself as a R-module.

18.2 Annihilators and Torsion

In this section we define torsion for an R-module and a very important subring of R called the annihilator.

Definition 18.2.1. Let M be an R-module. For a fixed $a \in M$ consider the map $\lambda_a : R \to M$, $\lambda_a(\alpha) := \alpha a$. λ_a is a module homomorphism considering R as an R-module. We call $\ker(\lambda_a)$ the *annihilator of* a denoted $\operatorname{Ann}(a)$, that is

$$\operatorname{Ann}(a) = \{\alpha \in R : \alpha a = 0\}.$$

Lemma 18.2.2. $\operatorname{Ann}(a)$ *is a submodule of* R *and the module isomorphism theorem* (1) *gives* $R/\operatorname{Ann}(a) \cong Ra$.

We next extend the annihilator to whole submodules of M.

Definition 18.2.3. Let U be a submodule of the R-module M. The *annihilator* $\operatorname{Ann}(U)$ is defined to be

$$\operatorname{Ann}(U) := \{\alpha \in R : \alpha u = 0 \text{ for all } u \in U\}.$$

As for single elements, since $\operatorname{Ann}(U) = \bigcap_{u \in U} \operatorname{Ann}(u)$, then $\operatorname{Ann}(U)$ is a submodule of R. If $\rho \in R$, $u \in U$, then $\rho u \in U$, that means, if $u \in \operatorname{Ann}(U)$ then also $\rho u \in \operatorname{Ann}(U)$ because $(\alpha\rho)u = \alpha(\rho u) = 0$. Hence, $\operatorname{Ann}(U)$ is an ideal in R.

Suppose that G is an abelian group. Then as mentioned G is a $\mathbb{Z}$-module. An element $g \in G$ is a *torsion element* or has *finite order* if $ng = 0$ for some $n \in \mathbb{N}$. The set $\operatorname{Tor}(G)$ consists of all the torsion elements in G. An abelian group is torsion-free if $\operatorname{Tor}(G) = \{0\}$.

Lemma 18.2.4. *Let* G *be an abelian group. Then* $\operatorname{Tor}(G)$ *is a subgroup of* G *and* $G/\operatorname{Tor}(G)$ *is torsion-free.*

We extend this concept now to general modules.

Definition 18.2.5. The R-module M is called *faithful* if $\operatorname{Ann}(M) = \{0\}$. An element $a \in M$ is called a *torsion element*, or element of finite order, if $\operatorname{Ann}(a) \neq \{0\}$. A module without torsion elements $\neq 0$ is called *torsion-free*. If the R-module M is torsion-free then R has no zero divisors $\neq 0$.

Theorem 18.2.6. *Let R be an integral domain and M an R-module (by our agreement M is unitary). Let $\operatorname{Tor}(M) = T(M)$ be the set of torsion elements of M. Then $\operatorname{Tor}(M)$ is a submodule of M and $M/\operatorname{Tor}(M)$ is torsion-free.*

Proof. If $m \in \operatorname{Tor}(M)$, $\alpha \in \operatorname{Ann}(m)$, $\alpha \neq 0$ and $\beta \in R$ then we get $\alpha(\beta m) = (\alpha\beta)m = (\beta\alpha)m = \beta(\alpha m) = 0$, that is, $\beta m \in \operatorname{Tor}(M)$, because $\alpha\beta \neq 0$ if $\beta \neq 0$ (R is an integral domain). Let m' another element of $\operatorname{Tor}(M)$ and $0 \neq \alpha' \in \operatorname{Ann}(m')$. Then $\alpha\alpha' \neq 0$ and $\alpha\alpha'(m + m') = \alpha\alpha' m + \alpha\alpha' m' = \alpha'(\alpha m) + \alpha(\alpha' m') = 0$, that is, $m + m' \in \operatorname{Tor}(M)$. Therefore $\operatorname{Tor}(M)$ is a submodule.

Now, let $m + \operatorname{Tor}(M)$ be a torsion element in $M/\operatorname{Tor}(M)$. Let $\alpha \in R$, $\alpha \neq 0$ with $\alpha(m + \operatorname{Tor}(M)) = \alpha m + \operatorname{Tor}(M) = \operatorname{Tor}(M)$. Then $\alpha m \in \operatorname{Tor}(M)$. Hence there exists a $\beta \in R$, $\beta \neq 0$, with $0 = \beta(\alpha m) = (\beta\alpha)m$. Since $\beta\alpha \neq 0$ we get that $m \in \operatorname{Tor}(M)$ and the torsion element $m + \operatorname{Tor}(M)$ is trivial. □

18.3 Direct Products and Direct Sums of Modules

Let M_i, $i \in I$, $I \neq \emptyset$, be a family of R-modules. On the direct product

$$P = \prod_{i \in I} M_i = \left\{ f : I \to \bigcup_{i \in I} M_i : f(i) \in M_i \text{ for all } i \in I \right\}$$

we define the module operations

$$+ : P \times P \to P \quad \text{and} \quad \cdot : R \times P \to P$$

via

$$(f + g)(i) := f(i) + g(i) \quad \text{and} \quad (\alpha f)(i) := \alpha f(i).$$

Together with this operations $P = \prod_{i \in I} M_i$ is an R-module, *the direct product of the M_i*. If we identify f with the I-tuple of the images $f = (f_i)_{i \in I}$ then the sum and the scalar multiplication are componentwise. If $I = \{1, \dots, n\}$ and $M_i = M$ for all $i \in I$ then we write, as usual, $M^n = \prod_{i \in I} M_i$.

We make the agreement that $\prod_{i \in I = \emptyset} M_i := \{0\}$.

$\bigoplus_{i \in I} M_i := \{f \in \prod_{i \in I} M_i : f(i) = 0$ for almost all $i\}$ ("for almost all i" means that there are at most finitely many i with $f(i) \neq 0$) is a submodule of the direct product, called the *direct sum of the M_i*. If $I = \{1, \dots, n\}$ then we write $\bigoplus_{i \in I} M_i = M_1 \oplus \cdots \oplus M_n$. Here $\prod_{i=1}^{n} M_i = \bigoplus_{i=1}^{n} M_i$ for finite I.

Theorem 18.3.1. (1) *If $\pi \in \mathrm{Per}(I)$ is a permutation of I then*

$$\prod_{i\in I} M_i \cong \prod_{i\in I} M_{\pi(i)}$$

and

$$\bigoplus_{i\in I} M_i \cong \bigoplus_{i\in I} M_{\pi(i)}.$$

(2) *If $I = \dot{\bigcup}_{j\in J} I_j$, the disjoint union, then*

$$\prod_{i\in I} M_i \cong \prod_{j\in J}\Big(\prod_{i\in I_j} M_i\Big)$$

and

$$\bigoplus_{i\in I} M_i \cong \bigoplus_{j\in J}\Big(\bigoplus_{i\in I_j} M_i\Big).$$

Proof. (1) Consider the map $f \mapsto f \circ \pi$.

(2) Consider the map $f \mapsto \bigcup_{j\in J} f_j$ where $f_j \in \prod_{i\in I_j} M_i$ is the restriction of f onto I_j, and $\bigcup_{j\in J} f_j$ is on J defined by $(\bigcup_{j\in J} f_j)(k) := f_k = f(k)$. □

Let $I \neq \emptyset$. If $M = \prod_{i\in I} M_i$ then we get in a natural manner module homomorphisms $\pi_i : M \to M_i$ via $f \mapsto f(i)$; π_i is called the *projection onto the i-th component*. In duality we define module homomorphisms $\delta_i : M_i \to \bigoplus_{i\in I} M_i \subset \prod_{i\in I} M_i$ via $\delta_i(m_i) = (n_j)_{j\in I}$ where $n_j = 0$ if $i \neq j$ and $n_i = m_i$. δ_i is called *the i-th canonical injection*. If $I = \{1,\dots,n\}$ then $\pi_i(a_1,\dots,a_i,\dots,a_n) = a_i$ and $\delta_i(m_i) = (0,\dots,0,m_i,0,\dots,0)$.

Theorem 18.3.2 (universal properties). *Let $A, M_i, i \in I \neq \emptyset$, be R-modules.*

(1) *If $\phi_i : A \to M_i$, $i \in I$, are module homomorphisms then there exists exactly one module homomorphism $\phi : A \to \prod_{i\in I} M_i$ such that for each i the following diagram commutes:*

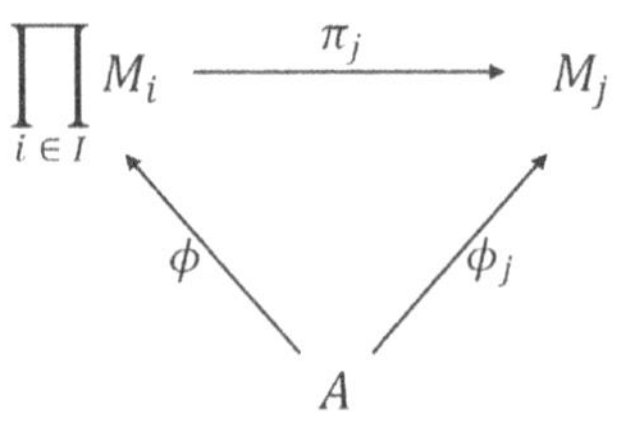

that is, $\phi_j = \pi_j \circ \phi$ where π_j is the j-th projection.

(2) *If $\Psi_i : M_i \to A$, $i \in I$, are module homomorphisms then there exists exactly one module homomorphism $\Psi : \bigoplus_{i \in I} M_i \to A$ such that for each $j \in J$ the following diagram commutes:*

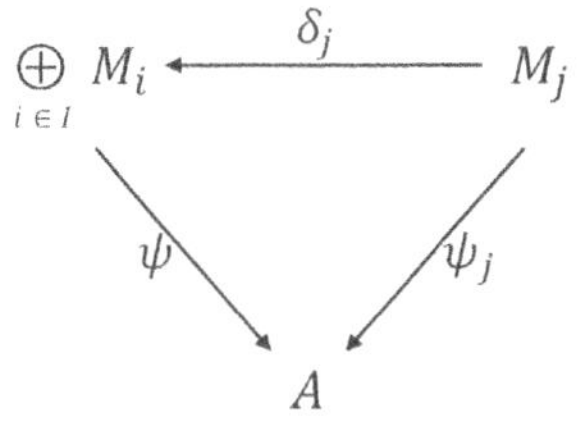

that is, $\Psi_j = \Psi \circ \delta_j$ where δ_j is the j-th canonical injection.

Proof. (1) If there is such ϕ then the j-th component of $\phi(a)$ is equal $\phi_j(a)$ because $\pi_j \circ \phi = \phi_j$. Hence, define $\phi(a) \in \prod_{i \in I} M_i$ via $\phi(a)(i) := \phi_i(a)$, and ϕ is the desired map.

(2) If there is such a Ψ with $\Psi \circ \alpha_j = \Psi_j$ then $\Psi(x) = \Psi((x_i)) = \Psi(\sum_{i \in I} \delta_i(x_i)) = \sum_{i \in I} \Psi \circ \delta_i(x_i) = \sum_{i \in I} \Psi_i(x_i)$. Hence define $\Psi((x_i)) = \sum_{i \in I} \Psi_i(x_i)$, and Ψ is the desired map (recall that the sum is well defined). □

18.4 Free Modules

If V is a vector space over a field F then V always has a basis over F which may be infinite. Despite the similarity to vector spaces, because the scalars may not have inverses this is not necessarily true for modules.

We now define a basis for a module and show that only free modules have bases.

Let R be a ring with identity 1, M be a unitary R-module and $S \subset M$. Each finite sum $\sum \alpha_i s_i$, the $\alpha_i \in R$ and the $s_i \in S$, is called a *linear combination in S*. Since M is unitary and $S \neq \emptyset$ then $\langle S \rangle$ is exactly the set of all linear combinations in S. In the following we assume that $S \neq \emptyset$. If $S = \emptyset$ then $\langle S \rangle = \langle \emptyset \rangle = \{0\}$, and this case is not interesting. For convention, in the following we always assume $m_i \neq m_j$ if $i \neq j$ in a finite sum $\sum \alpha_i m_i$ with all $\alpha_i \in R$ and all $m_i \in M$.

Definition 18.4.1. A finite set $\{m_1, \dots, m_n\} \subset M$ is called *linear independent* or *free (over R)* if a representation $0 = \sum_{i=1}^{n} \alpha_i m_i$ implies always $\alpha_i = 0$ for all $i \in \{1, \dots, n\}$, that is, 0 can be represented only trivially on $\{m_1, \dots, m_n\}$. A nonempty subset $S \subset M$ is called *free (over R)* if each finite subset of S is free.

Definition 18.4.2. Let M be a R-module (as above).

(1) $S \subset M$ is called a *basis* of M if

(a) $M = \langle S \rangle$ and

(b) S is free (over R).

(2) If M has a basis then M is called a *free R-module*. If S is a basis of M then M is called *free on S* or *free with basis S*.

In this sense we can consider $\{0\}$ as a free module with basis $\emptyset$.

Example 18.4.3. 1. $R \times R = R^2$, as R-module, is free with basis $\{(1,0),(0,1)\}$.

2. More general, let $I \neq \emptyset$. Then $\bigoplus_{i \in I} R_i$ with $R_i = R$ for all $i \in I$ is free with basis $\{\epsilon_i : I \to R : \epsilon_i(j) = \delta_{ij},\ i, j \in I\}$ where

$$\delta_{ij} = \begin{cases} 0 & \text{if } i \neq j, \\ 1 & \text{if } i = j. \end{cases}$$

Especially, if $I = \{1, \dots, n\}$ then $R^n = \{(a_1, \dots, a_n) : a_i \in R\}$ is free with basis $\{\epsilon_i = (\underbrace{0, \dots, 0}_{i-1}, 1, 0, \dots, 0); 1 \leq i \leq n\}$.

3. Let G be an abelian group. If G, as a $\mathbb{Z}$-module, is free on $S \subset G$, then G is called a *free abelian group with basis S*. If $|S| = n < \infty$ then $G \cong \mathbb{Z}^n$.

Theorem 18.4.4. *The R-module M is free on S if and only if each $m \in M$ can be written uniquely in the form $\sum \alpha_i s_i$ with $\alpha_i \in R$, $s_i \in S$. This is exactly the case when $M = \bigoplus_{s \in S} Rs$ is the direct sum of the cyclic submodules Rs, and each Rs is module isomorphic to R.*

Proof. If S is a basis then each $m \in M$ can be written as $m = \sum \alpha_i s_i$ because $M = \langle S \rangle$. This representation is unique because if $\sum \alpha_i s_i = \sum \beta_i s_i$ then $\sum(\alpha_i - \beta_i)s_i = 0$, that is $\alpha_i - \beta_i = 0$ for all i. If, on the other side, we assume that the representation is unique then we get from $\sum \alpha_i s_i = 0 = \sum 0 \cdot s_i$ that all $\alpha_i = 0$, and therefore M is free on S. The rest of the theorem essentially is a rewriting of the definition. If each $m \in M$ can be written as $m = \sum \alpha_i s_i$ then $M = \sum_{s \in S} Rs$. If $x \in Rs' \cap \sum_{s \in S, s \neq s'} Rs$ with $s' \in S$ then $x = \alpha' s' = \sum_{s_i \neq s', s_i \in S} \alpha_i s_i$ and $0 = \alpha' s' - \sum_{s_i \neq s', s_i \in S} \alpha_i s_i$ and therefore $\alpha' = 0$ and $\alpha_i = 0$ for all i. This gives $M = \bigoplus_{s \in S} Rs$. The cyclic modules Rs are isomorphic to $R/\operatorname{Ann}(s)$, and $\operatorname{Ann}(s) = \{0\}$ in the free modules. On the other side such modules are free on S. □

Corollary 18.4.5. (1) *M is free on $S \Leftrightarrow M \cong \bigoplus_{s \in S} R_s$, $R_s = R$ for all $s \in S$.*

(2) *If M is finitely generated and free then there exists an $n \in \mathbb{N}_0$ such that $M \cong R^n = \underbrace{R \oplus \cdots \oplus R}_{n\text{-times}}$.*

Proof. Part (1) is clear. We prove part (2). Let $M = \langle x_1, \dots, x_r \rangle$ and S a basis of M. Each x_i is uniquely representable on S as $x_i = \sum_{s_i \in S} \alpha_i s_i$. Since the x_i generate M we get $m = \sum \beta_i x_i = \sum_{i,j} \beta_i \alpha_j s_j$ for arbitrary $m \in M$, and we need only finitely many s_j to generate M. Hence S is finite. □

Theorem 18.4.6. *Let R be a commutative ring with identity 1 and M a free R-module. Then any two bases of M have the same cardinality.*

Proof. R contains a maximal ideal $\mathfrak{m}$, and $R/\mathfrak{m}$ is a field (see Theorem 2.3.2 and 2.4.2). Then $M/\mathfrak{m}M$ is a vector space over $R/\mathfrak{m}$. From $M \cong \bigoplus_{s\in S} Rs$ with basis S we get $\mathfrak{m}M \cong \bigoplus_{s\in S} \mathfrak{m}s$ and, hence,

$$M/\mathfrak{m}M \cong \Big(\bigoplus_{s\in S} Rs\Big)/\mathfrak{m}M \cong \bigoplus_{s\in S}(Rs/\mathfrak{m}M) \cong \bigoplus_{s\in S} R/\mathfrak{m}.$$

Hence the $R/\mathfrak{m}$-vector space $M/\mathfrak{m}M$ has a basis of the cardinality of S. This gives the result. □

Let R be a commutative ring with identity 1 and M a free R-module. The cardinality of a basis is an invariant of M, called the *rank of M* or *dimension of M*. If $\operatorname{rank}(M) = n < \infty$ then this means $M \cong R^n$.

Theorem 18.4.7. *Each R-module is a (module-)homomorphic image of a free R-module.*

Proof. Let M be a R-module. We consider $F := \bigoplus_{m\in M} R_m$ with $R_m = R$ for all $m \in M$. F is a free R-module. The map $f : F \to M$, $f((\alpha_m)_{m\in M}) = \sum \alpha_m m$ defines a surjective module homomorphism. □

Theorem 18.4.8. *Let F, M be R-modules, and let F be free. Let $f : M \to F$ be a module epimorphism. Then there exists a module homomorphism $g : F \to M$ with $f \circ g = \mathrm{id}_F$, and we have $M = \ker(f) \oplus g(F)$.*

Proof. Let S be a basis of F. By the axiom of choice there exists for each $s \in S$ an element $m_s \in M$ with $f(m_s) = s$ (f is surjective). We define the map $g : F \to M$ via $s \mapsto m_s$ linearly, that is, $g(\sum_{s_i\in S} \alpha_i s_i) = \sum_{s_i\in S} \alpha_i m_{s_i}$. Since F is free, the map g is well defined. Obviously $f\circ g(s) = f(m_s) = s$ for $s \in S$, that means $f\circ g = \mathrm{id}_F$ because F is free on S. For each $m \in M$ we have also $m = g\circ f(m) + (m - g\circ f(m))$ where $g \circ f(m) = g(f(m)) \in g(F)$, and since $f \circ g = \mathrm{id}_F$ the elements of the form $m - g \circ f(m)$ are in the kernel of f. Therefore $M = g(F) + \ker(f)$. Now let $x \in g(F) \cap \ker(f)$. Then $x = g(y)$ for some $y \in F$ and $0 = f(x) = f\circ g(y) = y$, and hence $x = 0$. Therefore the sum is direct: $M = g(F) \oplus \ker(f)$. □

Corollary 18.4.9. *Let M be an R-module and N a submodule such that M/N is free. Then there is a submodule N' of M with $M = N \oplus N'$.*

Proof. Apply the above theorem for the canonical map $\pi : M \to M/N$ with $\ker(\pi) = N$. □

18.5 Modules over Principal Ideal Domains

We now specialize to the case of modules over principal ideal domains. For the remainder of this section R is always a principal ideal domain $\neq \{0\}$. We now use the notation $(\alpha) := \alpha R$, $\alpha \in R$, for the principal ideal αR.

Theorem 18.5.1. *Let M be a free R-module of finite rank over the principal ideal domain R. Then each submodule U is free of finite rank, and* $\operatorname{rank}(U) \leq \operatorname{rank}(M)$.

Proof. We prove the theorem by induction on $n = \operatorname{rank}(M)$. The theorem certainly holds if $n = 0$. Now let $n \geq 1$ and assume that the theorem holds for all free R-modules of rank $< n$. Let M be a free R-module of rank n with basis $\{x_1, \dots, x_n\}$. Let U be a submodule of M. We represent the elements of U as linear combination of the basis elements $x_1, \dots, x_n$, and we consider the set of coefficients of x_1 for the elements of U:

$$\mathfrak{a} = \left\{\beta \in R : \beta x_1 + \sum_{i=2}^{n} \beta_i x_i \in U\right\}.$$

Certainly $\mathfrak{a}$ is an ideal in R. Since R is a principal ideal domain we have $\mathfrak{a} = (\alpha_1)$ for some $\alpha_1 \in R$. Let $u \in U$ be an element in U which has α_1 as its first coefficient, that is

$$u = \alpha_1 x_1 + \sum_{i=2}^{n} \alpha_i x_i \in U.$$

Let $v \in U$ be arbitrary. Then

$$v = \rho(\alpha_1 x_1) + \sum_{i=2}^{n} \rho_i x_i.$$

Hence $v - \rho u \in U' := U \cap M'$ where M' is the free R-module with basis $\{x_2, \dots, x_n\}$. By induction, U' is a free submodule of M' with a basis $\{y_1, \dots, y_t\}$, $t \leq n-1$. If $\alpha_1 = 0$ then $\mathfrak{a} = (0)$ and $U = U'$, and there is nothing to prove. Now let $\alpha_1 \neq 0$. We show that $\{u, y_1, \dots, y_t\}$ is a basis of U. $v - \rho u$ is a linear combination of the basis elements of U', that is, $v - \rho u = \sum_{i=1}^{t} \eta_i y_i$ uniquely. Hence $v = \rho u + \sum_{i=1}^{t} \eta_i y_i$ and $U = \langle u, y_1, \dots, y_t \rangle$. Now let be $0 = \gamma u + \sum_{i=1}^{t} \mu_i y_i$. We write u and the y_i as linear combinations in the basis elements $x_1, \dots, x_n$ of M. There is only an x_1-portion in γu. Hence

$$0 = \gamma \alpha_1 x_1 + \sum_{i=2}^{n} \mu_i' x_i.$$

Therefore first $\gamma \alpha_1 x_1 = 0$, that is, $\gamma = 0$ because R has no zero divisor $\neq 0$, and further $\mu_2' = \cdots = \mu_n' = 0$, that means, $\mu_1 = \cdots = \mu_t = 0$. □

Let R be a principal ideal domain. Then the annihilator $\mathrm{Ann}(x)$ in R-modules M has certain further properties. Let $x \in M$. By definition

$$\mathrm{Ann}(x) = \{\alpha \in R : \alpha x = 0\} \lhd R, \quad \text{an ideal in } R,$$

hence $\mathrm{Ann}(x) = (\delta_x)$. If $x = 0$ then $(\delta_x) = R$. δ_x is called the *order of* x and (δ_x) the *order ideal of* x. δ_x is uniquely determined up to units in R (that is, up to elements η with $\eta\eta' = 1$ for some $\eta' \in R$). For a submodule U of M we call $\mathrm{Ann}(U) = \bigcap_{u \in U}(\delta_u) = (\mu)$ the *order ideal* of U. In an abelian group G, considered as a $\mathbb{Z}$-module, this order for elements corresponds exactly with the order as group elements if we choose $\delta_x \geq 0$ for $x \in G$.

Theorem 18.5.2. *Let R be a principal ideal domain and M be a finitely generated torsion-free R-module. Then M is free.*

Proof. Let $M = \langle x_1, \dots, x_n \rangle$ torsion-free and R a principal ideal domain. Each submodule $\langle x_i \rangle = Rx_i$ is free because M is torsion-free. We call a subset $S \subset \langle x_1, \dots, x_n \rangle$ free if the submodule $\langle S \rangle$ is free. Since $\langle x_i \rangle$ is free there exist such nonempty subsets. Under all free subsets $S \subset \langle x_1, \dots, x_n \rangle$ we choose one with a maximal number of elements. We may assume that $\{x_1, \dots, x_s\}$, $1 \leq s \leq n$, is such a maximal set – after possible renaming. If $s = n$ then the theorem holds. Now, let $s < n$. By the choice of s the sets $\{x_1, \dots, x_s, x_j\}$ with $s < j \leq n$ are not free. Hence there are $\alpha_j \in R$ and $\alpha_i \in R$, not all 0, with

$$\alpha_j x_j = \sum_{i=1}^{s} \alpha_i x_i, \quad \alpha_j \neq 0,\ s < j \leq n.$$

For the product $\alpha := \alpha_{s+1} \cdots \alpha_n \neq 0$ we get $\alpha x_j \in Rx_1 \oplus \cdots \oplus Rx_s =: F$, $s < j \leq n$, because $\alpha x_i \in F$ for $1 \leq i \leq s$. Altogether we get $\alpha M \subset F$. αM is a submodule of the free R-module F of rank s. By Theorem 18.5.1 we have that αM is free. Since $\alpha \neq 0$ and M is torsion-free, the map $M \to \alpha M$, $x \mapsto \alpha x$, defines an (module) isomorphism, that is, $M \cong \alpha M$. Therefore, also M is free. □

We remind that for an integral domain R the set

$$\mathrm{Tor}(M) = T(M) = \{x \in M : \exists \alpha \in R, \alpha \neq 0, \text{ with } \alpha x = 0\}$$

of the torsion elements of an R-module M is a submodule with torsion-free factor module $M/T(M)$.

Corollary 18.5.3. *Let R be a principal ideal domain and M be a finitely generated R-module. Then $M = T(M) \oplus F$ with a free submodule $F \cong M/T(M)$.*

Proof. $M/T(M)$ is a finitely generated, torsion-free R-module, and, hence, free. By Corollary 18.4.9 we have $M = T(M) \oplus F$, $F \cong M/T(M)$. □

From now on we are interested in the case that $M \neq \{0\}$ is a torsion R-module, that is, $M = T(M)$. Let R be a principal ideal domain and $M = T(M)$ an R-module. Let $M \neq \{0\}$ and finitely generated. As above, let δ_x the order of $x \in M$ – unique up to units in R – and $(\delta_x) = \{\alpha \in R : \alpha x = 0\}$ the order ideal of x. Let $(\mu) = \bigcap_{x \in M}(\delta_x)$ the *order ideal of* M. Since $(\mu) \subset (\delta_x)$ we have $\delta_x | \mu$ for all $x \in M$. Since principal ideal domains are unique factorization domains, if $\mu \neq 0$ then there can not be many essentially different orders (that means different up to units). Since $M \neq \{0\}$ and finitely generated we have in any case $\mu \neq 0$, because if $M = \langle x_1, \dots, x_n \rangle, \alpha_i x_i = 0$ with $\alpha_i \neq 0$ then $\alpha M = \{0\}$ if $\alpha := \alpha_1 \cdots \alpha_n \neq 0$.

Lemma 18.5.4. *Let R be a principal ideal domain and $M \neq \{0\}$ be an R-module with $M = T(M)$.*

(1) *If the orders δ_x and δ_y of $x, y \in M$ are relatively prime, that is,* $\gcd(\delta_x, \delta_y) = 1$, *then $(\delta_{x+y}) = (\delta_x \delta_y)$.*

(2) *Let δ_z be the order of $z \in M$, $z \neq 0$. If $\delta_z = \alpha\beta$ with* $\gcd(\alpha, \beta) = 1$ *then there exist $x, y \in M$ with $z = x + y$ and $(\delta_x) = (\alpha)$, $(\delta_y) = (\beta)$.*

Proof. (1) Since $\delta_x \delta_y (x + y) = \delta_x \delta_y x + \delta_x \delta_y y = \delta_y \delta_x x + \delta_x \delta_y y = 0$ we get $(\delta_x \delta_y) \subset (\delta_{x+y})$. On the other side, from $\delta_x x = 0$ and $\delta_{x+y}(x + y) = 0$ we get $0 = \delta_x \delta_{x+y}(x + y) = \delta_x \delta_{x+y} y$, that means, $\delta_x \delta_{x+y} \in (\delta_y)$ and, hence, $\delta_y | \delta_x \delta_{x+y}$. Since $\gcd(\delta_x, \delta_y) = 1$ we have $\delta_y | \delta_{x+y}$. Analogously $\delta_x | \delta_{x+y}$. Hence, $\delta_x \delta_y | \delta_{x+y}$ and $(\delta_{x+y}) \subset (\delta_x \delta_y)$.

(2) Let $\delta_z = \alpha\beta$ with $\gcd(\alpha, \beta) = 1$. Then there are $\rho, \sigma \in R$ with $1 = \rho\alpha + \sigma\beta$. Therefore we get

$$z = 1 \cdot z = \underbrace{\rho\alpha z}_{=:y} + \underbrace{\sigma\beta z}_{=:x} = y + x = x + y.$$

Since $\alpha x = \alpha\sigma\beta z = \sigma\delta_z z = 0$ we get $\alpha \in (\delta_z)$, that means, $\delta_x | \alpha$. On the other side, from $0 = \delta_x x = \sigma\beta\delta_x z$ we get $\delta_z | \sigma\beta\delta_x$ and hence $\alpha\beta | \sigma\beta\delta_x$ because $\delta_z = \alpha\beta$. Therefore $\alpha | \sigma\delta_x$. From $\gcd(\alpha, \sigma) = 1$ we get $\alpha | \delta_x$. Therefore α is associated to δ_x, that is $\alpha = \delta_x \epsilon$ with ϵ a unit in R and further $(\alpha) = (\delta_x)$. Analogously $(\beta) = (\delta_y)$. □

In Lemma 18.5.4 we do not need $M = T(M)$. We only need $x, y, z \in M$ with $\delta_x \neq 0$, $\delta_y \neq 0$ and $\delta_z \neq 0$, respectively.

Corollary 18.5.5. *Let R be a principal ideal domain and $M \neq \{0\}$ be an R-module with $M = T(M)$.*

1. *Let $x_1, \dots, x_n \in M$ be pairwise different and pairwise relatively prime orders $\delta_{x_i} = \alpha_i$. Then $y = x_1 + \dots + x_n$ has order $\alpha := \alpha_1 \dots \alpha_n$.*

2. *Let* $0 \neq x \in M$ *and* $\delta_x = \epsilon \pi_1^{k_1} \dots \pi_n^{k_n}$ *be a prime decomposition of the order* δ_x *of* x *(*ϵ *a unit in* R *and the* π_i *pairwise nonassociate prime elements in* R*) where* $n > 0$, $k_i > 0$. *Then there exist* x_i, $i = 1, \dots, n$, *with* δ_{x_i} *associated to* $\pi_i^{k_i}$ *and* $x = x_1 + \dots + x_n$.

18.6 The Fundamental Theorem for Finitely Generated Modules

In Section 10.4 we described the following result called the basis theorem for finite abelian groups (in the following we give a complete proof in detail; an elementary proof is given in Chapter 19.).

Theorem 18.6.1 (Theorem 10.4.1, basis theorem for finite abelian groups). *Let* G *be a finite abelian group. Then* G *is a direct product of cyclic groups of prime power order.*

This allowed us, for a given finite order n, to present a complete classification of abelian groups of order n. In this section we extend this result to general modules over principal ideal domains. As a consequence we obtain the fundamental decomposition theorem for finitely generated (not necessarily finite) abelian groups, which finally proves Theorem 10.4.1. In the next chapter we present a separate proof of this in a slightly different format.

Definition 18.6.2. Let R be a principal ideal domain and M be an R-module. Let $\pi \in R$ be a prime element. $M_\pi := \{x \in M : \exists k \geq 0 \text{ with } \pi^k x = 0\}$ is called the *π-primary component* of M. If $M = M_\pi$ for some prime element $\pi \in R$ then M is called *π-primary*.

We certainly have the following.

1. M_π is a submodule of M.
2. The primary components correspond to the p-subgroup inabelian groups.

Theorem 18.6.3. *Let* R *be a principal ideal domain and* $M \neq \{0\}$ *be an* R*-module with* $M = T(M)$. *Then* M *is the direct sum of its* π*-primary components.*

Proof. $x \in M$ has finite order δ_x. Let $\delta_x = \epsilon \pi_1^{k_1} \cdots \pi_n^{k_n}$ be a prime decomposition of δ_x. By Corollary 18.5.5 we have that $x = \sum x_i$ with $x_i \in M_{\pi_i}$. That means $M = \sum_{\pi \in P} M_\pi$ where P is the set of the prime elements of R. Let $y \in M_\pi \cap \sum_{\sigma \in P, \sigma \neq \pi} M_\sigma$, that is, $\delta_y = \pi^k$ for some $k \geq 0$ and $y = \sum x_i$ with $x_i \in M_{\sigma_i}$, that means $\delta_{x_i} = \sigma^{l_i}$ for some $l_i \geq 0$. By Corollary 18.5.5 we get that y has the order $\prod_{\sigma_i \neq \pi} \sigma_i^{l_i}$, that means, π^k is associated to $\prod_{\sigma_i \neq \pi} \sigma_i^{l_i}$. Therefore $k = l_i = 0$ for all i, and the sum is direct. $\square$

If R is a principal ideal domain and $\{0\} \neq M = T(M)$ a finitely generated torsion R-module then there are only finitely many π-primary components, that is to say for the prime elements π with $\pi|\mu$ where (μ) is the order ideal of M.

Corollary 18.6.4. *Let R be a principal ideal domain and $\{0\} \neq M$ be a finitely generated torsion R-module. Then M has only finitely many nontrivial primary components $M_{\pi_1}, \dots, M_{\pi_n}$, and we have*

$$M = \bigoplus_{i=1}^{n} M_{\pi_i}.$$

Hence we have a reduction of the decomposition problem to the primary components.

Theorem 18.6.5. *Let R be a principal ideal domain, $\pi \in R$ a prime element and $M \neq \{0\}$ be a R-module with $\pi^k M = \{0\}$; further let $m \in M$ with $(\delta_m) = (\pi^k)$. Then there exists a submodule $N \subset M$ with $M = Rm \oplus N$.*

Proof. By Zorn's lemma the set $\{U : U$ submodule of M and $U \cap Rm = \{0\}\}$ has a maximal element N. This set is nonempty because it contains $\{0\}$. We consider $M' := N \oplus Rm \subset M$ and have to show that $M' = M$. Assume that $M' \neq M$. Then there exists a $x \in M$ with $x \notin M'$, especially $x \notin N$. Then N is properly contained in the submodule $Rx + N = \langle x, N \rangle$. By our choice of N we get $A := (Rx + N) \cap Rm \neq \{0\}$. If $z \in A$, $z \neq 0$, then $z = \rho m = \alpha x + n$ with ρ, $\alpha \in R$ and $n \in N$. Since $z \neq 0$ we have $\rho m \neq 0$; also $x \neq 0$ because otherwise $z \in Rm \cap N = \{0\}$. α is not a unit in R because otherwise $x = \alpha^{-1}(\rho m - n) \in M'$. Hence we have: If $x \in M$, $x \notin M'$ then there exist $\alpha \in R$, $\alpha \neq 0$, α not a unit in R, $\rho \in R$ with $\rho m \neq 0$ and $n \in N$ such that

$$\alpha x = \rho m + n. \qquad (\star)$$

Especially $\alpha x \in M'$.

Now let $\alpha = \epsilon \pi_1 \dots \pi_r$ be a prime decomposition. We consider one after the other the elements $x, \pi_r x, \pi_{r-1}\pi_r x, \dots, \epsilon\pi_1 \dots \pi_r x = \alpha x$. We have $x \notin M'$ but $\alpha x \in M'$; hence there exists an $y \notin M'$ with $\pi_i y \in N + Rm$.

1. $\pi_i \neq \pi$, π the prime element in the statement of the theorem. Then $\gcd(\pi_i, \pi^k) = 1$, hence there are $\sigma, \sigma' \in R$ with $\sigma\pi_i + \sigma'\pi^k = 1$, and we get $Rm = (R\pi_i + R\pi^k)m = \pi_i Rm$ because $\pi^k m = 0$. Therefore $\pi_i y \in M' = N \oplus Rm = N + \pi_i Rm$.
2. $\pi_i = \pi$. Then we write πy as $\pi y = n + \lambda m$ with $n \in N$ and $\lambda \in R$. This is possible because $\pi y \in M'$. Since $\pi^k M = \{0\}$ we get $0 = \pi^{k-1} \cdot \pi y = \pi^{k-1} n + \pi^{k-1}\lambda m$. Therefore $\pi^{k-1} n = \pi^{k-1}\lambda m = 0$ because $N \cap Rm = \{0\}$. Especially we get $\pi^{k-1}\lambda \in (\delta_m)$, that is, $\pi^k | \pi^{k-1}\lambda$ and, hence, $\pi|\lambda$. Therefore $\pi y = n + \lambda m = n + \pi\lambda' m \in N + \pi Rm$, $\lambda' \in R$.

Hence, in any case we have $\pi_i y \in N + \pi_i Rm$, that is, $\pi_i y = n + \pi_i z$ with $n \in N$ and $z \in Rm$. It follows $\pi_i(y - z) = n \in N$.

$y - z$ is not an element of M' because $y \notin M'$. By $(\star)$ we have therefore $\alpha, \beta \in R$, $\beta \neq 0$ not a unit in R with $\beta(y - z) = n' + \alpha m$, $\alpha m \neq 0$, $n' \in N$. We write $z' = \alpha m$, then $z' \in Rm$, $z' \neq 0$, and $\beta(y - z) = n' + z'$. So, we have the equations $\beta(y - z) = n' + z'$, $z' \neq 0$, and

$$\pi_i(y - z) = n. \qquad (\star\star)$$

We have $\gcd(\beta, \pi_i) = 1$ because otherwise $\pi_i | \beta$ and, hence, $\beta(y - z) \in N$ and $z' = 0$ because $N \cap Rm = \{0\}$. Then there exist γ, γ' with $\gamma\pi_i + \gamma'\beta = 1$. In $(\star\star)$ we multiply the first equation with γ' and the second with γ.

Addition gives $y - z \in N \oplus Rm = M'$, and hence $y \in M'$ which contradicts $y \notin M'$. Therefore $M = M'$. □

Theorem 18.6.6. *Let R be a principal ideal domain, $\pi \in R$ a prime element and $M \neq \{0\}$ a finitely generated π-primary R-module. Then there exist finitely many $m_1, \dots, m_s \in M$ with $M = \bigoplus_{i=1}^{s} Rm_i$.*

Proof. Let $M = \langle x_1, \dots, x_n \rangle$. Each x_i has an order π^{k_i}. We may assume that $k_1 = \max\{k_1, k_2, \dots, k_n\}$, possibly after renaming. We have $\pi^{k_i} x_i = 0$ for all i. Since $x_i^{k_1} = (x_i^{k_i})^{k_1 - k_i}$ we have also $\pi^{k_1} M = 0$, and also $(\delta_{x_1}) = (\pi^{k_1})$. Then $M = Rx_1 \oplus N$ for some submodule $N \subset M$ by Theorem 18.6.5.Now $N \cong M/Rx_1$ and M/Rx_1 is generated by the elements $x_2 + Rx_1, \dots, x_n + Rx_1$. Hence, N is finitely generated by $n - 1$ elements; and certainly N is π-primary. This proves the result by induction. □

Since $R_{m_i} \cong R/\operatorname{Ann}(m_i)$ and $\operatorname{Ann}(m_i) = (\delta_{m_i}) = (\pi^{k_i})$ we get the following extension of Theorem 18.6.6.

Theorem 18.6.7. *Let R be a principal ideal domain, $\pi \in R$ a prime element and $M \neq \{0\}$ a finitely generated π-primary R-module. Then there exist finitely many $k_1, \dots, k_s \in \mathbb{N}$ with*

$$M \cong \bigoplus_{i=0}^{s} R/(\pi^{k_i}),$$

and M is, up to isomorphism, uniquely determined by $(k_1, \dots, k_s)$.

Proof. The first part, that is, a description as $M \cong \bigoplus_{i=0}^{s} R/(\pi^{k_i})$ follows directly from Theorem 18.6.6. Now, let

$$M \cong \bigoplus_{i=0}^{n} R/(\pi^{k_i}) \cong \bigoplus_{i=0}^{m} R/(\pi^{l_i}).$$

We may assume that $k_1 \geq k_2 \geq \cdots \geq k_n > 0$ and $l_1 \geq l_2 \geq \cdots \geq l_m > 0$. We consider first the submodule $N := \{x \in M : \pi x = 0\}$. Let $M = \bigoplus_{i=1}^{n} R/(\pi^{k_i})$. If we write then $x = \sum(r_i + (\pi^{k_i}))$ we have $\pi x = 0$ if and only if $r_i \in (\pi^{k_i-1})$, that is, $N \cong \bigoplus_{i=1}^{n}(\pi^{k_i-1})/(\pi^{k_i}) \cong \bigoplus_{i=1}^{n} R/(\pi)$ because $\pi^{k-1}R/\pi^k R \cong R/\pi R$.

Since $(\alpha + (\pi))x = \alpha x$ if $\pi x = 0$ we get that N is an $R/(\pi)$-module, and, hence, a vector space over the field $R/(\pi)$. From the decompositions

$$N \cong \bigoplus_{i=1}^{n} R/(\pi) \quad \text{and, analogously,} \quad N \cong \bigoplus_{i=1}^{m} R/(\pi)$$

we get

$$n = \dim_{R/(\pi)} N = m. \tag{$\star\star\star$}$$

Assume that there is an i with $k_i < l_i$ or $l_i < k_i$. Without loss of generality assume that there is an i with $k_i < l_i$.

Let j be the smallest index for which $k_j < l_j$. Then (because of the ordering of the k_i)

$$M' := \pi^{k_j} M \cong \bigoplus_{i=1}^{n} \pi^{k_j} R/\pi^{k_i} R \cong \bigoplus_{i=1}^{j-1} \pi^{k_j} R/\pi^{k_i} R,$$

because if $i > j$ then $\pi^{k_j} R/\pi^{k_i} R = \{0\}$.

We now consider $M' = \pi^{k_j} M$ with respect to the second decomposition, that is, $M' \cong \bigoplus_{i=1}^{m} \pi^{k_j} R/\pi^{l_i} R$. By our choice of j we have $k_j < l_j \leq l_i$ for $1 \leq i \leq j$.

Therefore, in this second decomposition, the first j summands $\pi^{k_j} R/\pi^{l_i} R$ are unequal $\{0\}$, that is $\pi^{k_j} R/\pi^{l_i} R \neq \{0\}$ if $1 \leq i \leq j$. The remaining summands are $\{0\}$ or of the form $R/\pi^s R$. Hence, altogether, on the one side M' is a direct sum of $j-1$ cyclic submodules and on the other side a direct sum of $t \geq j$ nontrivial submodules. But this contradicts the above result $(\star\star\star)$ about the number of direct sums for finitely generated π-primary modules because, certainly, M' is also finitely generated and π-primary. Therefore $k_i = l_i$ for $i = 1, \dots, n$. This proves the theorem. □

Theorem 18.6.8 (fundamental theorem for finitely generated modules over principal ideal domains). *Let R be a principal ideal domain and $M \neq \{0\}$ be a finitely generated (unitary) R-module. Then there exist prime elements $\pi_1, \dots, \pi_r \in R$, $0 \leq r < \infty$ and numbers $k_1, \dots, k_r \in t\mathbb{N}$, $t \in \mathbb{N}_0$ such that*

$$M \cong R/(\pi_1^{k_1}) \oplus R/(\pi_2^{k_2}) \oplus \cdots \oplus R/(\pi_r^{k_r}) \oplus \underbrace{R \oplus \cdots \oplus R}_{t\text{-times}},$$

and M is, up to isomorphism, uniquely determined by $(\pi_1^{k_1}, \dots, \pi_r^{k_r}, t)$.

The prime elements π_i are not necessarily pairwise different (up to units in R), that means it can be $\pi_i = \epsilon \pi_j$ for $i \neq j$ where ϵ is a unit in R.

Proof. The proof is a combination of the preceding results. The free part of M is isomorphic to $M/T(M)$, and the rank of $M/T(M)$, which we call here t, is uniquely determined because two bases of $M/T(M)$ have the same cardinality. Therefore we may restrict ourselves on torsion modules. Here we have a reduction to π-primary modules because in a decomposition $M = \bigoplus_i R/(\pi_i^{k_i})$ is $M_\pi = \bigoplus_{\pi_i=\pi} R/(\pi_i^{k_i})$ the π-primary component of M (an isomorphism certainly maps a π-primary component onto a π-primary component). So it is only necessary, now, to consider π-primary modules M. The uniqueness statement now follows from Theorem 18.6.8. □

Since abelian groups can be considered as $\mathbb{Z}$-modules, and $\mathbb{Z}$ is a principal ideal domain, we get the following corollary. We will restate this result in the next chapter and prove a different version of it.

Theorem 18.6.9 (fundamental theorem for finitely generated abelian groups). *Let $\{0\} \neq G = (G, +)$ be a finitely generated abelian group. Then there exist prime numbers $p_1, \dots, p_r$, $0 \le r < \infty$, and numbers $k_1, \dots, k_r \in \mathbb{N}$, $t \in \mathbb{N}_0$ such that*

$$G \cong \mathbb{Z}/(p_1^{k_1}\mathbb{Z}) \oplus \cdots \oplus \mathbb{Z}/(p_r^{k_r}\mathbb{Z}) \oplus \underbrace{\mathbb{Z} \oplus \cdots \oplus \mathbb{Z}}_{t\text{-times}},$$

and G is, up to isomorphism, uniquely determined by $(p_1^{k_1}, \dots, p_r^{k_r}, t)$.

18.7 Exercises

1. Let M and N be isomorphic modules over a commutative ring R. Then $\operatorname{End}_R(M)$ and $\operatorname{End}_R(N)$ are isomorphic rings. ($\operatorname{End}_R(M)$ is the set of all R-modules endomorphisms of M.)
2. Let R be an integral domain and M an R-module with $M = \operatorname{Tor}(M)$ (torsion module). Show that $\operatorname{Hom}_R(M, R) = 0$. ($\operatorname{Hom}_R(M, R)$ is the set of all R-module homomorphisms from M to R.)
3. Prove the isomorphism theorems for modules (1), (2) and (3) in Theorem 18.1.11 in detail.
4. Let M, M', N be R-modules, R a commutative ring. Show:
   (i) $\operatorname{Hom}_R(M \oplus M', N) \cong \operatorname{Hom}_R(M, N) \times \operatorname{Hom}_R(M', N)$
   (ii) $\operatorname{Hom}_R(N, M \times M') \cong \operatorname{Hom}_R(N, M) \oplus \operatorname{Hom}_R(N, M')$.
5. Show that two free modules having bases whose cardinalities are equal are isomorphic.

6. Let M be an unitary R-module (R a commutative ring) and let $\{m_1, \dots, m_s\}$ be a finite subset of M. Show that the following are equivalent:

   (i) $\{m_1, \dots, m_s\}$ generates M freely.

   (ii) $\{m_1, \dots, m_s\}$ is linearly independent and generates M.

   (iii) Every element $m \in M$ is uniquely expressible in the form $m = \sum_{i=1}^{s} r_i m_i$ with $r_i \in R$.

   (iv) Each Rm_i is torsion-free, and $M = Rm_1 \oplus \cdots \oplus Rm_s$.

7. Let R be a principal domain and $M \neq \{0\}$ be an R-module with $M = T(M)$.

   (i) Let $x_1, \dots, x_n \in M$ be pairwise different and pairwise relatively prime orders $\delta_{x_i} = \alpha_i$. Then $y = x_1 + \cdots + x_n$ has order $\alpha := \alpha_1 \dots \alpha_n$.

   (ii) Let $0 \neq x \in M$ and $\delta_x = \epsilon \pi_1^{k_1} \cdots \pi_n^{k_n}$ be a prime decomposition of the order δ_x of x (ϵ a unit in R and the π_i pairwise nonassociate prime elements in R) where $n > 0$, $k_i > 0$. Then there exist x_i, $i = 1, \dots, n$, with δ_{x_i} associated to $\pi_i^{k_i}$ and $x = x_1 + \cdots + x_n$.

Chapter 19
Finitely Generated Abelian Groups

19.1 Finite Abelian Groups

In Chapter 10 we described the following theorem that completely provides the structure of finite abelian groups. As we saw in Chapter 18 this result is a special case of a general result on modules over principal ideal domains.

Theorem 19.1.1 (Theorem 10.4.1, basis theorem for finite abelian groups). *Let G be a finite abelian group. Then G is a direct product of cyclic groups of prime power order.*

We review two examples that show how this theorem leads to the classification of finite abelian groups. In particular this theorem allows us, for a given finite order n, to present a complete classification of abelian groups of order n.

Since all cyclic groups of order n are isomorphic to $(\mathbb{Z}_n, +)$, $\mathbb{Z}_n = \mathbb{Z}/n\mathbb{Z}$, we will denote a cyclic group of order n by $\mathbb{Z}_n$.

Example 19.1.2. Classify all abelian groups of order 60. Let G be an abelian group of order 60. From Theorem 10.4.1 G must be a direct product of cyclic groups of prime power order. Now $60 = 2^2 \cdot 3 \cdot 5$ so the only primes involved are 2, 3 and 5. Hence the cyclic groups involved in the direct product decomposition of G have order either 2, 4, 3 or 5 (by Lagrange's theorem they must be divisors of 60). Therefore G must be of the form

$$G \cong \mathbb{Z}_4 \times \mathbb{Z}_3 \times \mathbb{Z}_5$$

or

$$G \cong \mathbb{Z}_2 \times \mathbb{Z}_2 \times \mathbb{Z}_3 \times \mathbb{Z}_5.$$

Hence up to isomorphism there are only two abelian groups of order 60.

Example 19.1.3. Classify all abelian groups of order 180. Let G be an abelian group of order 180. Now $180 = 2^2 \cdot 3^2 \cdot 5$ so the only primes involved are 2, 3 and 5. Hence the cyclic groups involved in the direct product decomposition of G have order either 2, 4, 3, 9 or 5 (by Lagrange's theorem they must be divisors of 180). Therefore G

must be of the form

$$G \cong \mathbb{Z}_4 \times \mathbb{Z}_9 \times \mathbb{Z}_5$$
$$G \cong \mathbb{Z}_2 \times \mathbb{Z}_2 \times \mathbb{Z}_9 \times \mathbb{Z}_5$$
$$G \cong \mathbb{Z}_4 \times \mathbb{Z}_3 \times \mathbb{Z}_3 \times \mathbb{Z}_5$$
$$G \cong \mathbb{Z}_2 \times \mathbb{Z}_2 \times \mathbb{Z}_3 \times \mathbb{Z}_3 \times \mathbb{Z}_5.$$

Hence up to isomorphism there are four abelian groups of order 180.

The proof of Theorem 19.1.1 involves the following lemmas. We refer back to Chapter 10 or Chapter 18 for the proofs. Notice how these lemmas mirror the results for finitely generated modules over principal ideal domains considered in the last chapter.

Lemma 19.1.4. *Let G be a finite abelian group and let $p \| G|$ where p is a prime. Then all the elements of G whose orders are a power of p form a normal subgroup of G. This subgroup is called the p-primary component of G, which we will denote by G_p.*

Lemma 19.1.5. *Let G be a finite abelian group of order n. Suppose that $n = p_1^{e_1} \cdots p_k^{e_k}$ with $p_1, \dots, p_k$ distinct primes. Then*

$$G \cong G_{p_1} \times \cdots \times G_{p_k}$$

where G_{p_i} is the p_i-primary component of G.

Theorem 19.1.6 (basis theorem for finite abelian groups). *Let G be a finite abelian group. Then G is a direct product of cyclic groups of prime power order.*

19.2 The Fundamental Theorem: p-Primary Components

In this section we use the fundamental theorem for finitely generated modules over principal ideal domains to extend the basis theorem for finite abelian groups to the more general case of finitely generated abelian groups. In this section we consider the decomposition into p-primary components, mirroring our result in the finite case. In the next chapter we present a different form of the basis theorem with a more elementary proof.

In Chapter 18 we proved the following:

Theorem 19.2.1 (fundamental theorem for finitely generated modules over principal ideal domains). *Let R be a principal ideal domain and $M \neq \{0\}$ be a finitely generated (unitary) R-module. Then there exist prime elements $\pi_1, \dots, \pi_r \in R$, $0 \le r < \infty$ and numbers $k_1, \dots, k_r \in \mathbb{N}$, $t \in \mathbb{N}_0$ such that*

$$M \cong R/(\pi_1^{k_1}) \oplus R/(\pi_2^{k_2}) \oplus \cdots \oplus R/(\pi_r^{k_r}) \oplus \underbrace{R \oplus \cdots \oplus R}_{t\text{-times}},$$

and M is, up to isomorphism, uniquely determined by $(\pi_1^{k_1}, \dots, \pi_r^{k_r}, t)$.

The prime elements π_i are not necessarily pairwise different (up to units in R), that means it can be $\pi_i = \epsilon\pi_j$ for $i \neq j$ where ϵ is a unit in R.

Since abelian groups can be considered as $\mathbb{Z}$-modules, and $\mathbb{Z}$ is a principal ideal domain, we get the following corollary which is extremely important in its own right.

Theorem 19.2.2 (fundamental theorem for finitely generated abelian groups). *Let $\{0\} \neq G = (G, +)$ be a finitely generated abelian group. Then there exist prime numbers $p_1, \dots, p_r$, $0 \le r < \infty$, and numbers $k_1, \dots, k_r \in \mathbb{N}$, $t \in \mathbb{N}_0$ such that*

$$G \cong \mathbb{Z}/(p_1^{k_1}\mathbb{Z}) \oplus \cdots \oplus \mathbb{Z}/(p_r^{k_r}\mathbb{Z}) \oplus \underbrace{\mathbb{Z} \oplus \cdots \oplus \mathbb{Z}}_{t\text{-times}},$$

and G is, up to isomorphism, uniquely determined by $(p_1^{k_1}, \dots, p_r^{k_r}, t)$.

Notice that the number t of infinite components is unique. This is called the *rank* or *Betti number* of the abelian group G. This number plays an important role in the study of homology and cohomology groups in topology.

If $G = \mathbb{Z} \times \mathbb{Z} \times \cdots \times \mathbb{Z} = \mathbb{Z}^r$ for some r we call G a *free abelian* group of rank r. Notice that if an abelian group G is torsion-free then the p-primary components are just the identity. It follows that in this case G is a free abelian group of finite rank. Again using module theory it follows that subgroups of this must also be free abelian and of smaller or equal rank. Notice the distinction between free abelian groups and absolutely free groups (see Chapter 14). In the free group case a nonabelian free group of finite rank contains free subgroups of all possible countable ranks. In the free abelian case however the subgroups have smaller or equal rank. We summarize this comments.

Theorem 19.2.3. *Let $G \neq \{0\}$ be a finitely generated torsion-free abelian group. Then G is a free abelian group of finite rank r that is $G \cong \mathbb{Z}^r$. Further if H is a subgroup of G then H is also free abelian and the rank of H is smaller or equal than the rank of G.*

19.3 The Fundamental Theorem: Elementary Divisors

In this section we present the fundamental theorem of finitely generated abelian groups in a slightly different form and present an elementary proof of it.

In the following G is always a finitely generated abelian group. We use the addition "+" for the binary operation, that is,

$$+ : G \times G \to G, \quad (x, y) \mapsto x + y.$$

We also write ng instead of g^n and use 0 as the symbol for the identity element in G, that is, $0 + g = g$ for all $g \in G$. $G = \langle g_1, \ldots, g_t \rangle$, $0 \leq t < \infty$, that is, G is (finitely) generated by $g_1, \ldots, g_t$, is equivalent to the fact that each $g \in G$ can be written in the form $g = n_1 g_1 + n_2 g_2 + \cdots + n_t g_t$, $n_i \in \mathbb{Z}$. A *relation* between the g_i with coefficients $n_1, \ldots, n_t$ is then each equation of the form $n_1 g_1 + \cdots + n_t g_t = 0$. A relation is called *nontrivial* if $n_i \neq 0$ for at least one i. A system R of relations in G is called a *system of defining relations*, if each relation in G is a consequence of R. The elements $g_1, \ldots, g_t$ are called *integrally linear independent* if there are no nontrivial relations between them. A finite generating system $\{g_1, \ldots, g_t\}$ of G is called a *minimal generating system* if there is no generating system with $t - 1$ elements.

Certainly each finitely generated group has a minimal generating system. In the following we always assume that our finitely generated abelian group G is unequal $\{0\}$, that is, G is nontrivial.

As above, we may consider G as a finitely generated $\mathbb{Z}$-module, and in this sense, the subgroups of G are precisely the submodules. Hence, it is clear what we mean if we call G a direct product $G = U_1 \times \cdots \times U_s$ of its subgroups $U_1, \ldots, U_s$, namely, each $g \in G$ can be written as $g = u_1 + u_2 + \cdots + u_s$ with $u_i \in U_i$ and

$$U_i \cap \left(\prod_{j=1, j \neq i}^{s} U_j \right) = \{0\}.$$

To emphasize the little difference between abelian groups and $\mathbb{Z}$-modules we here use the notation "direct product" instead of "direct sum". Considered as $\mathbb{Z}$-modules, for finite index sets $I = \{1, \ldots, s\}$ we have anyway

$$\prod_{i=1}^{s} U_i = \bigoplus_{i=1}^{s} U_i.$$

Finally we use the notation $\mathbb{Z}_n$ instead of $\mathbb{Z}/n\mathbb{Z}$, $n \in \mathbb{N}$. In general, we use Z_n to be a cyclic group of order n.

The aim in this section is to prove the following:

Theorem 19.3.1 (basis theorem for finitely generated abelian groups). *Let $G \neq \{0\}$ be a finitely generated abelian group. Then G is a direct product*

$$G \cong Z_{k_1} \times \cdots \times Z_{k_r} \times U_1 \times \cdots \times U_s,$$

$r \geq 0$, $s \geq 0$, of cyclic subgroups with $|Z_{k_i}| = k_i$ for $i = 1, \ldots, r$, $k_i | k_{i+1}$ for $i = 1, \ldots, r-1$ and $U_j \cong \mathbb{Z}$ for $j = 1, \ldots, s$. Here the numbers $k_1, \ldots, k_r$, r and s are uniquely determined by G, that means, are $k_1', \ldots, k_r', r'$ and s' the respective numbers for a second analogous decomposition of G then $r = r'$, $k_1 = k_1', \ldots, k_r = k_r'$ and $s = s'$.

The numbers k_i are called the *elementary divisors* of G.

We can have $r = 0$ or $s = 0$ (but not both because $G \neq \{0\}$). If $s > 0$, $r = 0$ then G is a *free abelian group of rank s* (exactly the same rank if you consider G as a free $\mathbb{Z}$-module of rank s). If $s = 0$ then G is finite, in fact: $s = 0 \Leftrightarrow G$ is finite.

We first prove some preliminary results.

Lemma 19.3.2. *Let $G = \langle g_1, \ldots, g_t \rangle$, $t \geq 2$, an abelian group. Then also $G = \langle g_1 + \sum_{i=2}^{t} m_i g_i, g_2, \ldots, g_t \rangle$ for arbitrary $m_2, \ldots, m_t \in \mathbb{Z}$.*

Lemma 19.3.3. *Let G be a finitely generated abelian group. Among all nontrivial relations between elements of minimal generating systems of G we choose one relation*

$$m_1 g_1 + \cdots + m_t g_t = 0 \tag{$\star$}$$

with smallest possible positive coefficient, and let this smallest coefficient be m_1. Let

$$n_1 g_1 + \cdots + n_t g_t = 0 \tag{$\star\star$}$$

be another relation between the same generators $g_1, \ldots, g_t$. Then

(1) *$m_1 | n_1$ and*

(2) *$m_1 | m_i$ for $i = 1, 2, \ldots, t$.*

Proof. (1) Assume $m_1 \nmid n_1$. Then $n_1 = qm_1 + m_1'$ with $0 < m_1' < m_1$. If we multiply the relation ($\star$) with q and subtract the resulting relation from the relation ($\star\star$) then we get a relation with a coefficient $m_1' < m_1$ which contradicts the choice of m_1. Hence $m_1 | n_1$.

(2) Assume $m_1 \nmid m_2$. Then $m_2 = qm_1 + m_2'$ with $0 < m_2' < m_2$. $\{g_1 + qg_2, g_2, \ldots, g_t\}$ is a minimal generating system which satisfies the relation $m_1(g_1 + qg_2) + m_2' g_2 + m_3 g_3 + \cdots + m_t g_t = 0$, and this relation has a coefficient $m_2' < m_1$. This again contradicts the choice of m_1. Hence $m_1 | m_2$ and further $m_1 | m_i$ for $i = 1, \ldots, t$. □

Lemma 19.3.4 (invariant characterization of k_r for finite abelian groups G). *Let $G = Z_{k_1} \times \cdots \times Z_{k_r}$ and Z_{k_i} finite cyclic of order $k_i \geq 2$, $i = 1, \dots, r$, with $k_i | k_{i+1}$ for $i = 1, \dots, r-1$. Then k_r is the smallest natural number n such that $ng = 0$ for all $g \in G$. k_r is called the exponent or the maximal order of G.*

Proof. 1. Let $g \in G$ arbitrary, that is, $g = n_1 g_1 + \cdots + n_r g_r$ with $g_i \in Z_{k_i}$. Then $k_i g_i = 0$ for $i = 1, \dots, r$ by the theorem of Fermat. Since $k_i | k_r$ we get $k_r g = n_1 k_1 g_1 + \cdots + n_r k_r g_r = 0$.

2. Let $a \in G$ with $Z_{k_r} = \langle a \rangle$. Then the order of a is k_r and, hence, $na \neq 0$ for all $0 < n < k_r$. □

Lemma 19.3.5 (invariant characterization of s). *Let $G = Z_{k_1} \times \cdots \times Z_{k_r} \times U_1 \times \cdots \times U_s$, $s > 0$, where the Z_{k_i} are finite cyclic groups of order k_i and the U_j are infinite cyclic groups. Then, s is the maximal number of integrally linear independent elements of G; s is called the rank of G.*

Proof. 1. Let $g_i \in U_i$, $g_i \neq 0$, for $i = 1, \dots, s$. Then the $g_1, \dots, g_s$ are integrally linear independent because from $n_1 g_1 + \cdots + n_s g_s = 0$, the $n_i \in \mathbb{Z}$, we get $n_1 g_1 \in U_1 \cap (U_2 \times \cdots \times U_s) = \{0\}$, and, hence, $n_1 g_1 = 0$, that is $n_1 = 0$, because g_1 has infinite order. Analogously we get $n_2 = \cdots = n_s = 0$.

2. Let $g_1, \dots, g_{s+1} \in G$. We look for integers $x_1, \dots, x_{s+1}$, not all 0, such that a relation $\sum_{i=1}^{s+1} x_i g_i = 0$ holds. Let $Z_{k_i} \in \langle a_i \rangle$, $U_j = \langle b_j \rangle$. Then we may write each g_i as $g_i = m_{i1} a_1 + \cdots + m_{ir} a_r + n_{i1} b_1 + \cdots + n_{is} b_s$ for $i = 1, \dots, s+1$, where $m_{ij} a_j \in Z_{k_j}$ and $n_{il} b_l \in U_l$.

Case 1: all $m_{ij} a_j = 0$. Then $\sum_{i=1}^{s+1} x_i g_i = 0$ is equivalent to

$$\sum_{i=1}^{s+1} x_i \left(\sum_{j=1}^{s} n_{ij} b_j \right) = \sum_{j=1}^{s} \left(\sum_{i=1}^{s+1} n_{ij} x_i \right) b_j = 0.$$

The system $\sum_{i=1}^{s+1} n_{ij} x_i = 0$, $j = 1, \dots, s$, of linear equations has at least one nontrivial rational solution $(x_1, \dots, x_{s+1})$ because we have more unknown than equations. Multiplication with the common denominator gives a nontrivial integral solution $(x_1, \dots, x_{s+1}) \in \mathbb{Z}^{s+1}$. For this solution we get

$$\sum_{i=1}^{s+1} x_i g_i = 0.$$

Case 2: $m_{ij} a_j$ arbitrary. Let $k \neq 0$ be a common multiple of the orders k_j of the cyclic groups Z_{k_j}, $j = 1, \dots, r$. Then

$$k g_i = \underbrace{m_{i1} k a_1}_{=0} + \cdots + \underbrace{m_{ir} k a_r}_{=0} + n_{i1} k b_1 + \cdots + n_{is} k b_s$$

for $i = 1, \dots, s+1$. By case 1 the $kg_1, \dots, kg_{s+1}$ are integrally linear dependent, that is, we have integers $x_1, \dots, x_{s+1}$, not all 0, with $\sum_{i=1}^{s+1} x_i(kg_i) = 0 = \sum_{i=1}^{s+1}(x_i k)g_i$, and the $x_i k$ are not all 0. Hence, also $g_1, \dots, g_{s+1}$ are integrally linear dependent. □

Lemma 19.3.6. *Let* $G := Z_{k_1} \times \cdots \times Z_{k_r} \cong Z_{k'_1} \times \cdots \times Z_{k'_{r'}} =: G'$, *the* $Z_{k_i}, Z_{k'_j}$ *cyclic groups of orders* $k_i \neq 1$ *and* $k'_j \neq 1$, *respectively and* $k_i | k_{i+1}$ *for* $i = 1, \dots, r-1$ *and* $k'_j | k'_{j+1}$ *for* $j = 1, \dots, r'-1$. *Then* $r = r'$ *and* $k_1 = k'_1$, $k_2 = k'_2, \dots, k_r = k'_r$.

Proof. We prove this lemma by induction on the group order $|G| = |G'|$. Certainly, Lemma 19.3.6 holds if $|G| \leq 2$ because then either $G = \{0\}$, and here $r = r' = 0$ or $G \cong \mathbb{Z}_2$, and here $r = r' = 1$. Now let $|G| > 2$. Then especially $r \geq 1$. Inductively we assume that Lemma 19.3.6 holds for all finite abelian groups of order less than $|G|$. By Lemma 19.3.4 the number k_r is invariantly characterized, that is, from $G \cong G'$ follows $k_r = k'_{r'}$, that is especially, $Z_{k_r} \cong Z_{k'_{r'}}$. Then $G/Z_{k_r} \cong G/Z_{k'_{r'}}$, that is, $Z_{k_1} \times \cdots \times Z_{k_{r-1}} \cong Z_{k'_1} \times \cdots \times Z_{k'_{r'-1}}$. Inductively $r - 1 = r' - 1$, that is, $r = r'$, and $k_1 = k'_1, \dots, k_{r-1} = k'_{r'-1}$. This proves Lemma 19.3.6. □

We can now present the main result, which we state again, and its proof.

Theorem 19.3.7 (basis theorem for finitely generated abelian groups). *Let* $G \neq \{0\}$ *be a finitely generated abelian group. Then G is a direct product*

$$G \cong Z_{k_1} \times \cdots \times Z_{k_r} \times U_1 \times \cdots \times U_s, \quad r \geq 0,\ s \geq 0,$$

of cyclic subgroups with $|Z_{k_i}| = k_i$ *for* $i = 1, \dots, r$, $k_i | k_{i+1}$ *for* $i = 1, \dots, r-1$ *and* $U_j \cong \mathbb{Z}$ *for* $j = 1, \dots, s$. *Here the numbers* $k_1, \dots, k_r, r$ *and* s *are uniquely determined by G, that means, are* $k'_1, \dots, k'_r, r'$ *and* s' *the respective numbers for a second analogous decomposition of G then* $r = r'$, $k_1 = k'_1, \dots, k_r = k'_r$ *and* $s = s'$.

Proof. (a) We first prove the existence of the given decomposition. Let $G \neq \{0\}$ be a finitely generated abelian group. Let t, $0 < t < \infty$, be the number of elements in a minimal generating system of G. We have to show that G is decomposable as a direct product of t cyclic groups with the given description. We prove this by induction on t. If $t = 1$ then the basis theorem certainly is correct. Now let $t \geq 2$ and assume that the assertion holds for all abelian groups with less then t generators.

Case 1: There does not exist a minimal generating system of G which satisfies a nontrivial relation. Let $\{g_1, \dots, g_t\}$ be an arbitrary minimal generating system for G. Let $U_i = \langle g_i \rangle$. Then all U_i are infinite cyclic and we have $G = U_1 \times \cdots \times U_t$ because if, for instance, $U_1 \cap (U_2 + \cdots + U_t) \neq \{0\}$ then we must have a nontrivial relation between the $g_1, \dots, g_t$.

Case 2: There exist minimal generating systems of G which satisfy nontrivial relations. Among all nontrivial relations between elements of minimal generating systems of G we choose one relation

$$m_1 g_1 + \cdots + m_t g_t = 0 \qquad (\star)$$

with smallest possible positive coefficient. Without loss of generality, let m_1 be this coefficient. By Lemma 19.3.3 we get $m_2 = q_2 m_1, \dots, m_t = q_t m_1$. Now, $\{g_1 + \sum_{i=2}^t q_i g_i, g_2, \dots, g_t\}$ is a minimal generating system of G by Lemma 19.3.2. Define $h_1 = g_1 + \sum_{i=2}^t q_i g_i$, then $m_1 h_1 = 0$. If $n_1 h_1 + n_2 g_2 + \cdots + n_t g_t = 0$ is an arbitrary relation between $h_1, g_2, \dots, g_t$ then $m_1 | n_1$ by Lemma 19.3.3, and, hence $n_1 h_1 = 0$. Define $H_1 := \langle h_1 \rangle$ and $G' = \langle g_2, \dots, g_t \rangle$. Then $G = H_1 \times G'$. This we can see as follows: First, each $g \in G$ can be written as $g = m_1 h_1 + m_2 g_2 + \cdots + m_t g_t = m_1 h_1 + g'$ with $g' \in G'$. Also $H_1 \cap G' = \{0\}$ because $m_1 h_1 = g' \in G'$ implies a relation $n_1 h_1 + n_2 g_2 + \cdots + n_t g_t = 0$ and from this we get, as above, $n_1 h_1 = g' = 0$. Now, inductively, $G' = Z_{k_2} \times \cdots \times Z_{k_r} \times U_1 \times \cdots \times U_s$ with Z_{k_i} a cyclic group of order k_i, $i = 2, \dots, r$, $k_i | k_{i+1}$ for $i = 2, \dots, r-2$, $U_j \cong \mathbb{Z}$ for $j = 1, \dots, s$, and $(r-1) + s = t - 1$, that is, $r + s = t$. Further, $G = H_1 \times G'$ where H_1 is cyclic of order m_1. If $r \geq 2$ then we get a nontrivial relation

$$\underbrace{m_1 h_1}_{=0} + \underbrace{k_2 h_2}_{=0} = 0$$

since $k_2 \neq 0$. Again $m_1 | k_2$ by Lemma 19.3.3. This gives the desired decomposition.

(b) We now prove the uniqueness statement.

Case 1: G is finite abelian. Then the claim follows from Lemma 19.3.6

Case 2: G is arbitrary finitely generated and abelian. Let $T := \{x \in G : |x| < \infty\}$, that is the set of elements of G of finite order. Since G is abelian T is a subgroup of G, the so called *torsion subgroup of* G. If, as above, $G = Z_{k_1} \times \cdots \times Z_{k_r} \times U_1 \times \cdots \times U_s$ then $T = Z_{k_1} \times \cdots \times Z_{k_r}$ because an element $b_1 + \cdots + b_r + c_1 + \cdots + c_s$ with $b_i \in Z_{k_i}$, $c_j \in U_j$ has finite order if and only if all $c_j = 0$. That means: $Z_{k_1} \times \cdots \times Z_{k_r}$ is, independent of the special decomposition, uniquely determined by G, and hence, also the numbers $r, k_1, \dots, k_r$ by Lemma 19.3.6. Finally the number s, the rank of G, is uniquely determined by Lemma 19.3.5. This proves the basis theorem for finitely generated abelian groups. □

As a corollary we get the fundamental theorem for finitely generated abelian groups as given in Theorem 19.2.1.

Theorem 19.3.8. *Let $\{0\} \neq G = (G, +)$ be a finitely generated abelian group. Then there exist prime numbers $p_1, \dots, p_r$, $0 \leq r < \infty$, and numbers $k_1, \dots, k_r \in \mathbb{N}$, $t \in \mathbb{N}_0$ such that*

$$G \cong \mathbb{Z}_{p_1^{k_1}} \times \cdots \times \mathbb{Z}_{p_r^{k_r}} \times \underbrace{\mathbb{Z} \times \cdots \times \mathbb{Z}}_{t\text{-times}},$$

and G is, up to isomorphism, uniquely determined by $(p_1^{k_1}, \dots, p_r^{k_r}, t)$.

Proof. For the existence we only have to show that $\mathbb{Z}_{mn} \cong \mathbb{Z}_m \times \mathbb{Z}_n$ if $\gcd(m,n)=1$. For this we write $U_n = \langle m + mn\mathbb{Z}\rangle < \mathbb{Z}_{mn}$, $U_m = \langle n + nm\mathbb{Z}\rangle < \mathbb{Z}_{mn}$, and $U_n \cap U_m = \{mn\mathbb{Z}\}$ because $\gcd(m,n) = 1$. Further there are $h, k \in \mathbb{Z}$ with $1 = hm + kn$. Hence, $l + mn\mathbb{Z} = hlm + mn\mathbb{Z} + kln + mn\mathbb{Z}$, and therefore $\mathbb{Z}_{mn} = U_n \times U_m \cong \mathbb{Z}_n \times \mathbb{Z}_m$.

For the uniqueness statement we may reduce the problem to the case $|G| = p^k$ for a prime number p and $k \in \mathbb{N}$. But here the result follows directly from Lemma 19.3.6. □

From this proof we automatically get the Chinese remainder theorem for the case $\mathbb{Z}_n = \mathbb{Z}/n\mathbb{Z}$.

Theorem 19.3.9 (Chinese remainder theorem). *Let* $m_1, \dots, m_r \in \mathbb{N}$, $r \geq 2$, *with* $\gcd(m_i, m_j) = 1$ *for* $i \neq j$. *Define* $m := m_1 \cdots m_r$.

(1) $\pi : \mathbb{Z}_m \to \mathbb{Z}_{m_1} \times \cdots \times \mathbb{Z}_{m_r}$, $a + m\mathbb{Z} \mapsto (a + m_1\mathbb{Z}, \dots, a + m_r\mathbb{Z})$, *defines a ring isomorphism.*

(2) *The restriction of* π *on the multiplicative group of the prime residue classes defines a group isomorphism* $\mathbb{Z}_m^\star \to \mathbb{Z}_{m_1}^\star \times \cdots \times \mathbb{Z}_{m_r}^\star$.

(3) *For given* $a_1, \dots, a_r \in \mathbb{Z}$ *there exists modulo* m *exactly one* $x \in \mathbb{Z}$ *with* $x \equiv a_i \pmod{m_i}$ *for* $i = 1, \dots, r$.

Recall that for $k \in \mathbb{N}$ a prime residue class is defined by $a+k\mathbb{Z}$ with $\gcd(a,k) = 1$. The set of prime residue classes modulo k is certainly a multiplicative group.

Proof. By Theorem 19.3.1 we get that π is an additive group isomorphism which can be extended directly to a ring isomorphism via $(a + m\mathbb{Z})(b + m\mathbb{Z}) \mapsto (ab + m_1\mathbb{Z}, \dots, ab + m_r\mathbb{Z})$. The remaining statements are now obvious. □

Let $A(n)$ be the number of nonisomorphic finite abelian groups of order $n = p_1^{k_1} \cdots p_r^{k_r}$, $r \geq 1$, with pairwise different prime numbers $p_1, \dots, p_r$ and $k_1, \dots, k_r \in \mathbb{N}$. By Theorem 19.2.2 we have $A(n) = A(p_1^{k_1}) \cdots A(p_r^{k_r})$. Hence, to calculate $A(n)$, we have to calculate $A(p^m)$ for a prime number p $m \in \mathbb{N}$. Again, by Theorem 19.2.2, we get $G \cong \mathbb{Z}_{p^{m_1}} \times \cdots \times \mathbb{Z}_{p^{m_k}}$, all $m_i \geq 1$, if G is abelian of order p^m. If we compare the orders we get $m = m_1 + \cdots + m_k$. We may order the m_i by size. A k-tuple $(m_1, \dots, m_k)$ with $0 < m_1 \leq m_2 \leq \cdots \leq m_k$ and $m_1 + m_2 + \cdots + m_k = m$ is called a *partition of* m. From above each abelian group of order p^m gives a partition $(m_1, \dots, m_k)$ of m for some k with $1 \leq k \leq m$. On the other side, each partition $(m_1, \dots, m_k)$ of m gives an abelian group of order p^m, namely $\mathbb{Z}_{p^{m_1}} \times \cdots \times \mathbb{Z}_{p^{m_k}}$. Theorem 19.2.2 shows that different partitions give nonisomorphic groups. If we define $p(m)$ to be the number of partitions of m then we get the following: $A(p^m) = p(m)$ and $A(p_1^{k_1} \cdots p_r^{k_r}) = p(k_1) \cdots p(k_r)$.

19.4 Exercises

1. Let H be a finite generated abelian group, which is the homomorphic image of a torsion-free abelian group of finite rank n. Show that H is the direct sum of $\leq n$ cyclic groups.

2. Determine (up to isomorphism) all groups of order p^2 (p prime) and all abelian groups of order ≤ 15.

3. Let G be an abelian group with generating elements $a_1, \ldots, a_4$ and defining relations

$$\begin{aligned} 5a_1 + 4a_2 + a_3 + 5a_4 &= 0 \\ 7a_1 + 6a_2 + 5a_3 + 11a_4 &= 0 \\ 2a_1 + 2a_2 + 10a_3 + 12a_4 &= 0 \\ 10a_1 + 8a_2 - 4a_3 + 4a_4 &= 0. \end{aligned}$$

   Express G as a direct product of cyclic groups.

4. Let G be a finite abelian group and $u = \prod_{g \in G} g$ the product of all elements of G. Show: If G has an element a of order 2, then $u = a$, otherwise $u = e$. Conclude from this the theorem of Wilson:

$$(p-1)! \equiv -1 (\text{mod } p) \quad \text{for each prime } p.$$

5. Let p be a prime and G a finite abelian p-group, that is the order of all elements of G is finite and a power of p. Show that G is cyclic, if G has exactly one subgroup of order p. Is the statement still correct, if G is not abelian?

Chapter 20
Integral and Transcendental Extensions

20.1 The Ring of Algebraic Integers

Recall that a complex number α is an *algebraic number* if it is algebraic over the rational numbers $\mathbb{Q}$. That is α is a zero of a polynomial $p(x) \in \mathbb{Q}[x]$. If $\alpha \in \mathbb{C}$ is not algebraic then it is a *transcendental number.*

We will let $\mathcal{A}$ denote the totality of algebraic numbers within the complex numbers $\mathbb{C}$, and $\mathcal{T}$ the set of transcendentals so that $\mathbb{C} = \mathcal{A} \cup \mathcal{T}$. The set $\mathcal{A}$ is the algebraic closure of $\mathbb{Q}$ within $\mathbb{C}$.

The set $\mathcal{A}$ of algebraic numbers forms a subfield of $\mathbb{C}$ (see Chapter 5) and the subset $\mathcal{A}' = \mathcal{A} \cap \mathbb{R}$ of real algebraic numbers forms a subfield of $\mathbb{R}$. The field $\mathcal{A}$ is an algebraic extension of the rationals $\mathbb{Q}$, however the degree is infinite.

Since each rational is algebraic it is clear that there are algebraic numbers. Further there are irrational algebraic numbers, $\sqrt{2}$ for example, since it is a root of the irreducible polynomial $x^2 - 2$ over $\mathbb{Q}$. In Chapter 5 we proved that there are uncountably infinitely many transcendental numbers (Theorem 5.5.3). However it is very difficult to prove that any particular real or complex number is actually transcendental. In Theorem 5.5.4 we showed that the real number

$$c = \sum_{j=1}^{\infty} \frac{1}{10^{j!}}$$

is transcendental.

In this section we examine a special type of algebraic number called an *algebraic integer*. These are the algebraic numbers that are zeros of monic integral polynomials. The set of all such algebraic integers forms a subring of $\mathbb{C}$. The proofs in this section can be found in [35].

After we do this we extend the concept of an algebraic integer to a general context and define integral ring extensions. We then consider field extensions that are nonalgebraic – transcendental field extensions. Finally we will prove that the familiar numbers e and π are transcendental.

Definition 20.1.1. An *algebraic integer* is a complex number α that is a root of a *monic* integral polynomial. That is, $\alpha \in \mathbb{C}$ is an algebraic integer if there exists $f(x) \in \mathbb{Z}[x]$ with $f(x) = x^n + b_{n-1}x^{n-1} + \cdots + b_0$, $b_i \in \mathbb{Z}$, $n \geq 1$, and $f(\alpha) = 0$.

An algebraic integer is clearly an algebraic number. The following are clear.

Lemma 20.1.2. *If $\alpha \in \mathbb{C}$ is an algebraic integer, then all its conjugates, $\alpha_1, \dots, \alpha_n$, over $\mathbb{Q}$ are also algebraic integers.*

Lemma 20.1.3. *$\alpha \in \mathbb{C}$ is an algebraic integer if and only if $m_\alpha \in \mathbb{Z}[x]$.*

To prove the converse of this lemma we need the concept of a *primitive integral polynomial.* This is a polynomial $p(x) \in \mathbb{Z}[x]$ such that the GCD of all its coefficients is 1. The following can be proved (see exercises):

(1) If $f(x)$ and $g(x)$ are primitive then so is $f(x)g(x)$.

(2) If $f(x) \in \mathbb{Z}[x]$ is monic then it is primitive.

(3) If $f(x) \in \mathbb{Q}[x]$ then there exists a rational number c such that $f(x) = cf_1(x)$ with $f_1(x)$ primitive.

Now suppose $f(x) \in \mathbb{Z}[x]$ is a monic polynomial with $f(\alpha) = 0$. Let $p(x) = m_\alpha(x)$. Then $p(x)$ divides $f(x)$ so $f(x) = p(x)q(x)$.

Let $p(x) = c_1 p_1(x)$ with $p_1(x)$ primitive and let $q(x) = c_2 q_1(x)$ with $q_1(x)$ primitive. Then

$$f(x) = cp_1(x)q_1(x).$$

Since $f(x)$ is monic it is primitive and hence $c = 1$ so $f(x) = p_1(x)q_1(x)$.

Since $p_1(x)$ and $q_1(x)$ are integral and their product is monic they both must be monic. Since $p(x) = c_1 p_1(x)$ and they are both monic it follows that $c_1 = 1$ and hence $p(x) = p_1(x)$. Therefore $p(x) = m_\alpha(x)$ is integral.

When we speak of algebraic integers we will refer to the ordinary integers as *rational integers.* The next lemma shows the close ties between algebraic integers and rational integers.

Lemma 20.1.4. *If α is an algebraic integer and also rational then it is a rational integer.*

The following ties algebraic numbers in general to corresponding algebraic integers. Notice that if $q \in \mathbb{Q}$ then there exists a rational integer n such that $nq \in \mathbb{Z}$. This result generalizes this simple idea.

Theorem 20.1.5. *If θ is an algebraic number then there exists a rational integer $r \neq 0$ such that $r\theta$ is an algebraic integer.*

We saw that the set $\mathcal{A}$ of all algebraic numbers is a subfield of $\mathbb{C}$. In the same manner the set $\mathcal{I}$ of all algebraic integers forms a subring of $\mathcal{A}$. First an extension of the following result on algebraic numbers.

Lemma 20.1.6. *Suppose $\alpha_1, \dots, \alpha_n$ are the set of conjugates over $\mathbb{Q}$ of an algebraic integer α. Then any integral symmetric function of $\alpha_1, \dots, \alpha_n$ is a rational integer.*

Theorem 20.1.7. *The set $\mathcal{I}$ of all algebraic integers forms a subring of $\mathcal{A}$.*

We note that $\mathcal{A}$, the field of algebraic numbers, is precisely the quotient field of the ring of algebraic integers.

An *algebraic number field* is a finite extension of $\mathbb{Q}$ within $\mathbb{C}$. Since any finite extension of $\mathbb{Q}$ is a simple extension each algebraic number field has the form $K = \mathbb{Q}(\theta)$ for some algebraic number θ.

Let $K = \mathbb{Q}(\theta)$ be an algebraic number field and let $R_K = K \cap \mathcal{I}$. Then R_K forms a subring of K called the algebraic integers or just integers of K. An analysis of the proof of Theorem 20.1.5 shows that each $\beta \in K$ can be written as

$$\beta = \frac{\alpha}{r}$$

with $\alpha \in R_K$ and $r \in \mathbb{Z}$.

These rings of algebraic integers share many properties with the rational integers. While there may not be unique factorization into primes there is always prime factorization.

Theorem 20.1.8. *Let K be an algebraic number field and R_K its ring of integers. Then each $\alpha \in R_K$ is either* 0*, a unit or can be factored into a product of primes.*

We stress again that the prime factorization need not be unique. However from the existence of a prime factorization we can mimic Euclid's original proof of the infinitude of primes (see [35]) to obtain:

Corollary 20.1.9. *There exist infinitely many primes in R_K for any algebraic number ring R_K.*

Just as any algebraic number field is finite dimensional over $\mathbb{Q}$ we will see that each R_K is of finite degree over $\mathbb{Q}$. That is if K has degree n over $\mathbb{Q}$ we show that there exists $\omega_1, \dots, \omega_n$ in R_K such that each $\alpha \in R_K$ is expressible as

$$\alpha = m_1\omega_1 + \cdots + m_n\omega_n$$

where $m_1, \dots, m_n \in \mathbb{Z}$.

Definition 20.1.10. An *integral basis* for R_K is a set of integers $\omega_1, \dots, \omega_t \in R_K$ such that each $\alpha \in R_K$ can be expressed uniquely as

$$\alpha = m_1\omega_1 + \cdots + m_t\omega_t$$

where $m_1, \dots, m_t \in \mathbb{Z}$.

The finite degree comes from the following result that shows there does exist an integral basis (see [35]).

Theorem 20.1.11. *Let R_K be the ring of integers in the algebraic number field K of degree n over $\mathbb{Q}$. Then there exists at least one integral basis for R_K.*

20.2 Integral ring extensions

We now extend the concept of an algebraic integer to general ring extensions. We first need the idea of an R-algebra where R is a commutative ring with identity $1 \neq 0$.

Definition 20.2.1. Let R be a commutative ring with an identity $1 \neq 0$. An *R-algebra* or *algebra over R* is a unitary R-module A in which there is an additional multiplication such that

(1) A is a ring with respect to the addition and this multiplication

(2) $(rx)y = x(ry) = r(xy)$ for all $r \in R$ and $x, y \in A$.

As examples of R-algebras first consider $R = K$ where K is a field and let $A = M_n(K)$ the set of all $(n \times n)$-matrices over K. Then $M_n(K)$ is a K-algebra. Further the set of polynomials $K[x]$ is also a K-algebra.

We now define *ring extensions*. Let A be a ring, not necessarily commutative, with an identity $1 \neq 0$, and R be a commutative subring of A which contains 1. Assume that R is contained in the center of A, that is, $rx = xr$ for all $r \in R$ and $x \in A$. We then call A a *ring extension of R* and write $A|R$. If $A|R$ is a ring extension then A is an R-algebra in a natural manner.

Let A be an R-algebra with an identity $1 \neq 0$. Then we have the canonical ring homomorphism $\phi : R \to A$, $r \mapsto r \cdot 1$. The image $R' := \phi(R)$ is a subring of the center of A, and R' contains the identity element of A. Then $A|R'$ is a ring extension (in the above sense). Hence, if A is a R-algebra with an identity $1 \neq 0$ then we may consider R as a subring of A and $A|R$ as a ring extension.

We now will extend to the general context of ring extensions the ideas of integral elements and integral extensions. As above, let R be a commutative ring with an identity $1 \neq 0$ and let A be an R-algebra.

Definition 20.2.2. An element $a \in A$ is said to be *integral* over R or integrally dependent over R if there is a monic polynomial $f(x) = x^n + \alpha_{n-1}x^{n-1} + \cdots + \alpha_0 \in R[x]$ of degree $n \geq 1$ over R with $f(a) = a^n + \alpha_{n-1}a^{n-1} + \cdots + \alpha_0 = 0$. That is, a is integral over R if it is a root of a monic polynomial of degree ≥ 1 over R.

An equation that an integral element satisfies is called *integral equation of a over R*. If A has an identity $1 \neq 0$ then we may write $a^0 = 1$ and $\sum_{i=0}^{n} \alpha_i a^i$ with $\alpha_n = 1$.

Example 20.2.3. 1. Let $E|K$ be a field extension. $a \in E$ is integral over K if and only if a is algebraic over K. If K is the quotient field of an integral domain R and $a \in E$ is algebraic over K then there exists an $\alpha \in R$ with αa integral over R, because if $0 = \alpha_n a^n + \cdots + \alpha_0$ then $0 = (\alpha_n a)^n + \cdots + \alpha_n^{n-1}\alpha_0$.

2. The elements of $\mathbb{C}$ which are integral over $\mathbb{Z}$ are precisely the algebraic integers over $\mathbb{Z}$, that is, the roots of monic polynomials over $\mathbb{Z}$.

Theorem 20.2.4. *Let R be as above and A an R-algebra with an identity $1 \neq 0$. If A is, as an R-module, finitely generated then each element of A is integral over R.*

Proof. Let $\{b_1, \dots, b_n\}$ be a finite generating system of A, as an R-module. We may assume that $b_1 = 1$, otherwise add 1 to the system. As explained in the preliminaries, without loss of generality, we may assume that $R \subset A$. Let $a \in A$. For each $1 \leq j \leq n$ we have an equation $ab_j = \sum_{k=1}^{n} \alpha_{kj} b_k$ for some $\alpha_{kj} \in R$. In other words:

$$\sum_{k=1}^{n} (\alpha_{kj} - \delta_{jk} a) b_k = 0 \qquad (\star\star)$$

for $j = 1, \dots, n$, where

$$\delta_{jk} = \begin{cases} 0 & \text{if } j \neq k, \\ 1 & \text{if } j = k. \end{cases}$$

Define $\gamma_{jk} := \alpha_{kj} - \delta_{jk} a$ and $C = (\gamma_{jk})_{j,k}$. C is an $(n \times n)$-matrix over the commutative ring $R[a]$; recall that $R[a]$ has an identity element. Let $\tilde{C} = (\tilde{\gamma}_{jk})_{j,k}$ be the complimentary matrix of C. Then $\tilde{C} C = (\det C) E_n$. From $(\star\star)$ we get

$$0 = \sum_{j=1}^{n} \tilde{\gamma}_{ij} \left(\sum_{k=1}^{n} \gamma_{jk} b_k \right) = \sum_{k=1}^{n} \sum_{j=1}^{n} \tilde{\gamma}_{ij} \gamma_{jk} b_k = \sum_{k=1}^{n} (\det C) \delta_{ik} b_k = (\det C) b_i$$

for all $1 \leq i \leq n$. Since $b_1 = 1$ we have necessarily that $\det C = \det(\alpha_{jk} - \delta_{jk} a)_{j,k} = 0$ (recall that $\delta_{jk} = \delta_{kj}$). Hence a is a root of the monic polynomial $f(x) = \det(\delta_{jk} x - \alpha_{jk}) \in R[x]$ of degree $n \geq 1$. Hence a is integral over R. □

Definition 20.2.5. A ring extension $A|R$ is called an *integral extension* if each element of A is integral over R. A ring extension $A|R$ is called *finite* if A, as a R-module, is finitely generated.

Recall that finite field extensions are algebraic extensions. As an immediate consequence of Theorem 20.2.4 we get the corresponding result for ring extensions.

Theorem 20.2.6. *Each finite ring extension $A|R$ is an integral extension.*

Theorem 20.2.7. *Let A be an R-algebra with an identity $1 \neq 0$. If $a \in A$ then the following are equivalent:*

(1) *a is integral over R.*

(2) *The subalgebra $R[a]$ is, as an R-module, finitely generated.*

(3) *There exists a subalgebra A' of A which contains a and which is, as an R-module, finitely generated.*

A subalgebra of an algebra over R is a submodule which is also a subring.

Proof. (1) $\Rightarrow$ (2): We have $R[a] = \{g(a) : g \in R[x]\}$. Let $f(a) = 0$ be an integral equation of a over R. Since f is monic, by the division algorithm, for each $g \in R[x]$ there are $h, r \in R[x]$ with $g = h \cdot f + r$ and $r = 0$ or $r \neq 0$ and $\deg(r) < \deg(f) =: n$. Let $r \neq 0$. Since $g(a) = r(a)$, we get that $\{1, a, \dots, a^{n-1}\}$ is a generating system for the R-module $R[a]$.

(2) $\Rightarrow$ (3): Take $A' = R[a]$.

(3) $\Rightarrow$ (1): Use Theorem 20.2.4 for A$'$. □

For the remainder of this chapter all rings are commutative with an identity $1 \neq 0$.

Theorem 20.2.8. *Let $A|R$ and $B|A$ be finite ring extensions. Then also $B|R$ is finite.*

Proof. From $A = Re_1 + \cdots + Re_m$ and $B = Af_1 + \cdots + Af_n$ we get $B = Re_1 f_1 + \cdots + Re_m f_n$. □

Theorem 20.2.9. *Let $A|R$ be a ring extension. Then the following are equivalent:*

(1) *There are finitely many, over R integral elements $a_1, \dots, a_m$ in A such that $A = R[a_1, \dots, a_m]$.*

(2) *$A|R$ is finite.*

Proof. (2) $\Rightarrow$ (1): We only need to take for $a_1, \dots, a_m$ a generating system of A as an R-module, and the result holds because $A = Ra_1 + \cdots + Ra_m$, and each a_i is integral over R by Theorem 20.2.4.

(1) $\Rightarrow$ (2): We use induction for m. If $m = 0$ then there is nothing to prove. Now let $m \geq 1$, and assume that (1) holds. Define $A' = R[a_1, \dots, a_{m-1}]$. Then $A = A'[a_m]$, and a_m is integral over A$'$. $A|A'$ is finite by Theorem 20.2.7. By the induction assumption, $A'|R$ is finite. Then $A|R$ is finite by Theorem 20.2.8. □

Definition 20.2.10. Let $A|R$ be a ring extension. Then the subset $C = \{a \in A : a$ is integral over $R\} \subset A$ is called the *integral closure of R in A*.

Theorem 20.2.11. *Let $A|R$ be a ring extension. Then the integral closure of R in A is a subring of A with $R \subset A$.*

Proof. $R \subset C$ because $\alpha \in R$ is a root of the polynomial $x - \alpha$. Let $a, b \in C$. We consider the subalgebra $R[a, b]$ of the R-algebra A. $R[a, b]|R$ is finite by Theorem 20.2.9. Hence, by Theorem 20.2.4, all elements from $R[a, b]$ are integral over R, that is, $R[a, b] \subset C$. Especially, $a + b$, $a - b$ and ab are in C. □

We extend to ring extensions the idea of a closure.

Definition 20.2.12. Let $A|R$ a ring extension. R is called *integrally closed in A*, if R itself is its integral closure in R, that is, $R = C$, the integral closure of R in A.

Theorem 20.2.13. *For each ring extension $A|R$ the integral closure C of R in A is integrally closed in A.*

Proof. Let $a \in A$ be integral over C. Then $a^n + \alpha_{n-1}a^{n-1} + \cdots + \alpha_0 = 0$ for some $\alpha_i \in C$, $n \geq 1$. Then a is also integral over the R-subalgebra $A' = R[\alpha_0, \ldots, \alpha_{n-1}]$ of C; and $A'|R$ is finite. Further $A'[a]|A$ is finite. Hence $A'[a]|R$ is finite. By Theorem 20.2.4, then $a \in A'[a]$ is already integral over R, that is, $a \in C$. □

Theorem 20.2.14. *Let $A|R$ and $B|A$ be ring extensions. If $A|R$ and $B|A$ are integral extensions then also $B|R$ is an integral extension (and certainly vice versa).*

Proof. Let C be the integral closure of R in B. We have $A \subset C$ since $A|R$ is integral. Together with $B|A$ we also have that $B|C$ is integral. By Theorem 20.2.13 we get that C is integrally closed in B. Hence, $B = C$. □

We now consider integrally closed integral domains.

Definition 20.2.15. An integral domain R is called *integrally closed* if R is integrally closed in its quotient field K.

Theorem 20.2.16. *Each unique factorization domain R is integrally closed.*

Proof. Let $\alpha \in K$ and $\alpha = \frac{a}{b}$ with $a, b \in R$, $a \neq 0$. Since R is a unique factorization domain we may assume that a and b are relatively prime. Let α be integral over R. Then we have over R an integral equation $\alpha^n + a_{n-1}\alpha^{n-1} + \cdots + a_0 = 0$ for α. Multiplication with b^n gives $a^n + ba_{n-1} + \cdots + b^n a_0 = 0$. Hence b is a divisor of a^n. Since a and b are relatively prime in R, we have that b is a unit in R and, hence, $\alpha = \frac{a}{b} \in R$. □

Theorem 20.2.17. *Let R be an integral domain and K its quotient field. Let $E|K$ be a finite field extension. Let R be integrally closed, and $\alpha \in E$ be integral over R. Then the minimal polynomial $g \in K[x]$ of α over K has only coefficients of R.*

Proof. Let $g \in K[x]$ be the minimal polynomial of α over K (recall that g is monic by definition). Let $\bar{E}$ be an algebraic closure of E. Then $g(x) = (x - \alpha_1)\cdots(x - \alpha_n)$ with $\alpha_1 = \alpha$ over $\bar{E}$. There are K-isomorphisms $\sigma_i : K(\alpha) \to \bar{E}$ with $\sigma_i(\alpha) = \alpha_i$. Hence all α_i are also integral over R. Since all coefficients of g are polynomial expressions $C_j(\alpha_1, \ldots, \alpha_n)$ in the α_i we get that all coefficients of g are integral over R (see Theorem 20.2.11). Now $g \in R[x]$ because $g \in K[x]$ and R is integrally closed. □

Theorem 20.2.18. *Let R be an integrally closed integral domain and K be its quotient field. Let $f, g, h \in K[x]$ be monic polynomials over K with $f = gh$. If $f \in R[x]$ then also $g, h \in R[x]$.*

Proof. Let E be the splitting field of f over K. Over E we have $f(x) = (x - \alpha_1) \cdots (x - \alpha_n)$. Since f is monic all α_k are integral over R (see the proof of Theorem 20.2.17). Since $f = gh$ there are $I, J \subset \{1, \dots, n\}$ with $g(x) = \prod_{i \in I}(x - \alpha_i)$ and $h(x) = \prod_{j \in J}(x - \alpha_j)$. As polynomial expressions in the α_i, $i \in I$, and α_j, $j \in J$, respectively, the coefficients of g and h, respectively, are integral over R. On the other side all these coefficients are in K and R is integrally closed. Hence $g, h \in R[x]$. □

Theorem 20.2.19. *Let $E|R$ be an integral ring extension. If E is a field then also R is a field.*

Proof. Let $\alpha \in R \setminus \{0\}$. The element $\frac{1}{\alpha} \in E$ satisfies an integral equation $(\frac{1}{\alpha})^n + a_{n-1}(\frac{1}{\alpha})^{n-1} + \cdots + a_0 = 0$ over R. Multiplication with α^{n-1} gives $\frac{1}{\alpha} = -a_{n-1} - a_{n-2}\alpha - \cdots - a_0\alpha^{n-1} \in R$. Hence, R is a field. □

20.3 Transcendental field extensions

Recall that a transcendental number is an element of $\mathbb{C}$ that is not algebraic over $\mathbb{Q}$. More generally if $E|K$ is a field extension then an element $\alpha \in E$ is *transcendental* over K if it is not algebraic, that is, it is not a zero of any polynomial $f(x) \in K[x]$. Since finite extensions are algebraic clearly $E|K$ will contain transcendental elements only if $[E : K] = \infty$. However this is not sufficient. The field $\mathcal{A}$ of algebraic numbers is algebraic over $\mathbb{Q}$ but infinite dimensional over $\mathbb{Q}$. We now extend the idea of a transcendental number to that of a transcendental extension.

Let $K \subset E$ be fields, that is, $E|K$ is a field extension. Let M be a subset of E. The *algebraic cover* of M in E is defined to be the algebraic closure $H(M)$ of $K(M)$ in E, that is, $H_{K,E}(M) = H(M) = \{\alpha \in E : \alpha \text{ algebraic over } K(M)\}$. $H(M)$ is a field with $K \subset K(M) \subset H(M) \subset E$. $\alpha \in E$ is called *algebraically dependent on M (over K)* if $\alpha \in H(M)$, that is, if α is algebraic over $K(M)$.

The following are clear.

1. $M \subset H(M)$,
2. $M \subset M' \Rightarrow H(M) \subset H(M')$ and
3. $H(H(M)) = H(M)$.

Definition 20.3.1. (a) M is said to be *algebraically independent (over K)* if $\alpha \notin H(M \setminus \{\alpha\})$ for all $\alpha \in M$, that is, if each $\alpha \in M$ is transcendental over $K(M \setminus \{\alpha\})$.

(b) M is said to be *algebraically dependent (over K)* if M is not algebraically independent.

The proofs of the statements in the following lemma are straightforward.

Lemma 20.3.2. (1) *M is algebraically dependent if and only if there exists an $\alpha \in M$ which is algebraic over $K(M \setminus \{\alpha\})$.*

(2) *Let $\alpha \in M$. Then $\alpha \in H(M \setminus \{\alpha\}) \Leftrightarrow H(M) = H(M \setminus \{\alpha\})$.*

(3) *If $\alpha \notin M$ and α is algebraic over $K(M)$ then $M \cup \{\alpha\}$ is algebraically dependent.*

(4) *M is algebraically dependent if and only if there is a finite subset in M which is algebraically dependent.*

(5) *M is algebraically independent if and only if each finite subset of M is algebraically independent.*

(6) *M is algebraically independent if and only if the following holds: If $\alpha_1, \dots, \alpha_n$ are finitely many, pairwise different elements of M then the canonical homomorphism $\phi : K[x_1, \dots, x_n] \to E$, $f(x_1, \dots, x_n) \mapsto f(\alpha_1, \dots, \alpha_n)$ is injective; or in other words: for all $f \in K[x_1, \dots, x_n]$ we have that $f = 0$ if $f(\alpha_1, \dots, \alpha_n) = 0$, that is, there is no nontrivial algebraic relation between the $\alpha_1, \dots, \alpha_n$ over K.*

(7) *Let $M \subset E$, $\alpha \in E$. If M is algebraically independent and $M \cup \{\alpha\}$ algebraically dependent then $\alpha \in H(M)$, that is, α is algebraically dependent on M.*

(8) *Let $M \subset E$, $B \subset M$. If B is maximal algebraically independent, that is, if $\alpha \in M \setminus B$ then $B \cup \{\alpha\}$ is algebraically dependent, then $M \subset H(B)$, that is, each element of M is algebraic over $K(B)$.*

We will show that any field extension can be decomposed into a transcendental extension over an algebraic extension. We need the idea of a transcendence basis.

Definition 20.3.3. $B \subset E$ is called a *transcendence basis* of the field extension $E|K$ if the following two conditions are satisfied:

1. $E = H(B)$, that is, the extension $E|K(B)$ is algebraic.
2. B is algebraically independent over K.

Theorem 20.3.4. *If $B \subset E$ then the following are equivalent:*

(1) *B is a transcendence basis of $E|K$.*

(2) *If $B \subset M \subset E$ with $H(M) = E$, then B is a maximal algebraically independent subset of M.*

(3) *There exists a subset $M \subset E$ with $H(M) = E$ which contains B as a maximal algebraically independent subset.*

Proof. (1) $\Rightarrow$ (2): Let $\alpha \in M \setminus B$. We have to show that $B \cup \{\alpha\}$ is algebraically dependent. But this is clear because $\alpha \in H(B) = E$.

(2) $\Rightarrow$ (3): We just take $M = E$.

(3) $\Rightarrow$ (1): We have to show that $H(B) = E$. Certainly $M \subset H(B)$. Hence, $E = H(M) \subset H(H(B)) = H(B) \subset E$. □

We next show that any field extension does have a transcendence basis.

Theorem 20.3.5. *Each field extension $E|K$ has a transcendence basis. More concretely: If there is a subset $M \subset E$ such that $E|K(M)$ is algebraic and if there is a subset $C \subset M$ which is algebraically independent then there exists a transcendence basis B of $E|K$ with $C \subset B \subset M$.*

Proof. We have to extend C to a maximal algebraically independent subset B of M. By Theorem 20.3.4, such a B is a transcendence basis of $E|K$. If M is finite then such a B certainly exists. Now let M be not finite. We argue analogously as for the existence of a basis of a vector space, for instance with Zorn's lemma: If a partially ordered, nonempty set S is inductive, then there exist maximal elements in S. Here, a partially ordered, nonempty set S is said to be inductive if every totally ordered subset of S has an upper bound in S. The set N of all algebraically independent subsets of M which contain C is partially ordered with respect to "$\subset$", and $N \neq \emptyset$ because $C \in N$. Let $K \neq \emptyset$ be an ascending chain in N, that is, given an ascending chain $\emptyset \neq Y_1 \subset Y_2 \subset \cdots$. in N. The union $U = \bigcup_{Y \in K} Y$ is also algebraically independent. Hence, there exists a maximal algebraically independent subset $B \subset M$ with $C \subset B$. □

Theorem 20.3.6. *Let $E|K$ be a field extension and M be a subset of E for which $E|K(M)$ is algebraic. Let C be an arbitrary subset of E which is algebraically independent on K. Then there exists a subset $M' \subset M$ with $C \cap M' = \emptyset$ such that $C \cup M'$ is a transcendence basis of $E|K$.*

Proof. Take $M \cup C$ and define $M' := B \setminus C$ in Theorem 20.3.5. □

Theorem 20.3.7. *Let B, B' be two transcendence bases of the field extension $E|K$. Then there is a bijection $\phi : B \to B'$. In other words, any two transcendence bases of $E|K$ have the same cardinal number.*

Proof. (a) If B is a transcendental basis of $E|K$ and M is a subset of E such that $E|K(M)$ is algebraic then we may write $B = \bigcup_{\alpha \in M} B_\alpha$ with finite sets B_α. Especially, if B is infinite then the cardinal number of B is not bigger than the cardinal number of M.

(b) Let B and B' be two transcendence bases of $E|K$. If B and B' are both infinite then B and B' have the same cardinal number by (a) and the theorem by

Schroeder–Bernstein [5]. We now prove Theorem 20.3.7 for the case that $E|K$ has a finite transcendence basis. Let B be finite with n elements. Let C be an arbitrary algebraically independent subset in E over K with m elements. We show that $m \leq n$. Let $C = \{\alpha_1, \dots, \alpha_m\}$ with $m \geq n$. We show by induction that for each integer k, $0 \leq k \leq n$, there are subsets $B \supsetneq B_1 \supsetneq \cdots \supsetneq B_k$ of B such that $\{\alpha_1, \dots, \alpha_k\} \cup B_k$ is a transcendence basis of $E|K$ and $\{\alpha_1, \dots, \alpha_k\} \cap B_k = \emptyset$. For $k = 0$ we take $B_0 = B$, and the statement holds. Assume now that the statement is correct for $0 \leq k < n$. By Theorem 20.3.4 and 20.3.5 there is a subset B_{k+1} of $\{\alpha_1, \dots, \alpha_k\} \cup B_k$ such that $\{\alpha_1, \dots, \alpha_{k+1}\} \cup B_{k+1}$ is a transcendence basis of $E|K$ and $\{\alpha_1, \dots, \alpha_{k+1}\} \cap B_{k+1} = \emptyset$. Then necessarily $B_{k+1} \subset B_k$. Assume $B_k = B_{k+1}$. Then on one side, $B_k \cup \{\alpha_1, \dots, \alpha_{k+1}\}$ is algebraic independent because $B_k = B_{k+1}$. On the other side, also $B_k \cup \{\alpha_1, \dots, \alpha_k\} \cup \{a_{k+1}\}$ is algebraically dependent, which gives a contradiction. Hence, $B_{k+1} \subsetneq B_k$. Now B_k has at most $n - k$ elements, hence $B_n = \emptyset$, that is, $\{\alpha_1, \dots, \alpha_n\} = \{\alpha_1, \dots, \alpha_n\} \cup B_n$ is a transcendence basis of $E|K$. Because $C = \{\alpha_1, \dots, \alpha_m\}$ is algebraically independent, we cannot have $m > n$. Hence $m \leq n$; and B and B' have the same number of elements because B' must also be finite. □

Since the cardinality of any transcendence basis for a field extension $E|K$ is the same we can define the transcendence degree.

Definition 20.3.8. The *transcendence degree* $\operatorname{trgd}(E|K)$ of a field extension is the cardinal number of one (and hence of each) transcendence basis of $E|K$. A field extension $E|K$ is called *purely transcendental*, if $E|K$ has a transcendence basis B with $E = K(B)$.

We note the following facts:

(1) If $E|K$ is purely transcendental and $B = \{\alpha_1, \dots, \alpha_n\}$ is a transcendence basis of $E|K$ then E is K-isomorphic to the quotient field of the polynomial ring $K[x_1, \dots, x_n]$ of the independence indeterminates $x_1, \dots, x_n$.

(2) K is algebraically closed in E if $E|K$ is purely transcendental.

(3) By Theorem 20.3.4, the field extension $E|K$ has an intermediate field F, $K \subset F \subset E$, such that $F|K$ is purely transcendental and $E|F$ is algebraic. Certainly F is not uniquely determined.

For example take $\mathbb{Q} \subset F \subset \mathbb{Q}(i, \pi)$, and for F we may take $F = \mathbb{Q}(\pi)$ and also $F = \mathbb{Q}(i\pi)$, for instance.

(4) $\operatorname{trgd}(\mathbb{R}|\mathbb{Q}) = \operatorname{trgd}(\mathbb{C}|\mathbb{Q}) = \operatorname{card} \mathbb{R}$, the cardinal number of $\mathbb{R}$. This holds because the set of the algebraic numbers (over $\mathbb{Q}$) is countable.

Theorem 20.3.9. *Let $E|K$ a field extension and F an arbitrary intermediate field, $K \subset F \subset E$. Let B a transcendence basis of $F|K$ and B' a transcendence base of*

$E|F$. Then $B \cap B' = \emptyset$, and $B \cup B'$ is a transcendence basis of $E|K$. Especially: $\operatorname{trgd}(E|K) = \operatorname{trgd}(E|F) + \operatorname{trgd}(F|K)$.

Proof. (1) Assume $\alpha \in B \cap B'$. As an element of F, then α is algebraic over $F(B') \setminus \{\alpha\}$. But this gives a contradiction because $\alpha \in B'$, and B' is algebraically independent over F.

(2) $F|K(B)$ is an algebraic extension, and also $F(B')|K(B \cup B') = K(B)(B')$. Since the relation "algebraic extension" is transitive, we have that $E|K(B \cup B')$ is algebraic.

(3) Finally we have to show that $B \cup B'$ is algebraically independent over K. By Theorems 20.3.5 and 20.3.6 there is a subset B'' of $B \cup B'$ with $B \cap B'' = \emptyset$ such that $B \cup B''$ is a transcendence basis of $E|K$. We have $B'' \subset B'$, and have to show that $B' \subset B''$. Assume that there is an $\alpha \in B'$ with $\alpha \notin B''$. Then α is algebraic over $K(B \cup B'') = K(B)(B'')$ and, hence, algebraic over $F(B'')$. Since $B'' \subset B'$ we have that α is algebraically independent over F, which gives a contradiction. Hence $B'' = B'$. □

Theorem 20.3.10 (Noether's normalization theorem). *Let K be a field and $A = K[a_1, \dots, a_n]$. Then there exist elements $u_1, \dots, u_m$, $0 \le m \le n$, in A with the following properties:*

(1) *$K[u_1, \dots, u_m]$ is K-isomorphic to the polynomial ring $K[x_1, \dots, x_m]$ of the independent indeterminates $x_1, \dots, x_m$.*

(2) *The ring extension $A|K[u_1, \dots, u_m]$ is an integral extension, that is, for each $a \in A \setminus K[u_1, \dots, u_m]$ there exists a monic polynomial $f(x) = x^n + \alpha_{n-1}x^{n-1} + \dots + \alpha_0 \in K[u_1, \dots, u_m][x]$ of degree $n \ge 1$ with $f(a) = a^n + \alpha_{n-1}a^{n-1} + \dots + \alpha_0 = 0$. Especially $A|K[u_1, \dots, u_m]$ is finite.*

Proof. Without loss of generality, let the $a_1, \dots, a_n$ be pairwise different. We prove the theorem by induction on n. If $n = 1$ then there is nothing to show. Now, let $n \ge 2$, and assume that the statement holds for $n-1$. If there is no nontrivial algebraic relation $f(a_1, \dots, a_n) = 0$ over K between the $a_1, \dots, a_n$ then there is nothing to show. Hence, let there exists a polynomial $f \in K[x_1, \dots, x_n]$ with $f \ne 0$ and $f(a_1, \dots, a_n) = 0$. Let $f = \sum_{\nu=(\nu_1,\dots,\nu_n)} c_\nu x_1^{\nu_1} \cdots x_n^{\nu_n}$. Let $\mu_2, \mu_3, \dots, \mu_n$ be natural numbers which we specify later. Define $b_2 = a_2 - a_1^{\mu_2}$, $b_3 = a_3 - a_1^{\mu_3}$, $\dots, b_n = a_n - a_1^{\mu_n}$. Then $a_i = b_i + a_1^{\mu_i}$ for $2 \le i \le n$, hence, $f(a_1, b_2 + a_1^{\mu_2}, \dots, b_n + a_1^{\mu_n}) = 0$. We write $R := K[x_1, \dots, x_n]$ and consider the polynomial ring $R[y_2, \dots, y_n]$ of the $n-1$ independent indeterminates $y_2, \dots, y_n$ over R. In $R[y_2, \dots, y_n]$ we consider the polynomial $f(x_1, y_2 + x_1^{\mu_2}, \dots, y_n + x_1^{\mu_n})$. We may rewrite this polynomial as

$$\sum_{\nu=(\nu_1,\dots,\nu_n)} c_\nu x_1^{\nu_1 + \mu_2\nu_2 + \dots + \mu_n\nu_n} + g(x_1, y_2, \dots, y_n)$$

with a polynomial $g(x_1, y_2, \ldots, y_n)$ for which, as a polynomial in x_1 over $K[y_2, \ldots, y_n]$, the degree in x_1 is smaller than the degree of $\sum_{\nu=(\nu_1,\ldots,\nu_n)} c_\nu x_1^{\nu_1+\mu_2\nu_2+\cdots+\mu_n\nu_n}$, provided that we may choose the $\mu_2, \ldots, \mu_n$ in such a way that this really holds. We now specify the $\mu_2, \ldots, \mu_n$. We write $\mu := (1, \mu_2, \ldots, \mu_n)$ and define the scalar product $\mu\nu = 1 \cdot \nu_1 + \mu_2\nu_2 + \cdots + \mu_n\nu_n$. Choose $p \in \mathbb{N}$ with $p > \deg(f) = \max\{\nu_1 + \cdots + \nu_n : c_\nu \neq 0\}$. We now take $\mu = (1, p, p^2, \ldots, p^{n-1})$. If $\nu = (\nu_1, \ldots, \nu_n)$ with $c_\nu \neq 0$ and $\nu' = (\nu_1', \ldots, \nu_n')$ with $c_\nu' \neq 0$ are different n-tuple then indeed $\mu\nu \neq \mu\nu'$ because $\nu_i, \nu_i' < p$ for all i, $1 \leq i \leq n$. This follows from the uniqueness of the p-adic expression of a natural number. Hence, we may choose $\mu_2, \ldots, \mu_n$ such that $f(x_1, y_2 + x_1^{\mu_2}, \ldots, y_n + x_1^{\mu_n}) = cx_1^N + h(x_1, y_2, \ldots, y_n)$ with $c \in K$, $c \neq 0$, and $h \in K[y_2, \ldots, y_n][x_1]$ has in x_1 a degree $< N$. If we divide by c and take $a_1, b_2, \ldots, b_n$ for $x_1, y_2, \ldots, y_n$ then we get an integral equation of a_1 over $K[b_2, \ldots, b_n]$. Therefore, the ring extension $A = K[a_1, \ldots, a_n] | K[b_2, \ldots, b_n]$ is integral (see Theorem 20.2.9), $a_i = b_i + a_1^{\mu_i}$ for $2 \leq i \leq n$. By induction there exists elements $u_1, \ldots, u_m$ in $K[b_2, \ldots, b_n]$ with the following properties:

1. $K[u_1, \ldots, u_m]$ is a polynomial ring of the m independent indeterminates $u_1, \ldots, u_m$ and
2. $K[b_2, \ldots, b_n] | K[u_1, \ldots, u_m]$ is integral.

Hence, also $A | K[u_1, \ldots, u_m]$ is integral by Theorem 20.2.14. □

Corollary 20.3.11. *Let $E|K$ be a field extension. If $E = K[a_1, \ldots, a_n]$ for $a_1, \ldots, a_n \in E$ then $E|K$ is algebraic.*

Proof. By Theorem 20.3.10 we have that E contains a polynomial ring $K[u_1, \ldots, u_m]$, $0 \leq m \leq n$, of the m independent indeterminates $u_1, \ldots, u_m$ as a subring for which $E | K[u_1, \ldots, u_m]$ is integral. We claim that then already $K[u_1, \ldots, u_m]$ is a field. To prove that, let $a \in K[u_1, \ldots, u_m]$, $a \neq 0$. The element $a^{-1} \in E$ satisfies an integral equation $(a^{-1})^n + \alpha_{n-1}(a^{-1})^{n-1} + \cdots + \alpha_0 = 0$ over $K[u_1, \ldots, u_m] =: R$. Hence, $a^{-1} = -\alpha_{n-1} - \alpha_{n-2}a - \cdots - \alpha_0 a^{n-1} \in R$. Therefore R is a field which proves the claim. This is possible only for $m = 0$, and then $E|K$ is integral, that is here algebraic. □

20.4 The transcendence of e and π

Although we have shown that within $\mathbb{C}$ there are continuously many transcendental numbers we have only shown that one particular number is transcendental. In this section we prove that the numbers e and π are transcendental. We start with e.

Theorem 20.4.1. *e is a transcendental number, that is, transcendental over $\mathbb{Q}$.*

Proof. Let $f(x) \in \mathbb{R}[x]$ with the degree of $f(x) = m \geq 1$. Let $z_1 \in \mathbb{C}$, $z_1 \neq 0$, and $\gamma : [0, 1] \to \mathbb{C}$, $\gamma(t) = tz_1$. Let

$$I(z_1) = \int_\gamma e^{z_1 - z} f(z)dz = \left(\int_0^{z_1}\right)_\gamma e^{z_1 - z} f(z)dz.$$

By $(\int_0^{z_1})_\gamma$ we mean the integral from 0 to z_1 along γ. Recall that

$$\left(\int_0^{z_1}\right)_\gamma e^{z_1 - z} f(z)dz = -f(z_1) + e^{z_1} f(0) + \left(\int_0^{z_1}\right)_\gamma e^{z_1 - z} f'(z)dz.$$

It follows then by repeated partial integration that

(1) $I(z_1) = e^{z_1} \sum_{j=0}^m f^{(j)}(0) - \sum_{j=0}^m f^{(j)}(z_1)$.

Let $|f|(x)$ be the polynomial that we get if we replace the coefficients of $f(x)$ by their absolute values. Since $|e^{z_1 - z}| \leq e^{|z_1 - z|} \leq e^{|z_1|}$, we get

(2) $|I(z_1)| \leq |z_1| e^{|z_1|} |f|(|z_1|)$.

Now assume that e is an algebraic number, that is,

(3) $q_0 + q_1 e + \cdots + q_n e^n = 0$ for $n \geq 1$ and integers $q_0 \neq 0, q_1, \ldots, q_n$, and the greatest common divisor of $q_0, q_1, \ldots, q_n$, is equal to 1.

We consider now the polynomial $f(x) = x^{p-1}(x-1)^p \ldots (x-n)^p$ with p a sufficiently large prime number, and we consider $I(z_1)$ with respect to this polynomial. Let

$$J = q_0 I(0) + q_1 I(1) + \cdots + q_n I(n).$$

From (1) and (3) we get that

$$J = -\sum_{j=0}^{m} \sum_{k=0}^{n} q_k f^{(j)}(k),$$

where $m = (n+1)p - 1$ since $(q_0 + q_1 e + \cdots + q_n e^n)(\sum_{j=0}^m (f^{(j)}(0)) = 0$.

Now, $f^{(j)}(k) = 0$ if $j < p$, $k > 0$, and if $j < p - 1$ then $k = 0$, and hence $f^{(j)}(k)$ is an integer that is divisible by $p!$ for all j, k except for $j = p-1$, $k = 0$. Further, $f^{(p-1)}(0) = (p-1)!(-1)^{np}(n!)^p$, and hence, if $p > n$, then $f^{(p-1)}(0)$ is an integer divisible by $(p-1)!$ but not by $p!$.

It follows that J is a nonzero integer that is divisible by $(p-1)!$ if $p > |q_0|$ and $p > n$. So let $p > n$, $p > |q_0|$, so that $|J| \geq (p-1)!$.

Now, $|f|(k) \leq (2n)^m$. Together with (2) we then get that

$$|J| \leq |q_1| e |f|(1) + \cdots + |q_n| n e^n |f|(n) \leq c^p$$

for a number c independent of p. It follows that

$$(p-1)! \leq |J| \leq c^p,$$

that is,

$$1 \le \frac{|J|}{(p-1)!} \le c\frac{c^{p-1}}{(p-1)!}.$$

This gives a contradiction, since $\frac{c^{p-1}}{(p-1)!} \to 0$ as $p \to \infty$. Therefore, e is transcendental. □

We now move on to the transcendence of π. We first need the following lemma.

Lemma 20.4.2. *Suppose $\alpha \in \mathbb{C}$ is an algebraic number and $f(x) = a_n x^n + \cdots + a_0$, $n \ge 1$, $a_n \ne 0$, and all $a_i \in \mathbb{Z}$ ($f(x) \in \mathbb{Z}[x]$) with $f(\alpha) = 0$. Then $a_n\alpha$ is an algebraic integer.*

Proof.

$$\begin{aligned} a_n^{n-1} f(x) &= a_n^n x^n + a_n^{n-1} a_{n-1} x^{n-1} + \cdots + a_n^{n-1} a_0 \\ &= (a_n x)^n + a_{n-1}(a_n x)^{n-1} + \cdots + a_n^{n-1} a_0 \\ &= g(a_n x) = g(y) \in \mathbb{Z}[y] \end{aligned}$$

where $y = a_n x$ and $g(y)$ is monic. Then $g(a_n\alpha) = 0$, and hence $a_n\alpha$ is an algebraic integer. □

Theorem 20.4.3. *π is a transcendental number, that is, transcendental over $\mathbb{Q}$.*

Proof. Assume that π is an algebraic number. Then $\theta = i\pi$ is also algebraic. Let $\theta_1 = \theta, \theta_2, \ldots, \theta_d$ be the conjugates of θ. Suppose

$$p(x) = q_0 + q_1 x + \cdots + q_d x^d \in \mathbb{Z}[x], \quad q_d > 0, \quad \text{and} \quad \gcd(q_0, \ldots, q_d) = 1$$

is the entire minimal polynomial of θ over $\mathbb{Q}$. Then $\theta_1 = \theta, \theta_2, \ldots, \theta_d$ are the zeros of this polynomial. Let $t = q_d$. Then from Lemma 20.4.2, $t\theta_i$ is an algebraic integer for all i. From $e^{i\pi} + 1 = 0$ and from $\theta_1 = i\pi$ we get that

$$(1 + e^{\theta_1})(1 + e^{\theta_2}) \cdots (1 + e^{\theta_d}) = 0.$$

The product on the left side can be written as a sum of 2^d terms e^{ϕ}, where $\phi = \epsilon_1\theta_1 + \cdots + \epsilon_d\theta_d$, $\epsilon_j = 0$ or 1. Let n be the number of terms $\epsilon_1\theta_1 + \cdots + \epsilon_d\theta_d$ that are nonzero. Call these $\alpha_1, \ldots, \alpha_n$. We then have an equation

$$q + e^{\alpha_1} + \cdots + e^{\alpha_n} = 0$$

with $q = 2^d - n > 0$. Recall that all $t\alpha_i$, are algebraic integers and we consider the polynomial

$$f(x) = t^{np} x^{p-1} (x - \alpha_1)^p \cdots (x - \alpha_n)^p$$

with p a sufficiently large prime integer. We have $f(x) \in \mathbb{R}[x]$, since the α_i are algebraic numbers and the elementary symmetric polynomials in $\alpha_1, \dots, \alpha_n$ are rational numbers.

Let $I(z_1)$ be defined as in the proof of Theorem 20.4.1, and now let

$$J = I(\alpha_1) + \cdots + I(\alpha_n).$$

From (1) in the proof of Theorem 20.4.1 and (4) we get

$$J = -q \sum_{j=0}^{m} f^{(j)}(0) - \sum_{j=0}^{m} \sum_{k=1}^{n} f^{(j)}(\alpha_k),$$

with $m = (n+1)p - 1$.

Now, $\sum_{k=1}^{n} f^{(j)}(\alpha_k)$ is a symmetric polynomial in $t\alpha_1, \dots, t\alpha_n$ with integer coefficients since the $t\alpha_i$ are algebraic integers. It follows from the main theorem on symmetric polynomials that $\sum_{j=0}^{m} \sum_{k=1}^{n} f^{(j)}(\alpha_k)$ is an integer. Further, $f^{(j)}(\alpha_k) = 0$ for $j < p$. Hence $\sum_{j=0}^{m} \sum_{k=1}^{n} f^{(j)}(\alpha_k)$ is an integer divisible by $p!$.

Now, $f^{(j)}(0)$ is an integer divisible by $p!$ if $j \neq p-1$, and $f^{(p-1)}(0) = (p-1)!(-t)^{np}(\alpha_1 \dots \alpha_n)^p$ is an integer divisible by $(p-1)!$ but not divisible by $p!$ if p is sufficiently large. In particular, this is true if $p > |t^n(\alpha_1 \cdots \alpha_n)|$ and also $p > q$.

From (2) in the proof of Theorem 20.4.1 we get that

$$|J| \le |\alpha_1| e^{|\alpha_1|} |f|(|\alpha_1|) + \cdots + |\alpha_n| e^{|\alpha_n|} |f|(|\alpha_n|) \le c^p$$

for some number c independent of p.

As in the proof of Theorem 20.4.1, this gives us

$$(p-1)! \le |J| \le c^p,$$

that is,

$$1 \le \frac{|J|}{(p-1)!} \le c \frac{c^{p-1}}{(p-1)!}.$$

This, as before, gives a contradiction, since $\frac{c^{p-1}}{(p-1)!} \to 0$ as $p \to \infty$. Therefore, π is transcendental. □

20.5 Exercises

1. A polynomial $p(x) \in \mathbb{Z}[x]$ is primitive if the GCD of all its coefficients is 1. Prove the following:

   (i) If $f(x)$ and $g(x)$ are primitive then so is $f(x)g(x)$.

   (ii) If $f(x) \in \mathbb{Z}[x]$ is monic then it is primitive.

(iii) If $f(x) \in \mathbb{Q}[x]$ then there exists a rational number c such that $f(x) = cf_1(x)$ with $f_1(x)$ primitive.

2. Let d be a square-free integer and $K = \mathbb{Q}(\sqrt{d})$ be a quadratic field. Let R_K be the subring of K of the algebraic integers of K. Show that

   (i) $R_K = \{m + n\sqrt{d} : m, n \in \mathbb{Z}\}$ if $d \equiv 2 (\mathrm{mod}\ 4)$ or $d \equiv 3 (\mathrm{mod}\ 4)$. $\{1, \sqrt{d}\}$ is an integral basis for R_K.

   (ii) $R_K = \{m + n\frac{1+\sqrt{d}}{2} : m, n \in \mathbb{Z}\}$ if $d \equiv 1 (\mathrm{mod}\ 4)$. $\{1, \frac{1+\sqrt{d}}{2}\}$ is an integral basis for R_K.

   (iii) If $d < 0$ then there are only finitely many units in R_K.

   (iv) If $d > 0$ then there are infinitely many units in R_K.

3. Let $K = \mathbb{Q}(\alpha)$ with $\alpha^3 + \alpha + 1 = 0$ and R_K the subring of the algebraic integers in K. Show that

   (i) $\{1, \alpha, \alpha^2\}$ is an integral basis for R_K.

   (ii) $R_K = \mathbb{Z}[\alpha]$.

4. Let $A|R$ be an integral ring extension. If A is an integral domain and R a field then A is also a field.

5. Let $A|R$ be an integral extension. Let $\mathcal{P}$ be a prime ideal of A and $\mathfrak{p}$ be a prime ideal of R such that $\mathcal{P} \cap R = \mathfrak{p}$. Show that

   (i) If $\mathfrak{p}$ is maximal in R then $\mathcal{P}$ is maximal in A. (Hint: consider $A/\mathcal{P}$.)

   (ii) If $\mathcal{P}_0$ is another prime ideal of A with $\mathcal{P}_0 \cap R = \mathfrak{p}$ and $\mathcal{P}_0 \subset \mathcal{P}$ then $\mathcal{P} = \mathcal{P}_0$. (Hint: we may assume that A is an integral domain and $\mathcal{P} \cap R = \{0\}$, otherwise go to $A/\mathcal{P}$.)

6. Show that for a field extension $E|K$ the following are equivalent:

   (i) $[E : K(B)] < \infty$ for each transcendence basis B of $E|K$.

   (ii) $\mathrm{trgd}(E|K) < \infty$ and $[E : K(B)] < \infty$ for each transcendence basis B of $E|K$.

   (iii) There is a finite transcendence basis B of $E|K$ with $[E : K(B)] < \infty$.

   (iv) There are finitely many $x_1, \dots, x_n \in E$ with $E = K(x_1, \dots, x_n)$.

7. Let $E|K$ be a field extension. If $E|K$ is purely transcendental then K is algebraically closed in E.

Chapter 21
The Hilbert Basis Theorem and the Nullstellensatz

21.1 Algebraic Geometry

An extremely important application of abstract algebra and an application central to all of mathematics is the subject of *algebraic geometry*. As the name suggests this is the branch of mathematics that uses the techniques of abstract algebra to study geometric problems. Classically algebraic geometry involved the study of *algebraic curves* which roughly are the sets of zeros of a polynomial or set of polynomials in several variables over a field. For example, in two variables a real algebraic plane curve is the set of zeros in $\mathbb{R}^2$ of a polynomial $p(x, y) \in \mathbb{R}[x, y]$. The common planar curves such as parabolas and the other conic sections are all plane algebraic curves. In actual practice plane algebraic curves are usually considered over the complex numbers and are projectivized.

The algebraic theory that deals most directly with algebraic geometry is called *commutative algebra*. This is the study of commutative rings, ideals in commutative rings and modules over commutative rings. A large portion of this book has dealt with commutative algebra.

Although we will not consider the geometric aspects of algebraic geometry in general we will close the book by introducing some of the basic algebraic ideas that are crucial to the subject. These include the concept of an *algebraic variety* or *algebraic set* and its radical. We also state and prove two of the cornerstones of the theory as applied to commutative algebra – the Hilbert basis theorem and the Nullstellensatz.

In this chapter we consider a fixed field extension $C|K$ and the polynomial ring $K[x_1, \dots, x_n]$ of the n independent indeterminates $x_1, \dots, x_n$. Again, in this chapter we often use letters $\mathfrak{a}, \mathfrak{b}, \mathfrak{m}, \mathfrak{p}, \mathfrak{P}, \mathfrak{A}, \mathfrak{Q}, \dots$ for ideals in rings.

21.2 Algebraic Varieties and Radicals

We first define the concept of an algebraic variety.

Definition 21.2.1. If $M \subset K[x_1, \dots, x_n]$ then we define

$$\mathcal{N}(M) = \{(\alpha_1, \dots, \alpha_n) \in C^n : f(\alpha_1, \dots, \alpha_n) = 0 \ \forall f \in M\}.$$

$\alpha = (\alpha_1, \dots, \alpha_n) \in \mathcal{N}(M)$ is called a *zero (Nullstelle) of* M *in* C^n; and $\mathcal{N}(M)$ is called the *zero set* of M in C^n. If we want to mention C then we write $\mathcal{N}(M) =$

$\mathcal{N}_C(M)$. A subset $V \subset C^n$ of the form $V = \mathcal{N}(M)$ for some $M \subset K[x_1, \dots, x_n]$ is called an *algebraic variety* or *(affine) algebraic set* of C^n over K, or just an *algebraic K-set of C^n*.

For any subset N of C^n we can reverse the procedure and consider the set of polynomials whose zero set is N.

Definition 21.2.2. Suppose that $N \subset C^n$. Then

$$I(N) = \{f \in K[x_1, \dots, x_n] : f(\alpha_1, \dots, \alpha_n) = 0 \;\forall (\alpha_1, \dots, \alpha_n) \in N\}.$$

Instead of $f \in I(N)$ we also say that f *vanishes on N (over K)*. If we want to mention K then we write $I(N) = I_K(N)$.

What is important is that the set $I(N)$ forms an ideal. The proof is straightforward.

Theorem 21.2.3. *For any subset $N \subset C^n$ the set $I(N)$ is an ideal in $K[x_1, \dots, x_n]$; it is called the vanishing ideal of $N \subset C^n$ in $K[x_1, \dots, x_n]$.*

The following result examines the relationship between subsets in C^n and their vanishing ideals.

Theorem 21.2.4. *The following properties hold:*

(1) $M \subset M' \Rightarrow \mathcal{N}(M') \subset \mathcal{N}(M)$,

(2) *If $\mathfrak{a} = (M)$ is the ideal in $K[x_1, \dots, x_n]$ generated by M, then $\mathcal{N}(M) = \mathcal{N}(\mathfrak{a})$,*

(3) $N \subset N' \Rightarrow I(N') \subset I(N)$,

(4) *$M \subset I\mathcal{N}(M)$ for all $M \subset K[x_1, \dots, x_n]$,*

(5) *$N \subset \mathcal{N}I(N)$ for all $N \subset C^n$,*

(6) *If $(\mathfrak{a}_i)_{i \in I}$ is a family of ideals in $K[x_1, \dots, x_n]$ then $\bigcap_{i \in I} \mathcal{N}(\mathfrak{a}_i) = \mathcal{N}(\sum_{i \in I} \mathfrak{a}_i)$. Here $\sum_{i \in I} \mathfrak{a}_i$ is the ideal in $K[x_1, \dots, x_n]$, generated by the union $\bigcup_{i \in I} \mathfrak{a}_i$,*

(7) *If $\mathfrak{a}, \mathfrak{b}$ are ideals in $K[x_1, \dots, x_n]$ then $\mathcal{N}(\mathfrak{a}) \cup \mathcal{N}(\mathfrak{b}) = \mathcal{N}(\mathfrak{a}\mathfrak{b}) = \mathcal{N}(\mathfrak{a} \cap \mathfrak{b})$. Here $\mathfrak{a}\mathfrak{b}$ is the ideal in $K[x_1, \dots, x_n]$ generated by all products fg where $f \in \mathfrak{a}$ and $g \in \mathfrak{b}$,*

(8) *$\mathcal{N}(M) = \mathcal{N}I\mathcal{N}(M)$ for all $M \subset K[x_1, \dots, x_n]$,*

(9) *$V = \mathcal{N}I(V)$ for all algebraic K-sets V,*

(10) *$I(N) = I\mathcal{N}I(N)$ for all $N \subset C^n$.*

Proof. The proofs are straightforward. Hence, we prove only (7), (8) and (9). The rest can be left as exercise for the reader.

Proof of (7): Since $\mathfrak{a}\mathfrak{b} \subset \mathfrak{a} \cap \mathfrak{b} \subset \mathfrak{a}, \mathfrak{b}$ we have by (1) the inclusion $\mathcal{N}(\mathfrak{a}) \cup \mathcal{N}(\mathfrak{b}) \subset \mathcal{N}(\mathfrak{a} \cap \mathfrak{b}) \subset \mathcal{N}(\mathfrak{a}\mathfrak{b})$. Hence, we have to show that $\mathcal{N}(\mathfrak{a}\mathfrak{b}) \subset \mathcal{N}(\mathfrak{a}) \cup \mathcal{N}(\mathfrak{b})$.

Let $\alpha = (\alpha_1, \dots, \alpha_n) \in C^n$ be a zero of $\mathfrak{a}\mathfrak{b}$ but not a zero of $\mathfrak{a}$. Then there is an $f \in \mathfrak{a}$ with $f(\alpha) \neq 0$, and hence for all $g \in \mathfrak{b}$ we get $f(\alpha)g(\alpha) = (fg)(\alpha) = 0$ and, hence, $g(\alpha) = 0$. Therefore $\alpha \in \mathcal{N}(\mathfrak{b})$.

Proof of (8) and (9): Let $M \subset K[x_1, \dots, x_n]$. Then, on the one side, $M \subset I\mathcal{N}(M)$ by (5) and further $\mathcal{N}I\mathcal{N}(M) \subset \mathcal{N}(M)$ by (1). On the other side, $\mathcal{N}(M) \subset \mathcal{N}I\mathcal{N}(M)$ by (6). Therefore $\mathcal{N}(M) = \mathcal{N}I\mathcal{N}(M)$ for all $M \subset K[x_1, \dots, x_n]$.

Now, the algebraic K-sets of C^n are precisely the sets of the form $V = \mathcal{N}(M)$. Hence, $V = \mathcal{N}I(V)$. □

We make the following agreement: if $\mathfrak{a}$ is an ideal in $K[x_1, \dots, x_n]$ then we write

$$\mathfrak{a} \lhd K[x_1, \dots, x_n].$$

If $\mathfrak{a} \lhd K[x_1, \dots, x_n]$ then we do not have $\mathfrak{a} = I\mathcal{N}(\mathfrak{a})$ in general, that is, $\mathfrak{a}$ is in general not equal to the vanishing ideal of its zero set in C^n. The reason for this is that not each ideal $\mathfrak{a}$ occurs as a vanishing ideal of some $N \subset C^n$. If $\mathfrak{a} = I(N)$ then we must have:

$$f^m \in \mathfrak{a},\ m \geq 1 \Longrightarrow f \in \mathfrak{a}. \qquad (\star)$$

Hence, for instance, if $\mathfrak{a} = (x_1^2, \dots, x_n^2) \lhd K[x_1, \dots, x_n]$ then $\mathfrak{a}$ is not of the form $\mathfrak{a} = I(N)$ for some $N \subset C^n$. We now define the radical of an ideal.

Definition 21.2.5. Let R be a commutative ring, and $\mathfrak{a} \lhd R$ be an ideal in R. Then $\sqrt{\mathfrak{a}} = \{f \in R : f^m \in \mathfrak{a} \text{ for some } m \in \mathbb{N}\}$ is an ideal in R. $\sqrt{\mathfrak{a}}$ is called the *radical of* $\mathfrak{a}$ (in R). $\mathfrak{a}$ is said to be *reduced* if $\sqrt{\mathfrak{a}} = \mathfrak{a}$.

We note that the $\sqrt{0}$ is called the *nil radical* of R; it contains exactly the *nilpotent elements* of R, that is, the elements $a \in R$ with $a^m = 0$ for some $m \in \mathbb{N}$.

Let $\mathfrak{a} \lhd R$ be an ideal in R and $\pi : R \to R/\mathfrak{a}$ the canonical mapping. Then $\sqrt{\mathfrak{a}}$ is exactly the preimage of the nil radical of $R/\mathfrak{a}$.

21.3 The Hilbert Basis Theorem

In this section we show that if K is a field then each ideal $\mathfrak{a} \lhd K[x_1, \dots, x_n]$ is finitely generated. This is the content of the Hilbert basis theorem. This has as an important consequence that any algebraic variety of C^n is the zero set of only finitely many polynomials.

The Hilbert basis theorem follows directly from the following Theorem 21.3.2. Before we state this theorem we need a definition.

Definition 21.3.1. Let R be a commutative ring with an identity $1 \neq 0$. R is said to be *noetherian* if each ideal in R is generated by finitely many elements, that is, each ideal in R is finitely generated.

Theorem 21.3.2. *Let R be a noetherian ring. Then the polynomial ring $R[x]$ over R is also noetherian.*

Proof. Let $0 \neq f_k \in R[x]$. With $\deg(f_k)$ we denote the degree of f_k. Let $\mathfrak{a} \lhd R[x]$ be an ideal in $R[x]$. Assume that $\mathfrak{a}$ is not finitely generated. Then, especially, $\mathfrak{a} \neq 0$. We construct a sequence of polynomials $f_k \in \mathfrak{a}$ such that the highest coefficients a_k generate an ideal in R which is not finitely generated. This produces then a contradiction, and, hence, $\mathfrak{a}$ is in fact finitely generated. Choose $f_1 \in \mathfrak{a}$, $f_1 \neq 0$, so that $\deg(f_1) = n_1$ is minimal.

If $k \geq 1$ then choose $f_{k+1} \in \mathfrak{a}$, $f_{k+1} \notin (f_1, \dots, f_k)$ so that $\deg(f_{k+1}) = n_{k+1}$ is minimal for the polynomials in $\mathfrak{a} \setminus (f_1, \dots, f_k)$. This is possible because we assume that $\mathfrak{a}$ is not finitely generated. We have $n_k \leq n_{k+1}$ by our construction. Further $(a_1, \dots, a_k) \subsetneqq (a_1, \dots, a_k, a_{k+1})$.

Proof of this claim: Assume that $(a_1, \dots, a_k) = (a_1, \dots, a_k, a_{k+1})$. Then $a_{k+1} \in (a_1, \dots, a_k)$. Hence, there are $b_i \in R$ with $a_{k+1} = \sum_{i=1}^{k} a_i b_i$. Let $g(x) = \sum_{i=1}^{k} b_i f_i(x) x^{n_{k+1}-n_i}$, hence, $g \in (f_1, \dots, f_k)$ and $g = a_{k+1}x^{n_{k+1}} + \cdots$. Therefore $\deg(f_{k+1} - g) < n_{k+1}$ and $f_{k+1} - g \notin (f_1, \dots, f_k)$ which contradicts the choice of f_{k+1}. This proves the claim.

Hence $(a_1, \dots, a_k) \subsetneqq (a_1, \dots, a_k, a_{k+1})$ which contradicts the fact that R is noetherian. Hence $\mathfrak{a}$ is finitely generated. □

We now have the Hilbert basis theorem.

Theorem 21.3.3 (Hilbert basis theorem). *Let K be a field. Then each ideal $\mathfrak{a} \lhd K[x_1, \dots, x_n]$ is finitely generated, that is, $\mathfrak{a} = (f_1, \dots, f_m)$ for finitely many $f_1, \dots, f_m \in K[x_1, \dots, x_n]$.*

Corollary 21.3.4. *If $C|K$ is a field extension then each algebraic K-set V of C^n is already the zero set of only finitely many polynomials $f_1, \dots, f_m \in K[x_1, \dots, x_n]$:*

$$V = \{(\alpha_1, \dots, \alpha_n) \in C^n : f_i(\alpha_1, \dots, \alpha_n) = 0 \text{ for } i = 1, \dots, m\}.$$

Further we write $V = \mathcal{N}(f_1, \dots, f_m)$.

21.4 The Hilbert Nullstellensatz

Vanishing ideals of subsets of C^n are not necessarily reduced. For an arbitrary field C, the condition

$$f^m \in \mathfrak{a},\ m \geq 1 \Longrightarrow f \in \mathfrak{a}$$

is, in general, not sufficient for $\mathfrak{a} \lhd K[x_1, \dots, x_n]$ to be a vanishing ideal of a subset of C^n. For example let $n \geq 2$, $K = C = \mathbb{R}$ and $\mathfrak{a} = (x_1^2 + \cdots + x_n^2) \lhd \mathbb{R}[x_1, \dots, x_n]$. $\mathfrak{a}$ is a prime ideal in $\mathbb{R}[x_1, \dots, x_n]$ because $x_1^2 + \cdots + x_n^2$ is a prime element in

$\mathbb{R}[x_1, \ldots, x_n]$. Hence, $\mathfrak{a}$ is reduced. But, on the other side, $\mathcal{N}(\mathfrak{a}) = \{0\}$ and $I(\{0\}) = (x_1, \ldots, x_n)$. Therefore $\mathfrak{a}$ is not of the form $I(N)$ for some $N \subset C^n$. If this would be the case, then $\mathfrak{a} = I(N) = I\mathcal{N}I(N) = I\{0\} = (x_1, \ldots, x_n)$ because of Theorem 21.2.4(10), which gives a contradiction.

The Nullstellensatz of Hilbert which we give in two forms shows that if $\mathfrak{a}$ is reduced, that is, $\mathfrak{a} = \sqrt{\mathfrak{a}}$, then $I\mathcal{N}(\mathfrak{a}) = \mathfrak{a}$.

Theorem 21.4.1 (Hilbert's Nullstellensatz, first form). *Let $C|K$ be a field extension with C algebraically closed. If $\mathfrak{a} \lhd K[x_1, \ldots, x_n]$ then $I\mathcal{N}(\mathfrak{a}) = \sqrt{\mathfrak{a}}$. Moreover, if $\mathfrak{a}$ is reduced, that is, $\mathfrak{a} = \sqrt{\mathfrak{a}}$, then $I\mathcal{N}(\mathfrak{a}) = \mathfrak{a}$. Therefore $\mathcal{N}$ defines a bijective map between the set of reduced ideals in $K[x_1, \ldots, x_n]$ and the set of the algebraic K-sets in C^n; and I defines the inverse map.*

The proof follows from:

Theorem 21.4.2 (Hilbert's Nullstellensatz, second form). *Let $C|K$ be a field extension with C algebraically closed. Let $\mathfrak{a} \lhd K[x_1, \ldots, x_n]$ with $\mathfrak{a} \neq K[x_1, \ldots, x_n]$. Then there exists an $\alpha = (\alpha_1, \ldots, \alpha_n) \in C^n$ with $f(\alpha) = 0$ for all $f \in \mathfrak{a}$, that is, $\mathcal{N}_C(\mathfrak{a}) \neq \emptyset$.*

Proof. Since $\mathfrak{a} \neq K[x_1, \ldots, x_n]$ there exists a maximal ideal $\mathfrak{m} \lhd K[x_1, \ldots, x_n]$ with $\mathfrak{a} \subset \mathfrak{m}$. We consider the canonical map $\pi : K[x_1, \ldots, x_n] \to K[x_1, \ldots, x_n]/\mathfrak{m}$. Let $\beta_i = \pi(x_i)$ for $i = 1, \ldots, n$. Then $K[x_1, \ldots, x_n]/\mathfrak{m} = K[\beta_1, \ldots, \beta_n] =: E$. Since $\mathfrak{m}$ is maximal, E is a field. Moreover $E|K$ is algebraic by Corollary 20.3.11. Hence there exists a K-homomorphism $\sigma : K[\beta_1, \ldots, \beta_n] \to C$ (C is algebraically closed). Let $\alpha_i = \sigma(\beta_i)$; we have $f(\alpha_1, \ldots, \alpha_n) = 0$ for all $f \in \mathfrak{m}$. Since $\mathfrak{a} \subset \mathfrak{m}$ this holds also for all $f \in \mathfrak{a}$. Hence we get a zero $(\alpha_1, \ldots, \alpha_n)$ of $\mathfrak{a}$ in C^n. □

Proof of Theorem 21.4.1. Let $\mathfrak{a} \lhd K[x_1, \ldots, x_n]$, and let $f \in I\mathcal{N}(\mathfrak{a})$. We have to show that $f^m \in \mathfrak{a}$ for some $m \in \mathbb{N}$. If $f = 0$ then there is nothing to show.

Now, let $f \neq 0$. We consider $K[x_1, \ldots, x_n]$ as a subring of $K[x_1, \ldots, x_n, x_{n+1}]$ of the $n + 1$ independent indeterminates $x_1, \ldots, x_n, x_{n+1}$. In $K[x_1, \ldots, x_n, x_{n+1}]$ we consider the ideal $\bar{\mathfrak{a}} = (\mathfrak{a}, 1 - x_{n+1}f) \lhd K[x_1, \ldots, x_n, x_{n+1}]$, generated by $\mathfrak{a}$ and $1 - x_{n+1}f$.

Case 1: $\bar{\mathfrak{a}} \neq K[x_1, \ldots, x_n, x_{n+1}]$.

$\bar{\mathfrak{a}}$ then has a zero $(\beta_1, \ldots, \beta_n, \beta_{n+1})$ in C^{n+1} by Theorem 21.2.4. Hence, for $(\beta_1, \ldots, \beta_n, \beta_{n+1}) \in \mathcal{N}(\bar{\mathfrak{a}})$ we have the equations:

(1) $g(\beta_1, \ldots, \beta_n) = 0$ for all $g \in \mathfrak{a}$ and

(2) $f(\beta_1, \ldots, \beta_n)\beta_{n+1} = 1$.

From (1) we get $(\beta_1, \ldots, \beta_n) \in \mathcal{N}(\mathfrak{a})$. Hence, especially, $f(\beta_1, \ldots, \beta_n) = 0$ for our $f \in I\mathcal{N}(\mathfrak{a})$. But this contradicts (2). Therefore $\bar{\mathfrak{a}} \neq K[x_1, \ldots, x_n, x_{n+1}]$ is not possible. Therefore we have

Case 2: $\bar{\mathfrak{a}} = K[x_1, \dots, x_n, x_{n+1}]$, that is, $1 \in \bar{\mathfrak{a}}$. Then there exists a relation of the form

$$1 = \sum_i h_i g_i + h(1 - x_{n+1} f) \quad \text{for some } g_i \in \mathfrak{a} \text{ and } h_i, h \in K[x_1, \dots, x_n, x_{n+1}].$$

The map $x_i \mapsto x_i$ for $1 \leq i \leq n$ and $x_{n+1} \mapsto \frac{1}{f}$ defines a homomorphism $\phi : K[x_1, \dots, x_n, x_{n+1}] \to K(x_1, \dots, x_n)$, the quotient field of $K[x_1, \dots, x_n]$. From (3) we get a relation $1 = \sum_i h_i(x_1, \dots, x_n, \frac{1}{f}) g_i(x_1, \dots, x_n)$ in $K(x_1, \dots, x_n)$. If we multiply this with a suitable power f^m of f we get $f^m = \sum_i \tilde{h}_i(x_1, \dots, x_n) g_i(x_1, \dots, x_n)$ for some polynomials $\tilde{h} \in K[x_1, \dots, x_n]$. Since $g_i \in \mathfrak{a}$ we get $f^m \in \mathfrak{a}$. □

21.5 Applications and Consequences of Hilbert's Theorems

Theorem 21.5.1. *Each nonempty set of algebraic K-sets in C^n contains a minimal element. In other words: For each descending chain*

$$V_1 \supset V_2 \supset \cdots \supset V_m \supset V_{m+1} \supset \cdots \tag{1}$$

of algebraic K-sets V_i in C^n there exists an integer m such that $V_m = V_{m+1} = V_{m+2} = \cdots$, or equivalently, every strictly descending chain $V_1 \supsetneqq V_2 \supsetneqq \cdots$ of algebraic K-sets V_i in C^n is finite.

Proof. We apply the operator I, that is, we pass to the vanishing ideals. This gives an ascending chain of ideals

$$I(V_1) \subset I(V_2) \subset \cdots \subset I(V_m) \subset I(V_{m+1}) \subset \cdots. \tag{2}$$

The union of the $I(V_i)$ is an ideal in $K[x_1, \dots, x_n]$, and, hence, by Theorem 21.3.3 finitely generated. Hence, there is an m with $I(V_m) = I(V_{m+1}) = I(V_{m+2}) = \cdots$.

Now we apply the operator $\mathcal{N}$ and get the desired result because $V_i = \mathcal{N} I(V_i)$ by Theorem 21.2.4 (10). □

Definition 21.5.2. An algebraic K-set $V \neq \emptyset$ in C^n is called *irreducible* if it is not describable as a union $V = V_1 \cup V_2$ of two algebraic K-sets $V_i \neq \emptyset$ in C^n with $V_i \neq V$ for $i = 1, 2$. An irreducible algebraic K-set in C^n is also called a *K-variety* in C^n.

Theorem 21.5.3. *An algebraic K-set $V \neq \emptyset$ in C^n is irreducible if and only if its vanishing ideal $I_k(V) = I(V)$ is a prime ideal of $R = K[x_1, \dots, x_n]$ with $I(V) \neq R$.*

Proof. (1) Let V be irreducible. Let $fg \in I(V)$. Then $V = \mathcal{N} I(V) \subset \mathcal{N}(fg) = \mathcal{N}(f) \cup \mathcal{N}(g)$, hence $V = V_1 \cup V_2$ with the algebraic K-sets $V_1 = \mathcal{N}(f) \cap V$ and

$V_2 = \mathcal{N}(g) \cap V$. Now V is irreducible, hence, $V = V_1$ or $V = V_2$; say $V = V_1$. Then $V \subset \mathcal{N}(f)$, and therefore $f \in I\mathcal{N}(f) \subset I(V)$. Since $V \neq \emptyset$ we have further $1 \notin I(V)$, that is, $I(V) \neq R$.

(2) Let $I(V) \lhd R$ with $I(V) \neq R$ be a prime ideal. Let $V = V_1 \cup V_2$, $V_1 \neq V$, with algebraic K-sets V_i in C^n. First,

$$I(V) = I(V_1 \cup V_2) = I(V_1) \cap I(V_2) \supset I(V_1)I(V_2), \qquad (\star)$$

where $I(V_1)I(V_2)$ is the ideal generated by all products fg with $f \in I(V_1)$, $g \in I(V_2)$. We have $I(V_1) \neq I(V)$ because otherwise $V_1 = \mathcal{N}I(V_1) = \mathcal{N}I(V) = V$ contradicting $V_1 \neq V$. Hence, there is a $f \in I(V_1)$ with $f \notin I(V)$. Now, $I(V) \neq R$ is a prime ideal, and hence, necessarily $I(V_2) \subset I(V)$ by $(\star)$. It follows that $V \subset V_2$, and, hence, V is irreducible. □

Note that the affine space K^n is, as the zero set of the zero polynomial 0, itself an algebraic K-set in K^n. If K is infinite then $I(K^n) = \{0\}$ and, hence K^n is irreducible by Theorem 21.5.3. Moreover, if K is infinite then K^n can not be written as a union of finitely many proper algebraic K-subsets. If K is finite then K^n is not irreducible.

Further each algebraic K-set V in C^n is also an algebraic C-set in C^n. If V is an irreducible algebraic K-set in C^n then, in general, it is not an irreducible algebraic C-set in C^n.

Theorem 21.5.4. *Each algebraic K-set V in C^n can be written as a finite union $V = V_1 \cup V_2 \cup \cdots \cup V_r$ of irreducible algebraic K-sets V_i in C^n. If here $V_i \not\subset V_k$ for all pairs (i, k) with $i \neq k$ then this presentation is unique, up to the ordering of the V_i; and then the V_i are called the irreducible K-components of V.*

Proof. Let $\mathfrak{a}$ be the set of all algebraic K-sets in C^n which can not be presented as a finite union of irreducible algebraic K-sets in C^n.

Assume that $\mathfrak{a} \neq \emptyset$. By Theorem 21.4.1 there is a minimal element V in $\mathfrak{a}$. This V is not irreducible, otherwise we have a presentation as desired. Hence there exists a presentation $V = V_1 \cup V_2$ with algebraic K-sets V_i which are strictly smaller than V. By definition, both V_1 and V_2 have a presentation as desired, and hence V has one, too, which gives a contradiction. Hence, $\mathfrak{a} = \emptyset$.

Now suppose that $V = V_1 \cup \cdots \cup V_r = W_1 \cup \cdots \cup W_s$ be two presentations of the desired form. For each V_i we have a presentation $V_i = (V_i \cap W_1) \cup \cdots \cup (V_i \cap W_s)$. Each $V_i \cap W_j$ is a K-algebraic set (see Theorem 21.2.4). Since V_i is irreducible, we get that there is a W_j with $V_i = V_i \cap W_j$, that is, $V_i \subset W_j$. Analogously, for this W_j there is a V_k with $W_j \subset V_k$. Altogether $V_i \subset W_j \subset V_k$. But $V_p \not\subset V_q$ if $p \neq q$. Hence, from $V_i \subset W_j \subset V_k$ we get $i = k$ and therefore $V_i = W_j$, that means, for each V_i there is a W_j with $V_i = W_j$. Analogously, for each W_k there is a V_l with $W_k = V_l$. This proves the theorem. □

Example 21.5.5. 1. Let $M = \{gh\} \subset \mathbb{R}[x,y]$ with $g(x) = x^2 + y^2 - 1$ and $f(x) = x^2 + y^2 - 2$. Then $\mathcal{N}(M) = V = V_1 \cup V_2$ where $V_1 = \mathcal{N}(g)$ and $V_2 = \mathcal{N}(f)$, and V is not irreducible.

2. Let $M = \{f\} \subset \mathbb{R}[x,y]$ with $f(x,y) = xy - 1$; f is irreducible in $\mathbb{R}[x,y]$, therefore the ideal (f) is a prime ideal in $\mathbb{R}[x,y]$. Hence $V = \mathcal{N}(f)$ is irreducible.

Definition 21.5.6. Let V be an algebraic K-set in C^n. The residue class ring $K[V] = K[x_1,\dots,x_n]/I(V)$ is called the *(affine) coordinate ring* of V.

$K[V]$ can be identified with the ring of all those functions $V \to C$ which are given by polynomials from $K[x_1,\dots,x_n]$. As a homomorphic image of $K[x_1,\dots,x_n]$, we get that $K[V]$ can be described in the form $K[V] = K[\alpha_1,\dots,\alpha_n]$; therefore a K-algebra of the form $K[\alpha_1,\dots,\alpha_n]$ is often called an *affine K-algebra*. If the algebraic K-set V in C^n is irreducible – we can call V now an *(affine) K-variety in C^n* – then $K[V]$ is an integral domain with an identity because $I(V)$ is then a prime ideal with $I(V) \neq R$ by Theorem 21.4.2. The quotient field $K(V) = \text{Quot}\,K[V]$ is called the *field of rational functions on the K-variety V*.

We note the following:

1. If C is algebraically closed then $V = C^n$ is a K-variety and $K(V)$ is the field $K(x_1,\dots,x_n)$ of the rational functions in n variables over K.

2. Let the affine K-algebra $A = K[\alpha_1,\dots,\alpha_n]$ be an integral domain with an identity $1 \neq 0$. Then $A \cong K[x_1,\dots,x_n]/\mathfrak{p}$ for some prime ideal $\mathfrak{p} \neq K[x_1,\dots,x_n]$. Hence, if C is algebraically closed then A is isomorphic to the coordinate ring of the K-variety $V = \mathcal{N}(\mathfrak{p})$ in C^n (see Hilbert's Nullstellensatz, first form, Theorem 21.4.1).

3. If the affine K-algebra $A = K[\alpha_1,\dots,\alpha_n]$ is an integral domain with an identity $1 \neq 0$ then we define the transcendence degree $\text{trgd}(A|K)$ to the transcendence degree of the field extension $\text{Quot}(A)|K$, that is, $\text{trgd}(A|K) = \text{trgd}(\text{Quot}(A)|K)$, $\text{Quot}(A)$ the quotient field of A.

   In this sense $\text{trgd}(K[x_1,\dots,x_n]|K) = n$. Since $\text{Quot}(A) = K(\alpha_1,\dots,\alpha_n)$ we get $\text{trgd}(A|K) \leq n$ by Theorem 20.3.10.

4. An arbitrary affine K-algebra $K[\alpha_1,\dots,\alpha_n]$ is, as a homomorphic image of the polynomial ring $K[x_1,\dots,x_n]$, noetherian (see Theorem 21.2.4 and Theorem 21.2.3).

Example 21.5.7. Let $\omega_1, \omega_2 \in \mathbb{C}$ two elements which are linear independent over $\mathbb{R}$. An element $\omega = m_1\omega_1 + m_2\omega_2$ with $m_1, m_2 \in \mathbb{Z}$, is called a *period*. The periods describe an abelian group $\Omega = \{m_1\omega_1 + m_2\omega_2 : m_1, m_2 \in \mathbb{Z}\} \cong \mathbb{Z} \oplus \mathbb{Z}$ and give a lattice in $\mathbb{C}$.

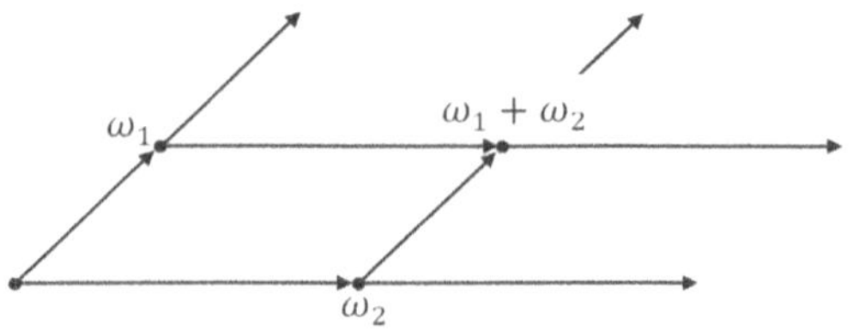

An elliptic function f (with respect to Ω) is a meromorphic function with period group Ω, that is, $f(z + w) = f(z)$ for all $z \in \mathbb{C}$. The Weierstrass $\wp$-function,

$$\wp(z) = \frac{1}{z^2} + \sum_{0 \neq w \in \Omega} \left(\frac{1}{(z - w)^2} - \frac{1}{w^2} \right).$$

is an elliptic function.

With $g_2 = 60 \sum_{0 \neq w \in \Omega} \frac{1}{w^4}$ and $g_3 = 140 \sum_{0 \neq w \in \Omega} \frac{1}{w^6}$ we get the differential equation $\wp'(z)^2 = 4\wp(z)^3 + g_2\wp(z) + g_3 = 0$. The set of elliptic functions is a field E, and each elliptic function is a rational function in $\wp$ and $\wp'$ (for details see, for instance, [27]).

The polynomial $f(t) = t^2 - 4s^3 + g_2 s + g_3 \in \mathbb{C}(s)[t]$ is irreducible over $\mathbb{C}(s)$. For the corresponding algebraic $\mathbb{C}(s)$-set V we get $K(V) = \mathbb{C}(s)[t]/(t^2 - 4s^3 + g_2 s + g_3) \cong E$ with respect to $t \mapsto \wp'$, $s \mapsto \wp$.

21.6 Dimensions

From now we assume that C is algebraically closed.

Definition 21.6.1. (1) The *dimension* $\dim(V)$ of an algebraic K-set V in C^n is said to be the supremum of all integers m for which there exists a strictly descending chain $V_0 \supsetneq V_1 \supsetneq \cdots \supsetneq V_m$ of K-varieties V_i in C^n with $V_i \subset V$ for all i.

(2) Let A be a commutative ring with an identity $1 \neq 0$. The *height* $h(\mathfrak{p})$ of a prime ideal $\mathfrak{p} \neq A$ of A is said to be the supremum of all integers m for which there exists a strictly ascending chain $\mathfrak{p}_0 \subsetneq \mathfrak{p}_1 \subsetneq \cdots \subsetneq \mathfrak{p}_m = \mathfrak{p}$ of prime ideals $\mathfrak{p}_i$ of A with $\mathfrak{p}_i \neq A$. The *dimension* (Krull dimension) $\dim(A)$ of A is said to be the supremum of the heights of all prime ideals $\neq A$ in A.

Theorem 21.6.2. *Let V be an algebraic K-set in C^n. Then* $\dim(V) = \dim(K[V])$.

Proof. By Theorem 21.2.4 and Theorem 21.4.2 we have a bijective map between the K-varieties W with $W \subset V$ and the prime ideals $\neq R = K[x_1, \dots, x_n]$ of R which contain $I(V)$ (the bijective map reverses the inclusion). But these prime ideals correspond exactly with the prime ideals $\neq K[V]$ of $K[V] = K[x_1, \dots, x_n]/I(V)$ which gives the statement. □

Suppose that V is an algebraic K-set in C^n and let $V_1, \dots, V_r$ the irreducible components of V. Then $\dim(V) = \max\{\dim V_1, \dots, \dim V_r\}$ because if V is a K-variety with $V' \subset V$ then $V' = (V' \cap V_1) \cup \dots \cup (V' \cap V_r)$. Hence, we may restrict ourselves on K-varieties V.

If we consider the special case of the K-variety $V = C^1 = C$ (recall that C is algebraically closed and, hence, especially C is infinite). Then $K[V] = K[x]$, the polynomial ring $K[x]$ in one indeterminate x. Now, $K[x]$ is a principal ideal domain, and hence, each prime ideal $\neq K[x]$ is either a maximal ideal or the zero ideal $\{0\}$ of $K[x]$. The only K-varieties in $V = C$ are therefore V itself and the zero set of irreducible polynomials in $K[x]$. Hence, if $V = C$ then $\dim(V) = \dim K[V] = 1 = \operatorname{trgd}(K[V]|K)$.

Theorem 21.6.3. *Let $A = K[\alpha_1, \dots, \alpha_n]$ be an affine K-algebra and let A be also an integral domain. Let $\{0\} = \mathfrak{p}_0 \subsetneq \mathfrak{p}_1 \subsetneq \cdots \subsetneq \mathfrak{p}_m$ be a maximal strictly ascending chain of prime ideals in A (such a chain exists since A is noetherian). Then $m =$* $\operatorname{trgd}(A|K) = \dim(A)$. *In other words:*

All maximal ideals of A have the same height, and this height is equal to the transcendence degree of A over K.

Corollary 21.6.4. *Let V be a K-variety in C^n. Then* $\dim(V) = \operatorname{trgd}(K[V]|K)$.

We prove Theorem 21.6.3 in several steps.

Lemma 21.6.5. *Let R be an unique factorization domain. Then each prime ideal $\mathfrak{p}$ with height $h(\mathfrak{p}) = 1$ is a principal ideal.*

Proof. $\mathfrak{p} \neq \{0\}$ since $h(\mathfrak{p}) = 1$. Hence there is an $f \in \mathfrak{p}$, $f \neq 0$. Since R is an unique factorization domain, f has a decomposition $f = p_1 \cdots p_s$ with prime elements $p_i \in R$. Now, $\mathfrak{p}$ is a prime ideal, hence some $p_i \in \mathfrak{p}$ because $f \in \mathfrak{p}$, say $p_1 \in \mathfrak{p}$. Then we have the chain $\{0\} \subsetneq (p_1) \subset \mathfrak{p}$, and (p_1) is a prime ideal of R. Since $h(\mathfrak{p}) = 1$ we get $(p_1) = \mathfrak{p}$. □

Lemma 21.6.6. *Let $R = K[y_1, \dots, y_r]$ be the polynomial ring of the r independent indeterminates $y_1, \dots, y_r$ over the field K (recall that R is a unique factorization domain). If $\mathfrak{p}$ is a prime ideal in R with height $h(\mathfrak{p}) = 1$ then the residue class ring $\bar{R} = R/\mathfrak{p}$ has transcendence degree $r - 1$ over K.*

Proof. By Lemma 21.6.5 we have that $\mathfrak{p} = (p)$ for some nonconstant polynomial $p \in K[y_1, \dots, y_r]$. Let the indeterminate $y = y_r$ occur in p, that is, $\deg_y(p) \geq 1$, the degree in y. If f is a multiple of p then also $\deg_y(f) \geq 1$. Hence, $\mathfrak{p} \cap K[y_1, \dots, y_r] \neq \{0\}$. Therefore the residue class mapping $R \to \bar{R} = K[\bar{y}_1, \dots, \bar{y}_r]$ induces an isomorphism $K[y_1, \dots, y_{r-1}] \to K[\bar{y}_1, \dots, \bar{y}_{r-1}]$ of the subring $K[y_1, \dots, y_{r-1}]$, that is, $\bar{y}_1, \dots, \bar{y}_{r-1}$ are algebraically independent over K. On the other side $p(\bar{y}_1, \dots,$

$\bar{y}_{r-1}, \bar{y}_r) = 0$ is a nontrivial algebraic relation for $\bar{y}_r$ over $K(\bar{y}_1, \ldots, \bar{y}_{r-1})$. Hence, altogether $\operatorname{trgd}(\bar{R}|K) = \operatorname{trgd}(K(\bar{y}_1, \ldots, \bar{y}_r)|K) = r - 1$ by Theorem 20.3.9. □

Before we describe the last technical lemma we need some preparatory theoretical material.

Let R, A be integral domains (with identity $1 \neq 0$) and let $A|R$ be a ring extension. We first consider only R.

(1) A subset $S \subset R \setminus \{0\}$ is called a multiplicative subset of R if $1 \in S$ for the identity 1 of R, and if $s, t \in S$ then also $st \in S$. $(x, s) \sim (y, t) :\Leftrightarrow xt - ys = 0$ defines an equivalence relation on $M = R \times S$. Let $\frac{x}{s}$ be the equivalence class of (x, s) and $S^{-1}R$ the set of all equivalence classes. We call $\frac{x}{s}$ a fraction. If we add and multiply fractions as usual we get that $S^{-1}R$ becomes an integral domain; it is called *the ring of fractions of R with respect to S*. If, especially, $S = R \setminus \{0\}$ then $S^{-1}R = \operatorname{Quot}(R)$, the quotient field of R.

Now, back to the general situation. $i : R \to S^{-1}R$, $i(r) = \frac{r}{1}$, defines an embedding of R into $S^{-1}R$. Hence, we may consider R as a subring of $S^{-1}R$. For each $s \in S \subset R \setminus \{0\}$ we have that $i(s)$ is an unit in $S^{-1}R$, that is, $i(s)$ is invertible, and that each element of $S^{-1}R$ has the form $i(s)^{-1}i(r)$ with $r \in R$, $s \in S$. Therefore $S^{-1}R$ is uniquely determined up to isomorphisms; and we have the following universal property:

If $\phi : R \to R'$ is a ring homomorphism (of integral domains) such that $\phi(s)$ is invertible for each $s \in S$ then there exist exactly one ring homomorphism $\lambda : S^{-1}R \to R'$ with $\lambda \circ i = \phi$. If $\mathfrak{a} \lhd R$ is an ideal in $\mathfrak{a}$ then we write $S^{-1}\mathfrak{a}$ for the ideal in $S^{-1}R$ generated by $i(\mathfrak{a})$. $S^{-1}\mathfrak{a}$ is the set of all elements of the form $\frac{a}{s}$ with $a \in \mathfrak{a}$ and $s \in S$; further $S^{-1}\mathfrak{a} = (1) \Leftrightarrow \mathfrak{a} \cap S \neq \emptyset$.

Vice versa, if $\mathfrak{A} \lhd S^{-1}R$ is an ideal in $S^{-1}R$ then we denote the ideal $i^{-1}(\mathfrak{A}) \lhd R$ with $\mathfrak{A} \cap R$, too. An ideal $\mathfrak{a} \lhd R$ is of the form $\mathfrak{a} = i^{-1}(\mathfrak{A})$ if and only if there is no $s \in S$ such that its image in $R/\mathfrak{a}$ under the canonical map $R \to R/\mathfrak{a}$ is a proper zero divisor in $R/\mathfrak{a}$. Under the mapping $\mathfrak{P} \to \mathfrak{P} \cap R$ and $\mathfrak{p} \mapsto S^{-1}\mathfrak{p}$ the prime ideals in $S^{-1}R$ correspond exactly to the prime ideals in R which do not contain an element of S.

We now identify R with $i(R)$.

(2) Now, let $\mathfrak{p} \lhd R$ be a prime ideal in R. Then $S = R \setminus \mathfrak{p}$ is multiplicative. In this case we write $R_{\mathfrak{p}}$ instead of $S^{-1}R$ and call $R_{\mathfrak{p}}$ the *quotient ring* of R with respect to $\mathfrak{p}$ or the localization of R of $\mathfrak{p}$. Put $\mathfrak{m} = \mathfrak{p}R_{\mathfrak{p}} = S^{-1}\mathfrak{p}$. Then $1 \notin \mathfrak{m}$. Each element of $R_{\mathfrak{p}}/\mathfrak{m}$ is a unit in $R_{\mathfrak{p}}$ and vice versa. In other words: Each ideal $\mathfrak{a} \neq (1)$ in $R_{\mathfrak{p}}$ is contained in $\mathfrak{m}$, or equivalently, $\mathfrak{m}$ is the only maximal ideal in $R_{\mathfrak{p}}$. A commutative ring with an identity $1 \neq 0$, which has exactly one maximal ideal, is called a *local ring*. Hence $R_{\mathfrak{p}}$ is a local ring. From part (1) we get further: the prime ideals of the local ring $R_{\mathfrak{p}}$ correspond bijectively to the prime ideals of R which are contained in $\mathfrak{p}$.

(3) Now we consider our ring extension $A|R$ as above. Let $\mathfrak{q}$ be a prime ideal in R.

Claim: If $\mathfrak{q}A \cap R = \mathfrak{q}$ then there exists a prime ideal $\mathfrak{Q} \lhd A$ with $\mathfrak{Q} \cap R = \mathfrak{q}$ (and vice versa).

Proof of the claim: If $S = R \setminus \mathfrak{q}$ then $\mathfrak{q}A \cap S = \emptyset$. Hence $\mathfrak{q}S^{-1}A$ is a proper ideal in $S^{-1}A$ and hence contained in a maximal ideal $\mathfrak{m}$ in $S^{-1}A$, here $\mathfrak{q}S^{-1}A$ is the ideal in $S^{-1}A$ which is generated by $\mathfrak{q}$. Define $\mathfrak{Q} = \mathfrak{m} \cap A$; $\mathfrak{Q}$ is a prime ideal in A, and $\mathfrak{Q} \cap R = \mathfrak{q}$ by part (1) because $\mathfrak{Q} \cap S = \emptyset$ where $S = R \setminus \mathfrak{q}$.

(4) Now let $A|R$ be an integral extension (A, R integral domains as above). Assume that R is integrally closed in its quotient field K. Let $\mathfrak{P} \lhd A$ be a prime ideal in A and $\mathfrak{p} = \mathfrak{P} \cap R$.

Claim: If $\mathfrak{q} \lhd R$ is a prime ideal in A with $\mathfrak{q} \subset \mathfrak{p}$ then $\mathfrak{q}A_{\mathfrak{p}} \cap R = \mathfrak{q}$.

Proof of the claim: An arbitrary $\beta \in \mathfrak{q}A_{\mathfrak{p}}$ has the form $\beta = \frac{\alpha}{s}$ with $\alpha \in \mathfrak{q}A$, $\mathfrak{q}A$ the ideal in A generated by $\mathfrak{q}$, and $s \in S = A \setminus \mathfrak{p}$. An integral equation for $\alpha \in \mathfrak{q}A$ over K is given a form $\alpha^n + a_{n-1}\alpha^{n-1} + \cdots + a_0 = 0$ with $a_i \in \mathfrak{q}$. This can be seen as follows: we have certainly a form $\alpha = b_1\alpha_1 + \cdots + b_m\alpha_m$ with $b_i \in \mathfrak{q}$ and $\alpha_i \in A$. The subring $A' = R[\alpha_1, \ldots, \alpha_m]$ is, as an R-module, finitely generated, and $\alpha A' \subset \mathfrak{q}A'$. Now, $a_i \in \mathfrak{q}$ follows with the same type of arguments as in the proof of Theorem 20.2.4.

Now, in addition, let $\beta \in R$. Then, for $s = \frac{\alpha}{\beta}$, we have an equation

$$s^n + \frac{a_{n-1}}{\beta}s^{n-1} + \cdots + \frac{a_0}{\beta^n} = 0$$

over K. But s is integral over R, and, hence, all $\frac{a_{n-i}}{\beta^i} \in R$.

We are now prepared to prove the last preliminary lemma which we need for the proof of Theorem 21.6.3.

Lemma 21.6.7 (Krull's going up lemma). *Let $A|R$ be an integral ring extension of integral domains and let R be integrally closed in its quotient field. Let $\mathfrak{p}$ and $\mathfrak{q}$ be prime ideals in R with $\mathfrak{q} \subset \mathfrak{p}$. Further let $\mathfrak{P}$ be a prime ideal in A with $\mathfrak{P} \cap R = \mathfrak{p}$. Then there exists a prime ideal $\mathfrak{Q}$ in A with $\mathfrak{Q} \cap R = \mathfrak{q}$ and $\mathfrak{Q} \subset \mathfrak{P}$.*

Proof. It is enough to show that there exists a prime ideal $\mathfrak{Q}$ in $A_{\mathfrak{p}}$ with $\mathfrak{Q} \cap R = \mathfrak{q}$. This can be seen from the preceding preparations. By part (1) and (2) such a $\mathfrak{Q}$ has the form $\mathfrak{Q} = \mathfrak{Q}'A_{\mathfrak{p}}$ with a prime ideal $\mathfrak{Q}'$ in A with $\mathfrak{Q}' \subset \mathfrak{P}$ and $\mathfrak{Q} \cap A = \mathfrak{Q}'$. It follows $\mathfrak{q} = \mathfrak{Q}' \cap R \subset \mathfrak{P} \cap R = \mathfrak{p}$. And the existence of such a $\mathfrak{Q}$ follows from parts (3) and (4). □

Proof of Theorem 21.6.3. Let first be $m = 0$. Then $\{0\}$ is a maximal ideal in A and, hence, $A = K[\alpha_1, \ldots, \alpha_n]$ a field. By Corollary 20.3.11 then $A|K$ is algebraic and, hence, $\operatorname{trgd}(A|K) = 0$. So, Theorem 21.3.3 holds for $m = 0$.

Now, let $m \geq 1$. We use Noether's normalization theorem. A has a polynomial ring $R = K[y_1, \ldots, y_r]$ of the r independent indeterminates $y_1, \ldots, y_r$ as a subring, and $A|R$ is an integral extension. As a polynomial ring over K the ring R is a unique factorization domain and hence, certainly, algebraically closed (in its quotient field).

Now, let

$$\{0\} = \mathfrak{P}_0 \subsetneq \mathfrak{P}_1 \subsetneq \cdots \subsetneq \mathfrak{P}_m \tag{1}$$

be a maximal strictly ascending chain of prime ideals in A. If we intersect with R we get a chain

$$\{0\} = \mathfrak{p}_0 \subset \mathfrak{p}_1 \subset \cdots \subset \mathfrak{p}_m \tag{2}$$

of prime ideals $\mathfrak{p}_i = \mathfrak{P}_i \cap R$ of R. Since $A|R$ is integral, the chain (2) is also a strictly ascending chain. This follows from Krull's going up lemma (Lemma 21.6.7) because if $\mathfrak{p}_i = \mathfrak{p}_j$ then $\mathfrak{P}_i = \mathfrak{P}_j$. If $\mathfrak{P}_m$ is a maximal ideal in A then also $\mathfrak{p}_m$ is a maximal ideal in R because $A|R$ is integral (consider $A/\mathfrak{P}_m$ and use Theorem 17.2.21). If the chain (1) is maximal and strictly then also the chain (2).

Now, let the chain (1) be maximal and strictly. If we pass to the residue class rings $\bar{A} = A/\mathfrak{P}_1$ and $\bar{R} = R/\mathfrak{p}_1$ then we get the chains of prime ideals $\{0\} = \bar{\mathfrak{P}}_1 \subset \bar{\mathfrak{P}}_2 \subset \cdots \subset \bar{\mathfrak{P}}_m$ and $\{0\} = \bar{\mathfrak{p}}_1 \subset \bar{\mathfrak{p}}_2 \subset \cdots \subset \bar{\mathfrak{p}}_m$ for the affine K-algebras $\bar{A}$ and $\bar{R}$, respectively, but with a 1 less length. By induction, we may assume that already $\operatorname{trgd}(\bar{A}|K) = m - 1 = \operatorname{trgd}(\bar{R}|K)$. On the other side, by construction we have $\operatorname{trgd}(A|K) = \operatorname{trgd}(R|K) = r$. To prove Theorem 21.3.3 finally, we have to show that $r = m$. If we compare both equations then $r = m$ follows if $\operatorname{trgd}(\bar{R}|K) = r - 1$. But this holds by Lemma 21.6.6. □

Theorem 21.6.8. *Let V be a K-variety in C^n. Then* $\dim(V) = n - 1$ *if and only if $V = (f)$ for some irreducible polynomial $f \in K[x_1, \ldots, x_n]$.*

Proof. (1) Let V be a K-variety in C^n with $\dim(V) = n - 1$. The corresponding ideal (in the sense of Theorem 21.2.4) is by Theorem 21.4.2 a prime ideal $\mathfrak{p}$ in $K[x_1, \ldots, x_n]$. By Theorem 21.3.3 and Corollary 21.3.4 we get $h(\mathfrak{p}) = 1$ for the height of $\mathfrak{p}$ because $\dim(V) = n - 1$ (see also Theorem 21.3.2). Since $K[x_1, \ldots, x_n]$ is a unique factorization domain we get that $\mathfrak{p} = (f)$ is a principal ideal by Lemma 21.6.5.

(2) Now let $f \in K[x_1, \ldots, x_n]$ be irreducible. We have to show that $V = \mathcal{N}(f)$ has dimension $n - 1$. For that, by Theorem 21.6.3, we have to show that the prime ideal $\mathfrak{p} = (f)$ has the height $h(\mathfrak{p}) = 1$. Assume that this is not the case. Then there exists a prime ideal $\mathfrak{q} \neq \mathfrak{p}$ with $\{0\} \neq \mathfrak{q} \subset \mathfrak{p}$. Choose $g \in \mathfrak{q}$, $g \neq 0$. Let $g = uf^{e_1}\pi_2^{e_2} \cdots \pi_r^{e_r}$ be its prime factorization in $K[x_1, \ldots, x_n]$. Now $g \in \mathfrak{q}$ and $f \notin \mathfrak{q}$ because $\mathfrak{q} \neq \mathfrak{p}$. Hence, there is a π_i in $\mathfrak{q} \subsetneq \mathfrak{p} = (f)$ which is impossible. Therefore $h(\mathfrak{p}) = 1$. □

21.7 Exercises

1. Let $A = K[a_1, \dots, a_n]$ and $C|K$ a field extension with C algebraically closed. Show that there is a K-algebra homomorphism $K[a_1, \dots, a_n] \to C$.

2. Let $K[x_1, \dots, x_n]$ be the polynomial ring of the n independent indeterminates $x_1, \dots, x_n$ over the algebraically closed field K. The maximal ideals of $K[x_1, \dots, x_n]$ are exactly the ideals of the form $\mathfrak{m}(\alpha) = (x_1 - \alpha_1, x_2 - \alpha_2, \dots, x_n - \alpha_n)$ with $\alpha = (\alpha_1, \dots, \alpha_n) \in K^n$.

3. The nil radical $\sqrt{0}$ of $A = K[a_1, \dots, a_n]$ corresponds with the Jacobson radical of A, that is, the intersection of all maximal ideals of A.

4. Let R be a commutative ring with $1 \neq 0$. If each prime ideal of R is finitely generated then R is noetherian.

5. Prove the theoretical preparations for Krull's going up lemma in detail.

6. Let $K[x_1, \dots, x_n]$ be the polynomial ring of the n independent indeterminates $x_1, \dots, x_n$. For each ideal $\mathfrak{a}$ of $K[x_1, \dots, x_n]$ there exists a natural number m with the following property: if $f \in K[x_1, \dots, x_n]$ vanishes on the zero set of $\mathfrak{a}$ then $f^m \in \mathfrak{a}$.

7. Let K be a field with char $K \neq 2$ and $a, b \in K^\star$. We consider the polynomial $f(x, y) = ax^2 + by^2 - 1 \in K[x, y]$, the polynomial ring of the independent indeterminates x and y. Let C be the algebraic closure of $K(x)$ and $\beta \in C$ with $f(x, \beta) = 0$. Show that:

   (i) f is irreducible over the algebraic closure C_0 of K (in C).

   (ii) $\operatorname{trgd}(K(x, \beta)|K) = 1$, $[K(x, \beta) : K(x)] = 2$, and K is algebraically closed in $K(x, \beta)$.

Chapter 22
Algebraic Cryptography

22.1 Basic Cryptography

As we have mentioned, much of mathematics has been algebraicized, that is uses the methods and techniques of abstract algebra. Throughout this book we have looked at various applications of the algebraic ideas. Many of these were to other areas of mathematics, such as the insolvability of the quintic. In this final chapter we move in a slightly different direction and look at applications of algebra to cryptography. This has become increasingly important because of the extensive use of cryptography and cryptosystems in modern commerce and communications. We first give a brief introduction to general cryptography and its history.

Cryptography refers to the science and/or art of sending and receiving coded messages. Coding and hidden ciphering is an old endeavor used by governments and militaries and between private individuals from ancient times. Recently it has become even more prominent because of the necessity of sending secure and private information, such as credit card numbers, over essentially open communication systems.

Traditionally *cryptography* is the science and or art of devising and implementing secret codes or *cryptosystems*. *Cryptanalysis* is the science and or art of breaking cryptosystems while *cryptology* refers to the whole field of cryptography plus cryptanalysis. In most modern literature cryptography is used synonymously with cryptology. Theoretically cryptography uses mathematics, computer science and engineering.

A *cryptosystem* or *code* is an algorithm to change a plain message, called the *plaintext message*, into a coded message, called the *ciphertext message*. In general both the *plaintext message* (uncoded message) and the *ciphertext message* (coded message) are written in some N letter alphabet which is usually the same for both plaintext and code. The method of coding or the encoding algorithm is then a transformation of the N letters. The most common way to perform this transformation is to consider the N letters as N integers modulo N and then perform a number theoretical function on them. Therefore most encoding algorithms use modular arithmetic and hence cryptography is closely tied to number theory. The subject is very broad, and as mentioned above, very current, due to the need for publically viewed but coded messages. There are many references to the subject. The book by Koblitz [60] gives an outstanding introduction to the interaction between number theory and cryptography. It also includes many references to other sources. The book by Stinson [68] describes the whole area.

Modern cryptography is usually separated into *classical cryptography* also called *symmetric key cryptography* and *public key cryptography*. In the former, both the encoding and decoding algorithms are supposedly known only to the sender and receiver, usually referred to as Bob and Alice. In the latter, the encryption method is public knowledge but only the receiver knows how to decode.

The message that one wants to send is written in *plaintext* and then converted into code. The coded message is written in *ciphertext*. The plaintext message and ciphertext message are written in some *alphabets* that are usually the same. The process of putting the plaintext message into code is called *enciphering* or *encryption* while the reverse process is called *deciphering* or *decryption*. Encryption algorithms break the plaintext and ciphertext message into *message units*. These are single letters or pairs of letters or more generally k-vectors of letters. The transformations are done on these message units and the encryption algorithm is a mapping from the set of plaintext message units to the set of ciphertext message units. Putting this into a mathematical formulation we let

$$\mathcal{P} = \text{set of all plaintext message units}$$
$$\mathcal{C} = \text{set of all ciphertext message units.}$$

The encryption algorithm is then the application of an invertible function

$$f : \mathcal{P} \to \mathcal{C}.$$

The function f is the *encryption map*. The inverse

$$f^{-1} : \mathcal{C} \to \mathcal{P}$$

is the *decryption* or *deciphering map*. The triple $\{\mathcal{P}, \mathcal{C}, f\}$, consisting of a set of plaintext message units, a set of cipertext message units and an encryption map, is called a *cryptosystem*.

Breaking a code is called *cryptanalysis*. An attempt to break a code is called an *attack*. Most cryptanalysis depends on a statistical frequency analysis of the plaintext language used (see exercises). Cryptanalysis depends also on a knowledge of the form of the code, that is, the type of cryptosystem used.

We now give some examples of cryptosystems and cryptanalysis.

Example 22.1.1. The simplest type of encryption algorithm is a *permutation cipher*. Here the letters of the plaintext alphabet are permuted and the plaintext message is sent in the permuted letters. Mathematically if the alphabet has N letters and σ is a permutation on $1, \dots, N$, the letter i in each message unit is replaced by $\sigma(i)$. For example suppose the plaintext language is English and the plaintext word is BOB and the permutation algorithm is

a	b	c	d	e	f	g	h	i	j	k	l	m
b	c	d	f	g	h	j	k	l	n	o	p	r

n	o	p	q	r	s	t	u	v	w	x	y	z
s	t	v	w	x	a	e	i	z	m	q	y	u

then BOB $\to$ CSC.

Example 22.1.2. A very straightforward example of a permutation encryption algorithm is a *shift algorithm*. Here we consider the plaintext alphabet as the integers $0, 1, \ldots, N-1 \mod N$. We choose a fixed integer k and the encryption algorithm is

$$f : m \to m + k \mod N.$$

This is often known as a *Caesar code* after Julius Caesar who supposedly invented it. It was used by the Union Army during the American Civil War. For example if both the plaintext and ciphertext alphabets were English and each message unit was a single letter then $N = 26$. Suppose $k = 5$ and we wish to send the message ATTACK. If $a = 0$ then ATTACK is the numerical sequence $0, 20, 20, 0, 2, 11$. The encoded message would then be FZZFIP.

Any permutation encryption algorithm which goes letter to letter is very simple to attack using a statistical analysis. If enough messages are intercepted and the plaintext language is guessed then a frequency analysis of the letters will suffice to crack the code. For example in the English language the three most commonly occurring letters are E, T and A with a frequency of occurrence of approximately 13% and 9% and 8% respectively. By examining the frequency of occurrences of letters in the ciphertext the letters corresponding to E, T and A can be uncovered.

Example 22.1.3. A variation on the Caesar code is the *Vignère code*. Here message units are considered as k-vectors of integers mod N from an N letter alphabet. Let $B = (b_1, \ldots, b_k)$ be a fixed k-vector in $\mathbb{Z}_n^k$. The Vignère code then takes a message unit

$$(a_1, \ldots, a_k) \to (a_1 + b_1, \ldots, a_k + b_k) \mod N.$$

From a cryptanalysis point of view a Vignère code is no more secure than a Caesar code and is susceptible to the same type of statistical attack.

The *Alberti Code* is a *polyalphabetic cipher* and can be often be used to thwart a statistical frequency attack. We describe it in the next example.

Example 22.1.4. Suppose we have an N letter alphabet. We then form an $N \times N$ matrix P where each row and column is a distinct permutation of the plaintext alphabet. Hence P is a permutation matrix on the integers $0, \ldots, N-1$. Bob and Alice decide on a *keyword*. The keyword is placed above the plaintext message and the intersection of the keyword letter and plaintext letter below it will determine which cipher alphabet to use. We will make this precise with an 9 letter alphabet A, B, C, D, E, O, S, T, U. Here for simplicity we will assume that each row is just a shift of the previous row, but any permutation can be used.

Key Letters

		A	B	C	D	E	O	S	T	U
a	A	a	b	c	d	e	o	s	t	u
l	B	b	c	d	e	o	s	t	u	a
p	C	c	d	e	o	s	t	u	a	b
h	D	d	e	o	s	t	u	a	b	c
a	E	e	o	s	t	u	a	b	c	d
b	O	o	s	t	u	a	b	c	d	e
e	S	s	t	u	a	b	c	d	e	o
t	T	t	u	a	b	c	d	e	o	s
s	U	u	a	b	c	d	e	o	s	t.

Suppose the plaintext message is STAB DOC and Bob and Alice have chosen the keyword BET. We place the keyword repeatedly over the message

B E T B E T B
S T A B D O C.

To encode we look at B which lies over S. The intersection of the B key letter and the S alphabet is a t so we encrypt the S with T. The next key letter is E which lies over T. The intersection of the E keyletter with the T alphabet is c. Continuing in this manner and ignoring the space we get the encryption

$$\text{STAB DOC} \to \text{TCTCTDD.}$$

Example 22.1.5. A final example, which is not number theory based, is the so-called *Beale Cipher*. This has a very interesting history which is related in the popular book *Archimedes Revenge* by Paul Hoffman (see [56]). Here letters are encrypted by numbering the first letters of each word in some document like the Declaration of Independence or the Bible. There will then be several choices for each letter and a Beale cipher is quite difficult to attack.

Until relatively recent times cryptography was mainly concerned with message confidentiality – that is sending secret messages so that interceptors or eavesdroppers cannot decipher them. The discipline was primarily used in military and espionage situations. This changed with the vast amount of confidential data that had to be transmitted over public airways so the field has expanded to many different types of cryptographic techniques such as digital signatures and message authentications.

Cryptography and encryption does have a long and celebrated history. In the Bible, in the book of Jeremiah, they use what is called an *Atabash Code*. In this code the letters of the alphabet – Hebrew in the Bible but can be used with any alphabet – are permuted first to last. That is, in the Latin alphabet, Z would go to A and so on.

The Kabbalists and the Kabbala believe that the Bible – written in Hebrew where each letter also stands for a number – is a code from heaven. They have devised elaborate ways to decode it. This idea has seeped into popular culture where the book "The Bible Code" became a bestseller.

In his military campaigns Julius Caesar would send out coded messages. His method, which we looked at in the last section, is now known as a Caesar code. It is a shift cipher. That is each letter is shifted a certain amount to the right. A shift cipher is a special case of an affine cipher that will be elaborated upon in the next section. The Caesar code was resurrected and used during the American Civil War.

Coded messages produced by most of the historical methods reveal statistical information about the plaintext. This could be used in most cases to break the codes. The discovery of frequency analysis was done by the Arab mathematician Al-Kindi in the ninth century and the basic classical substitution ciphers became more or less easily breakable. About 1470 Leon Alberti developed a method to thwart statistical analysis. His innovation was to use a polyalphabetic cipher where different parts of the message are encrypted with different alphabets. We looked at an example of an Alberti code in this section.

A different way to thwart statistical attacks is to use blank and neutral letters, that is meaningless letters within the message. Mary, Queen of Scots, used a random permutation cipher with neutrals in it, where a neutral was a random meaningless symbol. Unfortunately for her, her messages were decoded and she was beheaded.

There have been various physical devices and aids used to create codes. Prior to the widespread use of the computer the most famous cryptographic aid was the *Enigma machine* developed and used by the German military during the Second World War. This was a rotor machine using a polyalphabetic cipher. An early version was broken by Polish cryptographers early in the war so a larger system was built that was considered unbreakable. British cryptographers led by Alan Turing broke this and British knowledge of German secrets had a great effect on the latter part of the war.

The development of digital computers allowed for the development of much more complicated cryptosystems. Further this allowed for the encryption using anything that can be placed in binary formats whereas historical cryptosystems could only be rendered using language texts. This has revolutionized cryptography.

In 1976 Diffie and Hellman developed the first usable public key exchange protocol. This allowed for the transmission of secret data over open airways. A year later Rivest, Adelman and Shamir, developed the RSA algorithm, a second public key protocol. There are now many and we will discuss them later. In 1997 it became known that public key cryptography had been developed earlier by James Ellis working for British

Intelligence and that both the Diffie–Hellman and RSA protocols had been developed earlier by Malcom Williamson and Clifford Cocks respectively.

22.2 Encryption and Number Theory

Here we describe some basic number theoretically derived cryptosystems. In applying a cryptosystem to an N letter alphabet we consider the letters as integers mod N. The encryption algorithms then apply number theoretic functions and use modular arithmetic on these integers. One example of this was the shift, or Caesar cipher, described in the last section. In this encryption method a fixed integer k is chosen and the encryption map is given

$$f : m \to m + k \mod N.$$

The shift algorithm is a special case of an *affine cipher*. Recall that an *affine map* on a ring R is a function $f(x) = ax + b$ with $a, b, x \in R$. We apply such a map to the ring of integers modulo n, that is, $R = \mathbb{Z}_n$, as the encryption map. Specifically again suppose we have an N letter alphabet and we consider the letters as the integers $0, 1, \dots, N - 1 \mod N$, that is in the ring $\mathbb{Z}_N$. We choose integers $a, b \in \mathbb{Z}_N$ with $(a, N) = 1$ and $b \neq 0$. a, b are called the *keys* of the cryptosystem . The encryption map is then given by

$$f : m \to am + b \mod N.$$

Example 22.2.1. Using an affine cipher with the English language and keys $a = 3$, $b = 5$ encode the message EAT AT JOE'S. Ignore spaces and punctuation.

The numerical sequence for the message ignoring the spaces and punctuation is

$$4, 0, 19, 0, 19, 9, 14, 4, 18.$$

Applying the map $f(m) = 3m + 5 \mod 26$ we get

$$17, 5, 62, 5, 62, 32, 47, 17, 59 \to 17, 5, 10, 5, 10, 6, 21, 17, 7.$$

Now rewriting these as letters we get

$$\text{EAT AT JOE'S} \to \text{RFKFKGVRH}.$$

Since $(a, N) = 1$ the integer a has a multiplicative inverse $a^{-1} \mod N$. The decryption map for an affine cipher with keys a, b is then

$$f^{-1} : m \to a^{-1}(m - b) \mod N.$$

Since an affine cipher, as given above, goes letter to letter it is easy to attack using a statistical frequency approach. Further if an attacker can determine two letters and knows that it is an affine cipher the keys can be determined and the code broken. To give better security it is preferable to use k-vectors of letters as message units. The form then of an affine cipher becomes

$$f : v \to Av + B$$

where here v and B are k-vectors from $\mathbb{Z}_N^k$ and A is an invertible $k \times k$ matrix with entries from the ring $\mathbb{Z}_N$. The computations are then done modulo N. Since v is a k-vector and A is a $k \times k$ matrix the matrix product Av produces another k-vector from $\mathbb{Z}_N^k$. Adding the k-vector B again produces a k-vector so the ciphertext message unit is again a k-vector. The keys for this affine cryptosystem are the enciphering matrix A and the shift vector B. The matrix A is chosen to be invertible over $\mathbb{Z}_N$ (equivalent to the determinant of A being a unit in the ring $\mathbb{Z}_N$) so the decryption map is given by

$$v \to A^{-1}(v - B).$$

Here A^{-1} is the matrix inverse over $\mathbb{Z}_N$ and v is a k-vector. The *enciphering matrix* A and the shift vector B are now the keys of the cryptosystem.

A statistical frequency attack on such a cryptosystem requires knowledge, within a given language, of the statistical frequency of k-strings of letters. This is more difficult to determine than the statistical frequency of single letters. As for a letter to letter affine cipher, if $k + 1$ message units, where k is the message block length, are discovered, then the code can be broken.

Example 22.2.2. Using an affine cipher with message units of length 2 in the English language and keys

$$A = \begin{pmatrix} 5 & 1 \\ 8 & 7 \end{pmatrix}, \qquad B = (5, 3)$$

encode the message EAT AT JOE'S. Again ignore spaces and punctuation.

Message units of length 2, that is 2-vectors of letters are called *digraphs*. We first must place the plaintext message in terms of these message units. The numerical sequence for the message EAT AT JOE'S ignoring the spaces and punctuation is as before

$$4, 0, 19, 0, 19, 9, 14, 4, 18.$$

Therefore the message units are

$$(4, 0), (19, 0), (19, 9), (14, 4), (18, 18)$$

repeating the last letter to end the message.

The enciphering matrix A has determinant 1 which is a unit mod 26 and hence is invertible so it is a valid key.

Now we must apply the map $f(v) = Av + B$ mod 26 to each digraph. For example

$$A\begin{pmatrix}4\\0\end{pmatrix} + B = \begin{pmatrix}5 & 1\\8 & 7\end{pmatrix}\begin{pmatrix}4\\0\end{pmatrix} + \begin{pmatrix}5\\3\end{pmatrix} = \begin{pmatrix}20\\32\end{pmatrix} + \begin{pmatrix}5\\3\end{pmatrix} = \begin{pmatrix}25\\9\end{pmatrix}.$$

Doing this to the other message units we obtain

$$(25, 9), (22, 25), (5, 10), (1, 13), (9, 13).$$

Now rewriting these as digraphs of letters we get

(Z, J), (W, Z), (F, K), (B, N), (J, N).

Therefore the coded message is

EAT AT JOE'S → ZJWZFKBNJN.

Example 22.2.3. Suppose we receive the message ZJWZFKBNJN and we wish to decode it. We know that an affine cipher with message units of length 2 in the English language and keys

$$A = \begin{pmatrix}5 & 1\\8 & 7\end{pmatrix}, \quad B = (5, 3)$$

is being used.

The decryption map is given by

$$v \to A^{-1}(v - B).$$

so we must find the inverse matrix for A. For a 2×2 invertible matrix $\begin{pmatrix}a & b\\c & d\end{pmatrix}$ we have

$$\begin{pmatrix}a & b\\c & d\end{pmatrix}^{-1} = \frac{1}{ad - bc}\begin{pmatrix}d & -b\\-c & a\end{pmatrix}.$$

Therefore in this case recalling that multiplication is mod 26

$$A = \begin{pmatrix}5 & 1\\8 & 7\end{pmatrix} \Longrightarrow A^{-1} = \begin{pmatrix}7 & -1\\-8 & 5\end{pmatrix}.$$

The message ZJWZFKBNJN in terms of message units is

$$(25, 9), (22, 25), (5, 10), (1, 13), (9, 13).$$

We apply the decryption map to each digraph. For example

$$A^{-1}\left(\begin{pmatrix}20\\6\end{pmatrix} - B\right) = \begin{pmatrix}7 & -1\\-8 & 5\end{pmatrix}\left(\begin{pmatrix}25\\9\end{pmatrix} - \begin{pmatrix}5\\3\end{pmatrix}\right) = (4, 0).$$

Doing this to each we obtain

$$(4, 0), (19, 0), (19, 9), (14, 4), (18, 18)$$

and rewriting in terms of letters

$$(E, A), (T, A), (T, J), (O, E), (S, S).$$

This gives us

$$\text{ZJWZFKBNJN} \rightarrow \text{EATATJOESS}.$$

Modern cryptography is done via a computer. Hence all messages both plaintext and ciphertext are actually presented as binary strings. Important in this regard is the concept of a *hash function*.

A *cryptographic hash function* is a deterministic function

$$h : S \rightarrow \{0, 1\}^n$$

that returns for each arbitrary block of data, called a *message*, a fixed size bit string. It should have the property that a change in the data will change the hash value. The hash value is called the *digest*.

An ideal cryptographic hash function has the following properties:

(1) It is easy to compute the hash value for any given message.

(2) It is infeasible to find a message that has a given hash value (preimage resistant).

(3) It is infeasible to modify a message without changing its hash.

(4) It is infeasible to find two different messages with the same hash (collision resistant).

A cryptographic hash function can serve as a *digital signature*.

Hash functions can also be used with encryption. Suppose that Bob and Alice want to communicate openly. They have exchanged a secret key K that supposedly only they know. Let f_K be an encryption function or encryption algorithm based on the key K. Alice wants to send the message m to Bob and m is given as a binary bit string. Alice sends to Bob

$$f_K(m) \oplus h(K)$$

where $\oplus$ is addition modulo 2.

Bob knows the key K and hence its hash value $h(K)$. He now computes

$$f_K(m) \oplus h(K) \oplus h(K).$$

Since addition modulo 2 has order 2 we have

$$f_K(m) \oplus h(K) \oplus h(K) = f_k(M).$$

Bob now applies the decryption algorithm f_K^{-1} to decode the message.

Alice could have just as easily sent $f_K(m) \oplus K$. However sending the hash has two benefits. Usually the hash is shorter than the key and from the properties of hash functions it gives another level of security. As we will see, tying the secret key to the actual encryption in this manner is the basis for the El-Gamal and elliptic curve cryptographic methods.

The encryption algorithm f_K is usually a symmetric key encryption so that anyone knowing K can encrypt and decrypt easily. However it should be resistant to plaintext-ciphertext attacks. That is if an attacker gains some knowledge of a piece of plaintext together with the corresponding ciphertext it should not compromise the whole system.

The encryption algorithm can either be a *block cipher* or a *stream cipher*. In the former, blocks of fixed length k are transformed into blocks of fixed length n and there is a method to ties the encrypted blocks together. In the latter, a stream cipher, bits are transformed one by one into new bit strings by some procedure.

In 2001 the National Institute of Standards and Technology adopted a block cipher now called AES for *Advanced Encryption System* as the industry standard for a symmetric key encryption. Although not universally used it is the most widely used. This block cipher was a standardization of the Rijnadel cipher named after its inventor Rijmen and Daeman.

AES replaced DES or *Digital Encryption System* which had been the standard. Parts of DES were found to be insecure. AES proceeds with several rounds of encrypting blocks and then mixing blocks. The mathematics in AES is done over the finite field GF(2^8).

22.3 Public Key Cryptography

Presently there are many instances where secure information must be sent over open communication lines. These include for example banking and financial transactions, purchasing items via credit cards over the Internet and similar things. This led to the development of *public key cryptography*. Roughly, in classical cryptography only the sender and receiver know the encoding and decoding methods. Further it is a feature of such cryptosystems, such as the ones that we've looked at, that if the encrypting method is known then the decryption can be carried out. In public key cryptography the encryption method is public knowledge but only the receiver knows how to decode. More precisely in a classical cryptosystem once the encrypting algorithm is known the decryption algorithm can be implemented in approximately the same order of magnitude of time. In a public key cryptosystem, developed first by Diffie and

Hellman, the decryption algorithm is much more difficult to implement. This difficulty depends on the type of computing machinery used and as computers get better, new and more secure public key cryptosystems become necessary.

The basic idea in a public key cryptosystem is to have a *one-way function* or *trapdoor function*. That is a function which is easy to implement but very hard to invert. Hence it becomes simple to encrypt a message but very hard, unless you know the inverse, to decrypt.

The standard model for public key systems is the following. Alice wants to send a message to Bob. The encrypting map f_A for Alice is public knowledge as well as the encrypting map f_B for Bob. On the other hand the decryption algorithms f_A^{-1} and f_B^{-1} are secret and known only to Alice and Bob respectively. Let $\mathcal{P}$ be the message Alice wants to send to Bob. She sends $f_B f_A^{-1}(\mathcal{P})$. To decode Bob applies first f_B^{-1}, which only he knows. This gives him $f_B^{-1}(f_B f_A^{-1}(\mathcal{P})) = f_A^{-1}(\mathcal{P})$. He then looks up f_A which is publically available and applies this $f_A(f_A^{-1}(\mathcal{P})) = \mathcal{P}$ to obtain the message. Why not just send $f_B(\mathcal{P})$. Bob is the only one who can decode this. The idea is *authentication*, that is being certain from Bob's point of view that the message really came from Alice. Suppose $\mathcal{P}$ is Alice's verification; signature, social security number etc. If Bob receives $f_B(\mathcal{P})$ it could be sent by anyone since f_B is public. On the other hand since only Alice supposedly knows f_A^{-1} getting a reasonable message from $f_A(f_B^{-1} f_B f_A^{-1}(\mathcal{P}))$ would verify that it is from Alice. Applying f_B^{-1} alone should result in nonsense.

Getting a reasonable one way function can be a formidable task. The most widely used (at present) public key systems are based on difficult to invert number theoretic functions. The original public key system was developed by Diffie and Hellman in 1976. It was followed closely by a second public key system developed by Rivest, Shamir and Adeelman known as the RSA system. Although at present there are many different public key systems in use most are variations of these original two. The variations are attempts to make the systems more secure. We will discuss four such systems.

22.3.1 The Diffie–Hellman Protocol

Diffie and Hellman in 1976 developed the original public key idea using the *discrete log problem*. In modular arithmetic it is easy to raise an element to a power but difficult to determine, given an element, if it is a power of another element. Specifically if G is a finite group, such as the cyclic multiplicative group of $\mathbb{Z}_p$ where p is a prime, and $h = g^k$ for some k then the *discrete log* of h to the base g is any integer t with $h = g^t$.

The rough form of the Diffie–Hellman public key system is as follows. Bob and Alice will use a classical cryptosystem based on a key k with $1 < k < q - 1$ where q is a prime. It is the key k that Alice must share with Bob. Let g be a multiplicative generator of $\mathbb{Z}_q^\star$ the multiplicative group of $\mathbb{Z}_q$. The generator g is public. It is known that this group is cyclic if q is a prime.

Alice chooses an $a \in \mathbb{Z}_q$ with $1 < a < q - 1$. She makes public g^a. Bob chooses a $b \in \mathbb{Z}_q^\star$ and makes public g^b. The secret key is g^{ab}. Both Bob and Alice, but presumably none else, can discover this key. Alice knows her secret power a and the value g^b is public from Bob. Hence she can compute the key $g^{ab} = (g^b)^a$. The analogous situation holds for Bob. An attacker however only knows g^a and g^b and g. Unless the attacker can solve the discrete log problem the key exchange is secure.

Given q, g, g^a, g^b the problem of determining the secret key g^{ab} is called the *Diffie–Hellman problem*. At present the only known solution is to solve the discrete log problem which appears to be very hard. In choosing the prime q and the generator g it is assumed that the prime q is very large so that the order of g is very large. There are algorithms to solve the discrete log problem is q is too small.

One attack on the Diffie–Hellman key exchange is a *man in the middle attack*. Since the basic protocol involves no authentication an attacker can pretend to be Bob and get information from Alice and then pretend to be Alice and get information from Bob. In this way the attacker could get the secret shared key. To prevent this, digital signatures are often used (see [60] for a discussion of these).

The *decision Diffie–Hellman problem* is given a prime q and $g^a \bmod q$, $g^b \bmod q$ and $g^c \bmod q$ determine if $g^c = g^{ab}$.

In 1997 it became known that the ideas of public key cryptography were developed by British Intelligence Services prior to Diffie and Hellman.

22.3.2 The RSA Algorithm

In 1977 Rivest, Adelman and Shamir developed the *RSA algorithm*, which is presently (in several variations) the most widely used public key cryptosystem. It is based on the difficulty of factoring large integers and in particular on the fact that it is easier to test for primality than to factor very large integers.

In basic form the RSA algorithm works as follows. Alice chooses two large primes p_A, q_A and an integer e_A relatively prime to $\phi(p_A q_A) = (p_A - 1)(q_A - 1)$ where ϕ is the Euler phi-function. It is assumed that these integers are chosen randomly to minimize attack. Primality tests arise in the following manner. Alice first randomly chooses a large odd integer m and tests it for primality. If its prime it is used. If not, she tests $m + 2$, $m + 4$ and so on until she gets her first prime p_A. She then repeats the process to get q_A. Similarly she chooses another odd integer m and tests until she gets an e_A relatively prime to $\phi(p_A q_A)$. The primes she chooses should be quite large. Originally RSA used primes of approximately 100 decimal digits, but as computing and attack have become more sophisticated, larger primes have had to be utilized. Presently keys with 400 decimal digits are not uncommon. Once Alice has obtained p_A, q_A, e_A she lets $n_A = p_A q_A$ and computes d_A, the multiplicative inverse of e_A modulo $\phi(n_A)$. That is d_A satisfies $e_A d_A \equiv 1 \bmod (p_A - 1)(q_A - 1)$. She makes public the enciphering key $K_A = (n_A, e_A)$ and the encryption algorithm known to all is

$$f_A(\mathcal{P}) = \mathcal{P}^{e_A} \mod n_A$$

where $\mathcal{P} \in \mathbb{Z}_{n_A}$ is a message unit. It can be shown that if $(e_A, (p_A - 1)(q_A - 1)) = 1$ and $e_A d_A \equiv 1 \mod (p_A - 1)(q_A - 1)$ then $\mathcal{P}^{e_A d_A} \equiv \mathcal{P} \mod n_A$ (see exercises). Therefore the decryption algorithm is

$$f_A^{-1}(\mathcal{C}) = \mathcal{C}^{d_a} \mod n_A.$$

Notice then that $f_A^{-1}(f_A(\mathcal{P})) = \mathcal{P}^{e_A d_A} \equiv \mathcal{P} \mod n_A$ so it is the inverse.

Now Bob makes the same type of choices to obtain p_B, q_B, e_B. He lets $n_B = p_B q_B$ and makes public his key $K_B = (n_B, e_B)$.

If Alice wants to send a message to Bob that can be authenticated to be from Alice she sends $f_B(f_A^{-1}(\mathcal{P}))$. An attack then requires factoring n_A or n_B which is much more difficult than obtaining the primes p_A, q_A, p_B, q_B.

In practice suppose there is an N letter alphabet which is to be used for both plaintext and ciphertext. The plaintext message is to consist of k vectors of letters and the ciphertext message of l vectors of letters with $k < l$. Each of the k plaintext letters in a message unit $\mathcal{P}$ are then considered as integers mod N and the whole plaintext message is considered as a k digit integer written to the base N (see example below). The transformed message is then written as an l digit integer mod N and then the digits are then considered integers mod N from which encrypted letters are found. To ensure that the range of plaintext messages and ciphertext messages are the same $k < l$ are chosen so that

$$N^k < n_U < N^l$$

for each user U, that is $n_U = p_U q_U$. In this case any plaintext message $\mathcal{P}$ is an integer less than N^k considered as an element of $\mathbb{Z}_{n_U}$. Since $n_U < N^l$ the image under the power transformation corresponds to an l digit integer written to the base N and hence to an l letter block. We give an example with relatively small primes. In real world applications the primes would be chosen to have over a hundred digits and the computations and choices must be done using good computing machinery.

Example 22.3.1. Suppose $N = 26$, $k = 2$ and $l = 3$. Suppose further that Alice chooses $p_A = 29$, $q_A = 41$, $e_A = 13$. Here $n_A = 29 \cdot 41 = 1189$ so she makes public the key $K_A = (1189, 13)$. She then computes the multiplicative inverse d_A of 13 mod $1120 = 28 \cdot 40$. Now suppose we want to send her the message TABU. Since $k = 2$ the message units in plaintext are 2 vectors of letters so we separate the message into TA BU. We show how to send TA. First the numerical sequence for the letters TA mod 26 is (19,0). We then use these as the digits of a 2-digit number to the base 26. Hence

$$\text{TA} = 19 \cdot 26 + 0 \cdot 1 = 494.$$

We now compute the power transformation using her $e_A = 13$ to evaluate

$$f(19, 0) = 494^{13} \mod 1189.$$

This is evaluated as 320. Now we write 320 to the base 26. By our choices of k, l this can be written with a maximum of 3 digits to this base. Then

$$320 = 0 \cdot 26^2 + 12 \cdot 26 + 8.$$

The letters in the encoded message then correspond to $(0, 12, 8)$ and therefore the encryption of TA is AMI.

To decode the message Alice knows d_A and applies the inverse transformation.

Since we have assumed that $k < l$ this seems to restrict the direction in which messages can be sent. In practice to allow messages to go between any two users the following is done. Suppose Alice is sending an authenticated message to Bob. The keys $k_A = (n_A, e_A), k_B = (n_B, e_B)$ are public. If $n_A < n_B$ Alice sends $f_B f_A^{-1}(\mathcal{P})$. On the other hand if $n_A > n_B$ she sends $f_A^{-1} f_B(\mathcal{P})$.

There have been attacks on RSA for special types of primes so care must be chosen in choosing the primes.

The computations and choices used in real world implementations of the RSA algorithm must be done with computers. Similarly, attacks on RSA are done via computers. As computing machinery gets stronger and factoring algorithms get faster, RSA becomes less secure and larger and larger primes must be used. In order to combat this, other public key methods are in various stages of ongoing development. RSA and Diffie–Hellman and many related public key cryptosystems use properties in abelian groups. In recent years a great deal of work has been done to encrypt and decrypt using certain nonabelian groups such as linear groups or braid groups. We will discuss these later in the chapter.

22.3.3 The El-Gamal Protocol

The El-Gamal cryptosystem is a method to use the Diffie–Hellman key exchange method to do encryption. The method works as follows and uses the fact that hash functions can also be used with encryption. Suppose that Bob and Alice want to communicate openly. They have exchanged a secret key K that supposedly only they know. Let f_K be an encryption function or encryption algorithm based on the key K. Alice wants to send the message m to Bob and m is given as a binary bit string. Alice sends to Bob

$$f_K(m) \oplus h(K)$$

where $\oplus$ is addition modulo 2.

Bob knows the key K and hence its hash value $h(K)$. He now computes

$$f_K(m) \oplus h(K) \oplus h(K).$$

Since addition modulo 2 has order 2 we have

$$f_K(m) \oplus h(K) \oplus h(K) = f_k(M).$$

Bob now applies the decryption algorithm f_K^{-1} to decode the message.

Hence if K is a publicly exchanged secret key and f_K is a cryptosystem based on K then the above format allows an encryption algorithm to go with the key exchange. The El-Gamal system does this with the Diffie–Hellman key exchange protocol.

Suppose that Bob and Alice want to communicate openly. Alice chooses a prime q and a generator g of the multiplicative group $\mathbb{Z}_q$. q should be large enough to thwart the known discrete logarithm algorithms. Alice then chooses an integer a with $1 < a < q - 1$. She then computes

$$A = g^a \mod q.$$

Her public key is then (q, g, A). Bob wants to send a message M to Alice. He first encrypts the message an integer m mod q. For Bob to now send an encrypted message m to Alice he chooses a random integer b with $1 < b < q - 2$ and compute

$$B = g^b \mod q.$$

Bob then sends to Alice the integer

$$c = A^b m \mod q$$

that is Bob encrypts the whole message by multiplying it by the Diffie–Hellman shared key. The complete El-Gamal ciphertext is then the pair (B, c).

How does Alice decode the message. Given the message m she knows how to reconstruct the plaintext message M so she must recover the mod q integer m. As in the Diffie–Hellman key exchange she can compute the shared key $A^b = B^a$. She can then divide c by this Diffie–Hellman key g^{ab} to obtain m. To avoid having to find the inverse of B^a mod q which can be difficult she computes the exponent $x = p - 1 - a$. The inverse is then B^x mod q.

For each new El-Gamal encryption a new exponent b is chosen so that there is a random component of El-Gamal which improves the security.

Breaking the El-Gamal system is as difficult as breaking the Diffie–Hellman protocol and hence is based on the difficulty of the discrete log problem. However the El-Gamal has the advantage that the choice of primes is random. As mentioned the primes should be chosen large enough to not be susceptible to known discrete log algorithms. Presently the primes should be of binary length at least 512.

$$c = A^b m \mod q.$$

22.3.4 Elliptic Curves and Elliptic Curve Methods

A very powerful approach which has had wide ranging applications in cryptography is to use *elliptic curves*. If F is a field of characteristic not equal to 2 or 3 then an *elliptic curve* over F is the locus of points $(x, y) \in F \times F$ satisfying the equation

$$y^2 = x^3 + ax + b \quad \text{with } 4a^3 + 27b^2 \neq 0.$$

We denote by 0 a single point at infinity and let

$$E(F) = \{(x, y) \in F \times F : y^2 = x^3 + ax + b\} \cup \{0\}.$$

The important thing about elliptic curves from the viewpoint of cryptography is that a group structure can be placed on $E(F)$. In particular we define the operation $+$ on $E(F)$ by

1. $0 + P = P$ for any point $P \in E(F)$.
2. If $P = (x, y)$ then $-P = (x, -y)$ and $-0 = 0$.
3. $P + (-P) = 0$ for any point $P \in E(F)$.
4. If $P_1 = (x_1, y_1)$, $P_2 = (x_2, y_2)$ with $P_1 \neq -P_2$ then

$$P_1 + P_2 = (x_3, y_3) \quad \text{with}$$
$$x_3 = m^2 - (x_1 + x_2), \quad y_3 = -m(x_3 - x_1) - y_1$$

and

$$m = \frac{y_2 - y_1}{x_2 - x_1} \quad \text{if } x_2 \neq x_1$$

and

$$m = \frac{3x_1^2 + a}{2y_1} \quad \text{if } x_2 = x_1.$$

This operation has a very nice geometric interpretation if $F = \mathbb{R}$ the real numbers. It is known as the chord and tangent method. If $P_1 \neq P_2$ are two points on the curve then the line through P_1, P_2 intersects the curve at another point P_3. If we reflect P_3 through the x-axis we get $P_1 + P_2$. If $P_2 = P_2$ we take the tangent line at P_1.

With this operation $E(F)$ becomes an abelian group (due to Cassels) whose structure can be worked out.

Theorem 22.3.2. *$E(F)$ together with the operations defined above forms an abelian group. In F is a finite field of order p^k then $E(F)$ is either cyclic or has the structure*

$$E(F) = \mathbb{Z}_{m_1} \times \mathbb{Z}_{m_2}$$

with $m_1 | m_2$ and $m_1 | (p^k - 1)$.

A comprehensive description and discussion of elliptic curve methods can be found in Crandall and Pomerance [52].

The groups of elliptic curves can be used for cryptography as developed by Koblitz and others. If q is a prime and $a, b \in \mathbb{Z}_q$ then we can form the elliptic curve $E(p : a, b)$ and the corresponding elliptic curve abelian group. In this group the Diffie–Hellman key exchange protocol and the corresponding El-Gamal encryption system can be implemented. Care must be taken that the discrete log problem in $E(q; a, b)$ is difficult. The curve is then called a *cryptographically secure elliptic curve*.

Elliptic curve public-key cryptosystems are at present the most important commutative alternatives to the use of the RSA algorithm. There are several reasons for this. They are more efficient in many cases than RSA and keys in elliptic curve systems are much smaller than keys in RSA. It is felt that it is important to have good workable alternatives to RSA in the event that factoring algorithms become strong enough to compromise RSA encryption.

22.4 Noncommutative Group based Cryptography

The public key cryptosystems and public key exchange protocols that we have discussed, such as the RSA algorithm, Diffie–Hellman, El-Gamal and elliptic curve methods are number theory based and hence depend on the structure of abelian groups. Although there have been no overall successful attacks on the standard methods there is a feeling that the strength of computing machinery has made these techniques theoretically susceptible to attack. As a result of this, there has been a recent active line of research to develop cryptosystems and key exchange protocols using noncommutative cryptographic platforms. This line of investigation has been given the broad title of *noncommutative algebraic cryptography*. Since most of the cryptographic platforms are groups this is also known as *group based cryptography*. The book by Myasnikov, Shpilrain and Ushakov [63] provides an overview of group based cryptographic methods tied to complexity theory.

Up to this point the main sources for noncommutative cryptographic platforms has been nonabelian groups. In cryptosystems based on these objects algebraic properties of the platforms are used prominently in both devising cryptosystems and in cryptanalysis. In particular the nonsolvability of certain algorithmic problems in finitely presented groups, such as the conjugator search problem, has been crucial in encryption and decryption.

The main sources for nonabelian groups are combinatorial group theory and linear group theory. Braid group cryptography (see [53]), where encryption is done within the classical braid groups, is one prominent example. The one way functions in braid group systems are based on the difficulty of solving group theoretic decision problems such as the conjugacy problem and conjugator search problem. Although braid group cryptography had initial spectacular success, various potential attacks have been identified. Borovik, Myasnikov, Shpilrain [49] and others have studied the statistical aspects of these attacks and have identified what are termed black holes in the platform groups outside of which present cryptographic problems. Baumslag. Fine and Xu in [46] and [69] suggested potential cryptosystems using a combination of combinatorial group theory and linear groups and a general schema for the these types of cryptosystems was given. In [47] a public key version of this schema using the classical modular group as a platform was presented. A cryptosystem using the extended modular group $SL_2(\mathbb{Z})$ was developed by Yamamura [70] but was subsequently shown to have loopholes [67]. In [47] attacks based on these loopholes were closed.

The extension of the cryptographic ideas to noncommutative platforms involves the following ideas,

(1) General Algebraic Techniques for Developing Cryptosystems

(2) Potential Algebraic Platforms (Specific Groups, Rings, etc.) for implementing the Techniques

(3) Cryptanalysis and Security Analysis of the Resulting Systems.

The main source for noncommutative platforms are nonabelian groups and the main method for handling nonabelian groups in cryptography is combinatorial group theory which we discussed in detail in Chapter 14. The basic idea in using combinatorial group theory for cryptography is that elements of groups can be expressed as words in some alphabet. If there is an easy method to rewrite group elements in terms of these words and further the technique used in this rewriting process can be supplied by a secret key then a cryptosystem can be created.

One of the earliest descriptions of a free group cryptosystem was in a paper by W. Magnus in the early 1970s [61]. Recall that the classical modular group M is $M = PSL_2(\mathbb{Z})$. Hence M consists of the 2×2 projective integral matrices:

$$M = \left\{ \pm \begin{pmatrix} a & b \\ c & d \end{pmatrix} : ad - bc = 1, a, b, c, d \in \mathbb{Z} \right\}.$$

Equivalently M can be considered as the set of integral linear fractional transformations with determinant 1:

$$z' = \frac{az + b}{cz + d}, \quad ad - bc = 1, \quad a, b, c, d \in \mathbb{Z}.$$

Magnus proved the following theorem.

Theorem 22.4.1 ([46]). *The matrices*

$$\pm \begin{pmatrix} 1 & 1 \\ 1 & 2 \end{pmatrix}, \quad \pm \begin{pmatrix} 1 + 4t^2 & 2t \\ 2t & 1 \end{pmatrix}, \quad t = 1, 2, 3, \ldots$$

freely generate a free subgroup F of infinite index in M. Further distinct elements of F have distinct first columns.

Since the entries in the generating matrices are positive we can do the following. Choose a set

$$T_1, \dots, T_n$$

of projective matrices from the set above with n large enough to encode a desired plaintext alphabet $\mathcal{A}$. Any message would be encoded by a word

$$W(T_1, \dots, T_n)$$

with nonnegative exponents. This represents an element g of F. The two elements in the first column determine W and therefore g. Receiving W then determines the message uniquely.

The idea of using the difficulty of group theory decision problems in infinite nonabelian groups was first developed by Magyarik and Wagner in 1985. They devised a public key protocol based on the difficulty of the solution of the word problem (see Chapter 14). Although this was a seminal idea their basic cryptosystem was really unworkable and not secure in the form they presented. Wagner and Magyarik outlined a conceptual public key cryptosystem based on the hardness of the word problem for finitely presented groups. At the same time, they gave a specific example of such a system. González Vasco and Steinwandt proved that their approach is vulnerable to so-called reaction attacks. In particular, for the proposed instance it is possible to retrieve the private key just by watching the performance of a legitimate recipient.

The general scheme of the Wagner and Magyarik public-key cryptosystemis as follows. Let X be a finite set of generators, and let R and S be finite sets of relators such that the group $G_0 = \langle X; R \cup S \rangle$ has an easy word problem. That is the word problem can be solved in polynomial time while the $G = \langle X; R \rangle$ has a hard word problem (see Chapter 14 for terminology).

Choose two words W_0 and W_1 which are not equivalent in G_0 (and hence not equivalent in G). The public key is the presentation $\langle X; R \rangle$ and the chosen words W_0 and W_1. To encrypt a single bit $\in \{0, 1\}$, pick W_i and transform it into a ciphertext word W by repeatedly and randomly applying Tietze transformations to the presentation $\langle X; R \rangle$. To decrypt a word W, run the algorithm for the word problem of G_0 in order to decide which of $W_i W^{-1}$ is equivalent to the empty word for the presentation $\langle X; R \cup S \rangle$ The private key is the set S. Actually, this is not sufficient and Wagner and Magyarik are not clear on this point. The public key should be a deterministic polynomial-time algorithm for the word problem of $G_0 = \langle X; R \cup S \rangle$. Just knowing S does not automatically and explicitly give us an efficient algorithm (even if such an algorithm exists).

22.4.1 Free Group Cryptosystems

The simplest example of a nonabelian group based cryptosystem is perhaps a *free group cryptosystem*. This can be described in the following manner.

Consider a free group F on free generators $x_1, \ldots, x_r$. Then each element g in F has a unique expression as a word $W(x_1, \ldots, x_r)$. Let $W_1, \ldots, W_k$ with $W_i = W_i(x_1, \ldots, x_r)$ be a set of words in the generators $x_1, \ldots, x_r$ of the free group F. At the most basic level, to construct a cryptosystem, suppose that we have a plaintext alphabet $\mathcal{A}$. For example, suppose that $\mathcal{A} = \{a, b, \ldots\}$ are the symbols needed to construct meaningful messages in English. To encrypt, use a substitution ciphertext

$$\mathcal{A} \to \{W_1, \ldots, W_k\}.$$

That is

$$a \mapsto W_1, \quad b \mapsto W_2, \ldots.$$

Then given a word $W(a, b, \ldots)$ in the plaintext alphabet form the free group word $W(W_1, W_2, \ldots)$. This represents an element g in F. Send out g as the secret message.

In order to implement this scheme we need a concrete representation of g and then for decryption a way to rewrite g back in terms of $W_1, \ldots, W_k$. This concrete representation is the idea behind *homomorphic cryptosystems*.

The decryption algorithm in a free group cryptosystem then depends on the *Reidemeister–Schreier rewriting process*. As described in Chapter 14 this is a method to rewrite elements of a subgroup of a free group in terms of the generators of that subgroup. Recall that roughly it works as follows. Assume that $W_1, \ldots, W_k$ are free generators for some subgroup H of a free group F on $\{x_1, \ldots, x_n\}$. Each W_i is then a reduced word in the generators $\{x_1, \ldots, x_n\}$. A *Schreier transversal* for H is a set $\{h_1, \ldots, h_t, \ldots\}$ of (left) coset representatives for H in F of a special form (see Chapter 14). Any subgroup of a free group has a Schreier transversal. The Reidemeister–Schreier process allows one to construct a set of generators $W_1, \ldots, W_k$ for H by using a Schreier transversal. Further given the Schreier transversal from which the set of generators for H was constructed, the *Reidemeister–Schreier rewriting process* allows us to algorithmically rewrite an element of H. Given such an element expressed as a word $W = W(x_1, \ldots, x_r)$ in the generators of F this algorithm rewrites W as a word $W^\star(W_1, \ldots, W_k)$ in the generators of H.

The knowledge of a Schreier transversal and the use of Reidemeister–Schreier rewriting facilitates the decoding process in the free group case but is not essential. Given a known set of generators for a subgroup the Stallings Folding Method to develop a subgroup graph can also be utilized to rewrite in terms of the given generators. The paper by Kapovich and Myasnikov [58] is now a standard reference for this method in free groups. At present there is an ongoing study of the complexity of Reidemeister–Schreier being done by Baumslag, Brukhov, Fine and Troeger.

Pure free group cryptosystems are subject to various attacks and can be broken easily. However a public key free group cryptosystem using a free group representation in the Modular group was developed by Baumslag, Fine and Xu [46, 47]. The most successful attacks on free group cryptosystems are called *length based attacks*. Here an attacker multiplies a word in ciphertext by a generator to get a shorter word which could possibly be decoded.

Baumslag, Fine and Xu in [46] described the following general encryption scheme using free group cryptography. A further enhancement was discussed in the paper [47].

We start with a finitely presented group

$$G = \langle X \mid R \rangle$$

where $X = \{x_1, \ldots, x_n\}$ and a faithful representation

$$\rho : G \to \overline{G}.$$

$\overline{G}$ can be any one of several different kinds of objects – linear group, permutation group, power series ring etc.

We assume that there is an algorithm to re-express an element of $\rho(G)$ in $\overline{G}$ in terms of the generators of G. That is if $g = W(x_1, \ldots, x_n, \ldots) \in G$ where W is a word in the these generators and we are given $\rho(g) \in \overline{G}$ we can algorithmically find g and its expression as the word $W(x_1, \ldots, x_n)$.

Once we have G we assume that we have two free subgroups K, H with

$$H \subset K \subset G.$$

We assume that we have fixed Schreier transversals for K in G and for H in K both of which are held in secret by the communicating parties Bob and Alice. Now based on the fixed Schreier transversals we have sets of Schreier generators constructed from the Reidemeister–Schreier process for K and for H.

$$k_1, \ldots, k_m, \ldots \quad \text{for } K$$

and

$$h_1, \ldots, h_t, \ldots \quad \text{for } H.$$

Notice that the generators for K will be given as words in $x_1, \ldots, x_n$ the generators of G while the generators for H will be given as words in the generators $k_1, k_2, \ldots$ for K. We note further that H and K may coincide and that H and K need not in general be free but only have a unique set of normal forms so that the representation of an element in terms of the given Schreier generators is unique.

We will encode within H, or more precisely within $\rho(H)$. We assume that the number of generators for H is larger than the set of characters within our plaintext alphabet. Let $\mathcal{A} = \{a, b, c, \dots\}$ be our plaintext alphabet. At the simplest level we choose a starting point i, within the generators of H, and encode

$$a \mapsto h_i, \quad b \mapsto h_{i+1}, \dots \quad \text{etc.}$$

Suppose that Bob wants to communicate the message $W(a, b, c, \dots)$ to Alice where W is a word in the plaintext alphabet. Recall that both Bob and Alice know the various Schreier transversals which are kept secret between them. Bob then encodes $W(h_i, h_{i+1}, \dots)$ and computes in $\overline{G}$ the element $W(\rho(h_i), \rho(h_{i+1}), \dots)$ which he sends to Alice. This is sent as a matrix if $\overline{G}$ is a linear group or as a permutation if $\overline{G}$ is a permutation group and so on.

Alice uses the algorithm for $\overline{G}$ relative to G to rewrite $W(\rho(h_i), \rho(h_{i+1}), \dots)$ as a word $W^{\star}(x_1, \dots, x_n)$ in the generators of G. She then uses the Schreier transversal for K in G to rewrite using the Reidemeister–Schreier process $W^{\star}$ as a word $W^{\star\star}(k_1, \dots, k_s, \dots)$ in the generators of K. Since K is free or has unique normal forms this expression for the element of K is unique. Once she has the word written in the generators of K she uses the transversal for H in K to rewrite again, using the Reidemeister–Schreier process, in terms of the generators for H. She then has a word $W^{\star\star\star}(h_i, h_{i+1}, \dots)$ and using $h_i \mapsto a, h_{i+1} \mapsto b, \dots$ decodes the message.

In actual implementation an additional *random noise factor* is added.

In [46] and [47] an implementation of this process was presented that used for the base group G the classical modular group $M = \mathrm{PSL}_2(\mathbb{Z})$. Further it was a polyalphabetic cipher which was secure.

The system in the modular group M was presented as follows. A list of finitely generated free subgroups $H_1, \dots, H_m$ of M is public and presented by their systems of generators (presented as matrices). In a full practical implementation it is assumed that m is large. For each H_i we have a Schreier transversal

$$h_{1,i}, \dots, h_{t(i),i}$$

and a corresponding ordered set of generators

$$W_{1,i}, \dots, W_{m(i),i}$$

constructed from the Schreier transversal by the Reidemeister–Schreier process. It is assumed that each $m(i) \gg l$ where l is the size of the plaintext alphabet, that is each subgroup has many more generators than the size of the plaintext alphabet. Although Bob and Alice know these subgroups in terms of free group generators what is made public are generating systems given in terms of matrices.

The subgroups on this list and their corresponding Schreier transversals can be chosen in a variety of ways. For example the commutator subgroup of the Modular group is free of rank 2 and some of the subgroups H_i can be determined from homomorphisms of this subgroup onto a set of finite groups.

Suppose that Bob wants to send a message to Alice. Bob first chooses three integers (m, q, t) where

$$\begin{aligned} m &= \text{choice of the subgroup } H_m \\ q &= \text{starting point among the generators of } H_m \\ &\quad \text{for the substitution of the plaintext alphabet} \\ t &= \text{size of the message unit.} \end{aligned}$$

We clarify the meanings of q and t. Once Bob chooses m, to further clarify the meaning of q, he makes the substitution

$$a \mapsto W_{m,q}, \quad b \mapsto W_{m,q+1}, \ldots.$$

Again the assumption is that $m(i) \gg l$ so that starting almost anywhere in the sequence of generators of H_m will allow this substitution. The message unit size t is the number of coded letters that Bob will place into each coded integral matrix.

Once Bob has made the choices (m, q, t) he takes his plaintext message $W(a, b, \ldots)$ and groups blocks of t letters. He then makes the given substitution above to form the corresponding matrices in the Modular group;

$$T_1, \ldots, T_s.$$

We now introduce a *random noise factor*. After forming $T_1, \ldots, T_s$, Bob then multiplies on the right each T_i by a random matrix in M say R_{T_i} (different for each T_i). The only restriction on this random matrix R_{T_i} is that there is no free cancellation in forming the product $T_i R_{T_i}$. This can be easily checked and ensures that the freely reduced form for $T_i R_{T_i}$ is just the concatenation of the expressions for T_i and R_{T_i}. Next he sends Alice the integral key (m, q, t) by some public key method (RSA, Anshel–Anshel–Goldfeld etc.). He then sends the message as s random matrices

$$T_1 R_{T_1}, T_2 R_{T_2}, \ldots, T_s R_{T_s}.$$

Hence what is actually being sent out are not elements of the chosen subgroup H_m but rather elements of random right cosets of H_m in M. The purpose of sending coset elements is two-fold. The first is to hinder any geometric attack by masking the subgroup. The second is that it makes the resulting words in the modular group generators longer – effectively hindering a brute force attack.

To decode the message Alice first uses public key decryption to obtain the integral keys (m, q, t). She then knows the subgroup H_m, the ciphertext substitution from the generators of H_m and how many letters t each matrix encodes. She next uses the algorithms described in Section 14.4 to express each $T_i R_{T_i}$ in terms of the free group generators of M say $W_{T_i}(y_1, \ldots, y_n)$. She has knowledge of the Schreier transversal, which is held secretly by Bob and Alice, so now uses the Reidemeister–Schreier rewriting process to start expressing this freely reduced word in terms of the generators of H_m. The Reidemeister–Schreier rewriting is done letter by letter from left to right (see Chapter 14). Hence when she reaches t of the free generators she stops. Notice that the string that she is rewriting is longer than what she needs to rewrite in order to decode as a result of the random polynomial R_{T_i}. This is due to the fact that she is actually rewriting not an element of the subgroup but an element in a right coset. This presents a further difficulty to an attacker. Since these are random right cosets it makes it difficult to pick up statistical patterns in the generators even if more than one message is intercepted. In practice the subgroups should be changed with each message.

The initial key (m, q, t) is changed frequently. Hence as mentioned above this method becomes a type of polyalphabetic cipher. Polyalphabetic ciphers have historically been very difficult to decode.

A further variation of this method using a formal power series ring in noncommuting variables over a field was described in [43].

There have been many cryptosystems based on the difficulty of solving hard group theoretic problems. The book by Myasnikov, Shpilrain and Ushakov [63] describes many of these in detail.

22.5 Ko–Lee and Anshel–Anshel–Goldfeld Methods

After the initial attempt by Wagner and Magyarik to develop a cryptosystem based on a hard group theoretic problem there have been many developments using nonabelian groups in cryptography. Among the first were the cryptographic schemes of Anshel, Anshel and Goldfeld [42] and Ko and Lee [59]. Both sets of authors, at about the same time, proposed using nonabelian groups and combinatorial group theory for public key exchange. The security of these systems depended on the difficulty of solving certain "hard" group theoretic problems.

The methods of both Anshel–Anshel–Goldfeld and Ko–Lee can be considered as group theoretic analogs of the number theory based Diffie–Hellman method. The basic underlying idea is the following. If G is a group and $g, h \in G$ we let g^h denote the conjugate of g by h, that is $g^h = h^{-1}gh$. The simple observation is that this behaves like ordinary exponentiation in that $(g^{h_1})^{h_2} = g^{h_1h_2}$. From this straightforward idea one can exactly mimic the Diffie–Hellman protocol within a nonabelian group.

Both the Anshel–Anshel–Goldfeld protocol and the Ko–Lee protocol start with a platform group G given by a group presentation. A major assumption in both protocols is that the elements of G have nice unique normal forms that are easy to compute for given group elements. However it is further assumed that given normal forms for $x, y \in G$ the normal form for the product xy does not reveal x or y.

22.5.1 The Ko–Lee Protocol

Ko and Lee [59] developed a public key exchange system that is a direct translation of the Diffie–Hellman protocol to a nonabelian group theoretic setting. Its security is based on the difficulty of the *conjugacy problem*. We again assume that the platform group has nice unique normal forms that are easy to compute given a group element but hard to recover the group element. Recall again that g^h means the conjugate of g by h.

In the Ko–Lee protocol, Alice and Bob choose commuting subgroups A and B of the platform group G. A is Alice's subgroup while Bob's subgroup is B and these are secret. Now they completely mimic the classical Diffie–Hellman technique. There is a public element $g \in G$, Alice chooses a random secret element $a \in A$ and makes public g^a. Bob chooses a random secret element $b \in B$ and makes public g^b. The secret shared key is g^{ab}. Notice that $ab = ba$ since the subgroups commute. It follows then, that $(g^a)^b = g^{ab} = g^{ba} = (g^b)^a$ just as if these were exponents. Hence both Bob and Alice can determine the common secret. The difficulty is in the difficulty of the conjugacy problem.

The *conjugacy problem* for a group G, or more precisely for a group presentation for G is given $g, h \in G$ to determine algorithmically if they are conjugates. As with the conjugator search problem it is known that the conjugacy is undecidable in general but there are groups where it is but hard. These groups then become the target platform groups for the Ko–Lee protocol. As with the Anshel–Anshel–Goldfeld protocol, Ko and Lee suggest the use of the braid groups.

As with the standard Diffie–Hellman key exchange protocol using number theory the Ko–Lee protocol can be changed to an encryption system via the El-Gamal method. Their are several different variants of noncommutative El-Gamal systems.

22.5.2 The Anshel–Anshel–Goldfeld Protocol

We now describe the Anshel–Anshel–Goldfeld public key exchange protocol. Let G be the platform group given by a finite presentation and with the assumptions on normal forms as described above.

Alice and Bob want to communicate a shared secret. First, Alice and Bob choose random finitely generated subgroups of G by giving a set of generators for each.

$$A = \{a_1, \dots, a_n\}, \quad B = \{b_1, \dots, b_m\}$$

and make them public. The subgroup A is Alice's subgroup while the subgroup B is Bob's subgroup.

Alice chooses a secret group word $a = W(a_1, \dots, a_n)$ in her subgroup while Bob chooses a secret group word $b = V(b_1, \dots, b_m)$ in his subgroup. For an element $g \in G$ we let $\mathrm{NF}(g)$ denote the normal form for g. Alice knows her secret word a and knows the generators b_i of Bob's subgroup. She makes public the normal forms of the conjugates

$$\mathrm{NF}(b_i^a), \quad i = 1, \ldots, m.$$

Bob knows his secret word b and the generators a_i of Alice's subgroup and makes public the normal forms of the conjugates

$$\mathrm{NF}(a_j^b), \quad j = 1, \ldots, n.$$

The common shared secret is the commutator

$$[a, b] = a^{-1}b^{-1}ab = a^{-1}a^b = (b^a)^{-1}b.$$

Notice that Alice knows a^b since she knows a in terms of generators a_i of her subgroup and she knows the conjugates by b since Bob has made the conjugates of the generators of A by b public. Since Alice knows a^b she knows $[a, b] = a^{-1}a^b$.

In an analogous manner Bob knows $[a, b] = (b^a)^{-1}b$. An attacker would have to know the corresponding *conjugator*, that is the element that conjugates each of the generators. Given elements g, h in a group G where it is known that $g^k = k^{-1}gk = h$ the *conjugator search problem* is to determine the conjugator k. It is known that this problem is undecidable in general, that is there are groups where the conjugator cannot be determined algorithmically. On the other hand there are groups where the conjugator search problem is solvable but "difficult", that is the complexity of solving the conjugator search problem is hard. Such groups become the ideal platform groups for the Anshel–Anshel–Goldfeld protocol.

The security in this system is then in the difficulty of the *conjugator search problem*. Anshel, Anshel and Goldfeld suggested the Braid Groups as potential platforms, use for example B_{80} with 12 or more generators in the subgroups. Their suggestion and that of Ko and Lee led to development of *braid group cryptography*. There have been various attacks on the Braid group system. However some have been handled by changing the parameters. In general the ideas remain valid despite the attacks.

The Anshel–Anshel–Goldfeld key exchange can be developed into a cryptosystem again by the El-Gamal method.

There have been many other public key exchange protocols developed using non-abelian groups. A large number of them are described in the book of Myasnikov, Shpilrain and Ushakov [63]. The authors of that book themselves have developed many of these methods. They use different "hard" group theoretic decision problems and many have been broken. On the other hand the security of many of them is still open and they perhaps can be used as viable alternatives to commutative methods.

22.6 Platform Groups and Braid Group Cryptography

Given a group based encryption scheme, such as Ko–Lee or Anshel–Anshel–Goldfeld a *platform group* is a group G in which the encryption is to take place. In general, platform groups for the noncommutative protocols that we have discussed require certain properties. The first is the existence of a normal form for elements in the group. Normal forms provide an effective method of disguising elements. Without this, one can determine a secret key simply by inspection of group elements. Further if $N(x)$, $N(y)$ are the normal forms for x, y respectively then it should difficult to determine $N(x)$ and $N(y)$ from $N(xy)$. The existence of a normal form in a group implies that the group has solvable word problem, which is essential for these protocols. For purposes of practicality, the group also needs an efficiently computable normal form, which ensures an efficiently solvable word problem.

In addition to the platform group having normal form, ideally, it would also be large enough so that a brute force search for the secret key is infeasible.

Currently, there are many potential platform groups that have been suggested. The following are some of the proposals. We refer to [63] for a discussion of many of these.

- Braid groups (Ko–Lee, Anshel–Anshel–Goldfeld),
- Thompson Groups (Shpilrain–Ushakov) [65],
- Polycyclic Groups (Eick–Kahrobaei) [54],
- Linear Groups (Baumslag–Fine–Xu) [46, 47],
- Free metabelian Groups (Shpilrain–Zapata) [66],
- Artin Groups (Shpilrain–Zapata) [66],
- Grigorchuk Groups (Petrides) [64],
- Groups of Matrices (Grigoriev–Ponomarenko) [55],
- Surface Braid Groups (Camps) [51].

Most of these are discussed in detail in [63].

As platform groups for their respective protocols, both Ko–Lee and Anshel–Anshel–Goldfeld suggested the braid groups B_n (see [50]). The groups in this class of groups possess the desired properties for the key exchange and key transport protocols; they have nice presentations with solvable word problems and conjugacy problems; the solution to the conjugacy and conjugator search problem is "hard"; there are several possibilities for normal forms for element and they have many choices for large commuting subgroups. Initially the braid groups were considered so ideal as platforms that many other cryptographic applications were framed within the braid group setting. These included *authentication*, that is identifying over a public airwave that a message received was from the correct sender and *digital signature*, that is sending an encrypted message with an included authentication. There was so much enthusiasm about using these groups that the whole area of study was named *braid group cryptography*. A comprehensive and well-written article by Dehornoy [31] provides a detailed overview of the subject and we refer the reader to that for technical details.

After the initial successes with braid group cryptographic schemes there were some surprisingly effective attacks. There were essentially three types of attacks; an attack using solutions to the conjugacy and conjugator search problems, an attack using heuristic probability within B_n and an attack based on the fact that there are faithful linear representations of each B_n (see [31]). What is most surprising is that the Anshel–Anshel–Goldfeld method was susceptible to a length based attack. In the Anshel–Anshel–Goldfeld method the *parameters* are the specific braid group B_n and the rank of the secret subgroups for Bob and Alice. A length based attack essentially broke the method for the initial parameters suggested by AAG. The parameters were then made larger and attacks by this method were less successful. However this led to research on why these attacks on the conjugator search problem within B_n were successful. What was discovered was that *generically* a random subgroup of B_n is a free group and hence length based attacks are essentially attacks on free group cryptography and therefore successful. What this indicated was that although randomness is important in cryptography in using the braid groups as platforms subgroups cannot be chosen purely randomly.

Braid groups arise in several different areas of mathematics and have several equivalent formulations. We close this chapter and the book with a brief introduction to braid groups. A complete topological and algebraic description can be found in the book of Joan Birman [50].

A *braid* on n strings is obtained by starting with n parallel strings and intertwining them. We number the strings at each vertical position and keep track of where each individual string begins and ends. We say that two braids are equivalent if it is possible to move the strings of one of the braids in space without moving the endpoints or moving through a string and obtain the other braid. A braid with no crossings is called a *trivial braid.* We form a product of braids in the following manner. If u is the first braid and v is the second braid then uv is the braid formed by placing the starting points for the strings in v at the endpoints of the strings in u. The inverse of a braid is the mirror image in the horizontal plane. It is clear that if we form the product of a braid and its mirror image we get a braid equivalent to the trivial braid. With these definitions the set of all equivalence classes braids on n strings forms a group B_n. We let σ_i denote the braid that has a single crossing from string i over string $i+1$. Since a general braid is just a series of crossings it follows that B_n is generated by the set σ_i; $i = 1, \dots, n-1$.

There is an equivalent algebraic formulation of the braid group B_n. Let F_n be a free on the n generators $x_1, \dots, x_n$ with $n > 2$. Let σ_i, $i = 1, \dots, n-1$ be the automorphism of F_n given by

$$\sigma_i : x_i \mapsto x_{i+1},\ x_{i+1} \mapsto x_{i+1}^{-1} x_i x_{i+1}$$
$$\sigma_i : x_j \mapsto x_j, \quad j \neq i, i+1.$$

Then each σ_i corresponds precisely to the basic crossings in B_n. Therefore B_n can be considered as the subgroup of $\mathrm{Aut}(F_n)$ generated by the automorphisms σ_i, Artin proved [28] (see also [24]) that a finite presentation for B_n is given by

$$B_n = \langle \sigma_1, \ldots, \sigma_{n-1}; [\sigma_i, \sigma_j] = 1 \text{ if } |i - j| > 1,$$
$$x_{i+1}x_i x_{i+1} = x_i x_{i+1} x_i, i = 1, \ldots, n-1\rangle.$$

This is now called the Artin presentation. The fact that B_n is contained in $\mathrm{Aut}(F_n)$ provides an elementary solution to the word problem in B_n since one can determine easily if an automorphism of F_n is trivial on all the generators. We note that although the braid groups B_n are linear (the Lawrence-Krammer representation is faithful (see [31]) it is known that $\mathrm{Aut}(F_n)$ is not linear (see [34]).

From the commuting relations in the Artin presentation it is clear that each B_n has the requisite collection of commuting subgroups.

The conjugacy problem for B_n was originally solved by Garside and it was assumed that it was hard in the complexity sense. Recently there has been significant research on the complexity of the solution to the conjugacy problem (see [63] and [31]).

There are several possibilities for normal forms for elements of B_n. The two most commonly used are the *Garside normal form* and the *Dehornoy handle form*. These are described in [31] and [63].

For braid group cryptography one must be careful in using more than one normal form in an encryption scheme. The second may expose what the first is hiding and vice versa (see [31]).

We describe first the Dehornoy handle form. Let W be a word in the generators of the braid group B_n. An x_i-handle is a subword of W of the form

$$x_i^{-\epsilon} V x_i^{\epsilon}$$

with $\epsilon = \pm 1$ and where the word V does not involve x_i. If V does not contain any x_{i+1}-handles then the x_i-handle is called *permitted*

A braid word W is obtained from a braid word W' by a *one step handle reduction* if some subword of W is a permitted x_i-handle $x_i^{-\epsilon} V x_i^{\epsilon}$ and W' is obtained from W by applying the following substitutions for all letters in the x_i-handle

$$\begin{aligned} x_j^{\pm 1} &\to 1 && \text{if } j = i \\ x_j^{\pm 1} &\to x_{i+1}^{-\epsilon} x_i^{\pm 1} x_{i+1}^{\epsilon} && \text{if } j = i + 1 \\ x_j^{\pm 1} &\to x_j^{\pm 1} && \text{if } j < i \text{ or } j > i + 1. \end{aligned}$$

W can be obtained from W' by an *m-step handle reduction* if W can be obtained from W' by a sequence of m one-step handle reductions.

A word is *handle free* if it has no handles. The handle free braid words provide normal forms for the elements of B_n.

Theorem 22.6.1. *Let W be a braid word. Then the following holds:*

(1) *Any sequence of handle reductions applied to W will eventually stop and produce a handle free braid word V representing the same element as W.*

(2) *The word W represents the identity in B_n if and only if any sequence of handle reductions applied to W produces the trivial word or equivalently the handle free form of W is trivial.*

The handle free reduction process is very efficient and most of the time works in polynomial time on the length of the braid word to produce the handle free form. However there is no known theoretical complexity estimate (see [31]).

Garside solved the conjugacy problem using a different type of normal form for B_n. Let S_n be the symmetric group on n letters and for each $s \in S_n$ let ζ_s be the shortest positive braid such that $\pi(\zeta_s) = s$. The elements

$$S = \{\zeta_s : s \in S_n\} \subset B_n$$

are called *simple elements.* We order the simple elements so that $\zeta_s < \zeta_t$ if there exists $r \in S_n$ such that $\zeta_t = \zeta_s \zeta_r$. This produces a lattice structure on S.

The trivial braid is the smallest element of S while the greatest element of S is the half-twist braid

$$\Delta = \zeta_{(n,n-1,\dots,2,1)}.$$

The *Garside left normal form* of a braid $a \in B_n$ is a pair $(p, (s_1, \dots, s_t))$ where $p \in \mathbb{Z}$ and $s_1, \dots, s_t$ is a sequence of permutations in $S_n \setminus \{1, \Delta\}$ satisfying for each $i = 1, \dots, t-1$

$$\zeta_1 = \gcd(\zeta_{s_i^{-1}\Delta}, \zeta_{s_{i+1}})$$

where

$$\gcd(\zeta_s, \zeta_t) = \max\{\zeta_r : \zeta_r < \zeta_s \text{ and } \zeta_r < \zeta_t\}.$$

A normal form $(p, (s_1, \dots, s_t))$ represents the element

$$\zeta_\Delta^p \zeta_{s_1} \cdots \zeta_{s_n}.$$

Theorem 22.6.2. *There exists an algorithm which computes the normal form of the corresponding braid for any braid word $W = w(x_1, \dots, x_n)$.*

22.7 Exercises

1. Show that if p, q are primes and e, d are positive integers with $(e, (p-1)(q-1)) = 1$ and $ed \equiv 1 \bmod (p-1)(q-1)$ then $a^{ed} \equiv a \bmod pq$ for any integer a. (This is the basis if the decryption function used in the RSA algorithm.

2. The following table gives the approximate statistical frequency of occurrence of letters in the English language. The passage below is encrypted with a simple permutation cipher without punctuation. Use a frequency analysis to try to decode it.

letter	frequency	letter	frequency	letter	frequency
A	.082	B	.015	C	.028
D	.043	E	.127	F	.022
G	.020	H	.061	I	070
J	.002	K	.008	L	.040
M	.024	N	.067	O	.075
P	.019	Q	.001	R	.060
S	.063	T	.091	U	.028
V	.010	W	.023	X	.001
Y	.020	Z	.001		

ZKIRNVMFNYVIRHZKLHRGREVRMGVTVIDSR
XSSZHZHGHLMOBKLHRGREVWRERHLIHLMVZ
MWRGHVOUKIRNVMFNYVIHKOZBZXIFXRZOI
LOVRMMFNYVIGSVLIBZMWZIVGSVYZHRHUL
IGHSHVMLGVHGSVIVZIVRMURMRGVOBNZMB
KIRNVHZMWGSVBHVIEVZHYFROWRMTYOLXP
HULIZOOGSVKLHRGREVRMGVTVIH

3. Encrypt the message NO MORE WAR using an affine cipher with single letters keys $a = 7, b = 5$.

4. Encrypt the message NO MORE WAR using an affine cipher on 2 vectors of letters and an encrypting keys

$$A = \begin{pmatrix} 5 & 2 \\ 1 & 1 \end{pmatrix}, \quad B = (3, 7).$$

5. What is the decryption algorithm for the affine cipher given in the last problem.

6. How many different affine enciphering transformations are there on single letters with an N letter alphabet.

7. Let $N \in \mathbb{N}$ with $N \geq 2$ and $n \to an + b$ with $(a, N) = 1$ is an affine cipher on an N letter alphabet. Show that if any two letters are guessed $n_1 \to m_1, n_2 \to m_2$ with $(n_1 - n_2, N) = 1$ then the code can be broken.

8. Let F be a free group of rank 3 with generators x, y, z. Code the English alphabet by $a \mapsto 0, b \mapsto 1, \ldots$. Consider the free group cryptosystem given by

$$i \mapsto W_i$$

where $W_i = x^i y^{i+1} z^{i+2} x^{-i+1}$. Code the message EAT AT JOES with this system.

9. In the Anshel–Anshel–Goldfeld protocol verify that both Bob and Alice will know the commutator.

Bibliography

General Abstract Algebra

[1] M. Artin, *Algebra*, Prentice-Hall. 1991.

[2] C. Curtis and I. Reiner, *Representation Theory of Finite Groups and Associative Algebras*, Wiley Interscience, 1966.

[3] B. Fine and G. Rosenberger, *The Fundamental Theorem of Algebra*, Springer-Verlag, 2000.

[4] J. Fraleigh, *A First Course in Abstract Algebra*, 7th ed., Addison-Wesley, 2003.

[5] P. R. Halmos, *Naive Set Theory*, Springer-Verlag, 1998.

[6] I. Herstein, *Topics in Algebra*, Blaisdell, 1964.

[7] M. Kreuzer and S. Robiano, *Computational Commutative Algebra I and II*, Springer-Verlag, 1999.

[8] S. Lang, *Algebra*, Addison-Wesley, 1965.

[9] S. MacLane and G. Birkhoff, *Algebra*, Macmillan, 1967.

[10] N. McCoy, *Introduction to Modern Algebra*. Allyn and Bacon, 1960.

[11] N. McCoy, *The Theory of Rings*, Macmillan, 1964.

[12] G. Stroth, *Algebra. Einführung in die Galoistheorie*, De Gruyter, 1998.

Group Theory and Related Topics

[13] G. Baumslag, *Topics in Combinatorial Group Theory*, Birkhäuser, 1993.

[14] O. Bogopolski, *Introduction to Group Theory*, European Mathematical Society, 2008.

[15] T. Camps, V. große Rebel and G. Rosenberger, *Einführung in die kombinatorische und die geometrische Gruppentheorie*, Heldermann Verlag, 2008.

[16] T. Camps, S. Künling and G. Rosenberger, *Einführung in die mengenteoretische und die algebraische Topologie*, Heldermann Verlag, 2006.

[17] B. Fine and G. Rosenberger, *Algebraic Generalizations of Discrete Groups*, Marcel Dekker, 2001.

[18] D. Gorenstein, *Finite Simple Groups. An Introduction to their Classification*, Plenum Press, 1982.

[19] D. Johnson, *Presentations of Groups*, Cambridge University Press, 1990.

[20] S. Katok, *Fuchsian Groups*, Univ. of Chicago Press, 1992.

[21] G. Kern-Isberner and G. Rosenberger. Normalteiler vom Geschlecht eins in freien Produkten endlicher zyklischer Gruppen, *Results in Math.*, 11, 1987, 272–288.

[22] R. C. Lyndon, *Groups and Geometry*, LMS Lecture Note Series 101, Cambridge University Press, 1985.

[23] R. C. Lyndon and P. Schupp, *Combinatorial Group Theory*, Springer-Verlag 1977.

[24] W. Magnus, A. Karrass and D. Solitar *Combinatorial Group Theory*, Wiley, 1966.

[25] D. J. S. Robinson, *A Course in the Theory of Groups*, Springer-Verlag, 1982.

[26] J. Rotman, *Group Theory*, 3rd ed., Wm. C. Brown, 1988.

Number Theory

[27] L. Ahlfors, *Introduction to Complex Analysis*, Springer-Verlag, 1968.

[28] T. M. Apostol, *Introduction to Analytic Number Theory*, Springer-Verlag, 1976.

[29] A. Baker, *Transcendental Number Theory*, Cambridge University Press, 1975.

[30] H. Cohn, *A Classical Invitation to Algebraic Numbers and Class Fields*, Springer-Verlag, 1978.

[31] L. E. Dickson, *History of the Theory of Numbers*, Chelsea, 1950.

[32] B. Fine, A note on the two-square theorem, *Can. Math. Bulletin*, 20, 1977, 93–94.

[33] B. Fine, Sums of squares rings, *Can. J. Math.*, 29, 1977, 155–160.

[34] B. Fine, *The Algebraic Theory of the Bianchi Groups*, Marcel Dekker, 1989.

[35] B. Fine and G. Rosenberger, *Number Theory: An Introduction via the Distribution of Primes*, Birkhäuser, 2006.

[36] G. H. Hardy and E. M. Wright, 5th ed., *An Introduction to the Theory of Numbers*. Clarendon Press, 1979.

[37] E. Landau, *Elementary Number Theory*, Chelsea, 1958.

[38] M. Newman, *Integral Matrics*, Academic Press, 1972.

[39] I. Niven and H. S. Zuckerman, *The Theory of Numbers*, 4th ed., John Wiley, 1980.

[40] O. Ore, *Number Theory and its History*, McGraw-Hill, 1949.

[41] H. Pollard and H. Diamond *The Theory of Algebraic Numbers*, Carus Mathematical Monographs, 9, Math. Assoc. of America, 1975.

Cryptography

[42] I. Anshel, M. Anshel and D. Goldfeld, An algebraic method for public key cryptography, *Math. Res. Lett.*, 6, 1999, 287–291.

[43] G. Baumslag, Y. Brjukhov, B. Fine and G. Rosenberger, Some cryptoprimitives for non-commutative algebraic cryptography, *Aspects of Infinite Groups*, World Scientific Press, 26–44, 2009.

[44] G. Baumslag, Y. Brjukhov, B. Fine and D. Troeger, Challenge response password security using combinatorial group theory, *Groups Complex. Cryptol.*, 2, 2010, 67–81.

[45] G. Baumslag, T. Camps, B. Fine, G. Rosenberger and X. Xu, Designing key transport protocols using combinatorial group theory, *Cont. Math.* 418, 2006, 35–43.

[46] G. Baumslag, B. Fine and X. Xu, Cryptosystems using linear groups, *Appl. Algebra Eng. Commun. Comput.* 17, 2006, 205–217.

[47] G. Baumslag, B. Fine and X. Xu, A proposed public key cryptosystem using the modular group, *Cont. Math.* 421, 2007, 35–44.

[48] J. Birman, *Braids, Links and Mapping Class Groups*, Annals of Math Studies, Vol. 82, Princeton University Press, 1975.

[49] A. V. Borovik, A. G. Myasnikov and V. Shpilrain, Measuring sets in infinite groups, *Computational and Statistical Group Theory*, Contemp. Math. 298, 2002, 21–42.

[50] J. A. Buchmann, *Introduction to Cryptography*, Springer 2004.

[51] T. Camps, *Surface Braid Groups as Platform Groups and Applications in Cryptography*, Ph.D. thesis, Universität Dortmund 2009.

[52] R. E. Crandall and C. Pomerance, *Prime Numbers. A Computational Perspective*, 2nd ed., Springer-Verlag, 2005.

[53] P. Dehornoy, Braid-based cryptography, *Cont. Math.*, 360, 2004, 5–34.

[54] B. Eick and D. Kahrobaei, Polycyclic groups: A new platform for cryptology? *math.GR/0411077* (2004), 1–7.

[55] D. Grigoriev and I. Ponomarenko, Homomorphic public-key cryptosystems over groups and rings, *Quaderni di Matematica*, 2005.

[56] P. Hoffman, *Archimedes' Revenge*, W. W. Norton & Company, 1988.

[57] D. Kahrobaei and B. Khan, A non-commutative generalization of the El-Gamal key exchange using polycyclic groups, *Proceeding of IEEE*, 2006, 1–5.

[58] I. Kapovich and A. Myasnikov, Stallings foldings and subgroups of free groups, *J. Algebra* 248, 2002, 608–668.

[59] K. H. Ko, S. J. Lee, J. H. Cheon, J. H. Han, J. S. Kang and C. Park, New public-key cryptosystems using Braid groups, *Advances in Cryptography*, Proceedings of Crypto 2000, Lecture Notes in Computer Science 1880, 2000, 166–183.

[60] N. Koblitz, *Algebraic Methods of Cryptography*, Springer, 1998.

[61] W. Magnus, Rational representations of fuchsian groups and non-parabolic subgroups of the modular group, *Nachrichten der Akad. Göttingen*, 1973, 179–189.

[62] A. G. Myasnikov, V. Shpilrain and A. Ushakov, A practical attack on some braid group based cryptographic protocols, *CRYPTO 2005*, Lecture Notes in Computer Science 3621, 2005, 86–96.

[63] A. G. Myasnikov, V. Shpilrain and A. Ushakov, *Group-Based Cryptography*, Advanced Courses in Mathematics, CRM Barcelona, 2007.

[64] G. Petrides, Cryptoanalysis of the public key cryptosystem based on the word problem on the Grigorchuk groups, *Cryptography and Coding*, Lecture Notes in Computer Science 2898, 2003, 234–244.

[65] V. Shpilrain and A. Ushakov, The conjugacy search problem in public key cryptography; unnecessary and insufficient, *Applicable Algebra in Engineering, Communication and computing*, 17, 2006 285–289.

[66] V. Shpilrain and A. Zapata, Using the subgroup memberhsip problem in public key cryptography, *Cont. Math.*, 418, 2006, 169–179.

[67] R. Steinwandt, Loopholes in two public key cryptosystems using the modular groups, *preprint, University of Karlsruhe*, 2000.

[68] R. Stinson, *Cryptography; Theory and Practice*, Chapman and Hall, 2002.

[69] X. Xu, *Cryptography and Infinite Group Theory*, Ph.D. thesis, CUNY, 2006.

[70] A. Yamamura, Public key cryptosystems using the modular group, *Public Key Cryptography*, Lecture Notes in Computer Sciences 1431, 1998, 203–216.

Index

Q

R

S

T

U

V

W

Z

Made in United States
Cleveland, OH
11 November 2026

48115028R00207